ISBN 978-1-330-16533-1
PIBN 10042213

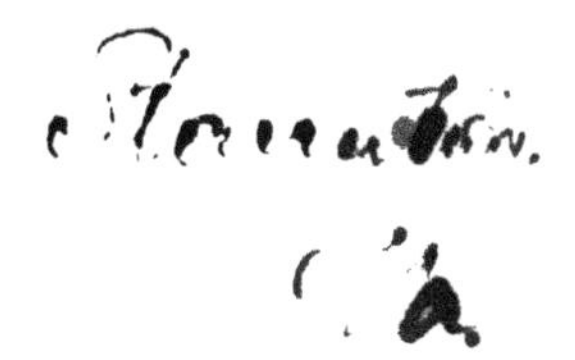

TREATISE

OF

PLANE TRIGONOMETRY.

TO WHICH IS PREFIXED

A SUMMARY VIEW OF THE NATURE AND USE OF

LOGARITHMS;

BEING

THE SECOND PART

OF

A COURSE OF MATHEMATICS.

ADAPTED TO THE METHOD OF INSTRUCTION IN THE AMERICAN COLLEGES.

BY JEREMIAH DAY, D.D. LL.D

President of Yale College

THE SECOND EDITION,

WITH ADDITIONS AND ALTERATIONS

NEW-HAVEN:

PUBLISHED BY HOWE & SPALDING

S. CONVERSE, Printer.

1824.

THE plan upon which this work was originally commenced, is continued in this second part of the course. As the single object is to provide for *a class in college*, such matter as is not embraced by this design is excluded. The mode of treating the subjects, for the reasons mentioned in the preface to Algebra, is, in a considerable degree, diffuse. It was thought better to err on this extreme, than on the other, especially in the early part of the course.

The section on right angled triangles will probably be considered as needlessly minute. The solutions might, in all cases, be effected by the theorems which are given for oblique angled triangles. But the applications of rectangular trigonometry are so numerous, in navigation, surveying, astronomy, &c. that it was deemed important, to render familiar the various methods of stating the relations of the sides and angles; and especially to bring distinctly into view the principle on which most trigonometrical calculations are founded, the proportion between the parts of the given triangle, and a similar one formed from the sines, tangents, &c. in the tables.

CONTENTS.

LOGARITHMS.

TRIGONOMETRY.

LOGARITHMS.

SECTION I.

NATURE OF LOGARITHMS.*

Art. 1. THE operations of Multiplication and Division, when they are to be often repeated, become so laborious, that it is an object of importance to substitute, in their stead, more simple methods of calculation, such as Addition and Subtraction. If these can be made to perform, in an expeditious manner, the office of multiplication and division, a great portion of the time and labour which the latter processes require, may be saved.

Now it has been shown, (Algebra, 233, 237,) that *powers* may be multiplied, by adding their *exponents*, and divided, by subtracting their exponents. In the same manner, *roots* may be multiplied and divided, by adding and subtracting their fractional exponents. (Alg. 280, 286.) When these exponents are arranged in tables, and applied to the general purposes of calculation, they are called *Logarithms*.

2. LOGARITHMS, THEN, ARE THE EXPONENTS OF A SERIES OF POWERS AND ROOTS.†

In forming a system of logarithms, some particular number is fixed upon, as the *base*, *radix*, or first power, whose logarithm is always 1. From this, a series of powers is raised; and the exponents of these are arranged in tables for use. To explain this, let the number which is chosen for the first

* Maskalyne's Preface to Taylor's Logarithms. Introduction to Hutton's Tables. Keil on Logarithms. Maseres Scriptores Logarithmici. Briggs' Logarithms. Dodson's Anti-logarithmic Canon. Euler's Algebra.

† See note A

power, be represented by a. Then taking a series of powers, both direct and reciprocal, as in Alg. 207;

$$a^4,\ a^3,\ a^2,\ a^1,\ a^0,\ a^{-1},\ a^{-2},\ a^{-3},\ a^{-4},\ \&c.$$

The logarithm of a^3 is 3, and the logarithm of a^{-1} is -1,
of a^1 is 1, of a^{-2} is -2,
of a^0 is 0, of a^{-3} is -3, &c.
Universally, the logarithm of a^x is x.

3. In the system of logarithms in common use, called *Briggs'* logarithms, the number which is taken for the radix or base is 10. The above series then, by substituting 10 for a, becomes

$$10^4,\ 10^3,\ 10^2,\ 10^1,\ 10^0,\ 10^{-1},\ 10^{-2},\ 10^{-3},\ \&c.$$
Or $10000,\ 1000,\ 100,\ 10,\ 1,\ \frac{1}{10},\ \frac{1}{100},\ \frac{1}{1000}$, &c.
Whose logarithms are
$4,\ 3,\ 2,\ 1,\ 0,\ -1,\ -2,\ -3$, &c.

4. The fractional exponents of *roots*, and of powers of roots, are converted into *decimals*, before they are inserted in the logarithmic tables. See Alg. 255.

The logarithm of $a^{\frac{1}{3}}$, or $a^{0.3333}$, is 0.3333,
of $a^{\frac{2}{3}}$, or $a^{0.6666}$, is 0.6666,
of $a^{\frac{3}{7}}$, or $a^{0.4285}$, is 0.4285,
of $a^{\frac{11}{3}}$, or $a^{3.6666}$, is 3.6666, &c.

These decimals are carried to a greater or less number of places, according to the degree of accuracy required.

5. In forming a system of logarithms, it is necessary to obtain the logarithm of each of the numbers in the natural series 1, 2, 3, 4, 5, &c.; so that the logarithm of any number may be found in the tables. For this purpose, the *radix* of the system must first be determined upon; and then every other number may be considered as some power or root of this. If the radix is 10, as in the common system, every other number is to be considered as some power of 10.

That a power or root of 10 may be found, which shall be equal to any other number whatever, or, at least, a very near approximation to it, is evident from this, that the *exponent* may be endlessly varied; and if this be increased or diminished, the *power* will be increased or diminished.

If the exponent is a fraction, and the *numerator* be increased, the power will be increased: but if the *denominator* be increased the power will be diminished.

6. To obtain then the logarithm of any number, according to Briggs' system, we have to find a power or root of 10 which shall be equal to the proposed number. The *exponent* of that power or root is the logarithm required. Thus

$$\left.\begin{array}{r} 7=10^{0.8451} \\ 20=10^{1.3010} \\ 30=10^{1.4771} \\ 400=10^{2.6020} \end{array}\right\} \text{therefore the logarithm} \left\{\begin{array}{l} \text{of } 7 \text{ is } 0.8451 \\ \text{of } 20 \text{ is } 1.3010 \\ \text{of } 30 \text{ is } 1.4771 \\ \text{of } 400 \text{ is } 2.6020, \text{\&c.} \end{array}\right.$$

7. A logarithm generally consists of two parts, an *integer* and a *decimal*. Thus the logarithm 2.60206, or, as it is sometimes written, 2+.60206, consists of the integer 2, and the decimal .60206. The integral part is called the *characteristic* or *index** of the logarithm; and is frequently omitted, in the common tables, because it can be easily supplied, whenever the logarithm is to be used in calculation.

By art. 3d, the logarithms of

	10000,	1000,	100,	10,	1,	.1,	.01,	.001,	&c.
are	4,	3,	2,	1,	0,	−1,	−2,	−3,	&c.

As the logarithms of 1 and of 10 are 0 and 1, it is evident, that, if any given number be between 1 and 10, its logarithm will be between 0 and 1, that is, it will be greater than 0, but less than 1. It will therefore have 0 for its index, with a decimal annexed.

Thus the logarithm of 5 is 0.69897.

For the same reason, if the given number be between

$$\left.\begin{array}{r} 10 \text{ and } 100, \\ 100 \text{ and } 1000, \\ 1000 \text{ and } 10000, \end{array}\right\} \begin{array}{c} \text{the log.} \\ \text{will be} \\ \text{between} \end{array} \left\{\begin{array}{ll} 1 \text{ and } 2, \text{ i. e.} & 1+\text{the dec. part.} \\ 2 \text{ and } 3, & 2+\text{the dec. part.} \\ 3 \text{ and } 4, & 3+\text{the dec. part.} \end{array}\right.$$

We have, therefore, when the logarithm of an integer or mixed number is to be found, this general rule.

* The term *index*, as it is used here, may possibly lead to some confusion in the mind of the learner. For the logarithm itself is the index or exponent of a power. The characteristic, therefore, is the index of an index.

8. *The index of the logarithm is always one less, than the number of integral figures, in the natural number whose logarithm is sought:* or, the index shows how far the first figure of the natural number is removed from the place of units.

Thus the logarithm of 37 is 1.56820.

Here, the number of figures being *two*, the index of the logarithm is 1.

The logarithm of 253 is 2.40312.

Here, the proposed number 253 consists of *three* figures, the first of which is in the second place from the unit figure. The index of the logarithm is therefore 2.

The logarithm of 62.8 is 1.79796.

Here it is evident that the mixed number 62.8 is between 10 and 100. The index of its logarithm must, therefore, be 1.

9. As the logarithm of 1 is 0, the logarithm of a number *less* than 1, that is, of any proper *fraction*, must be *negative*.

Thus by art. 3d
The logarithm of $\frac{1}{10}$ or .1 is -1,
of $\frac{1}{100}$ or .01 is -2,
of $\frac{1}{1000}$ or .001 is -3, &c.

10. If the proposed number is *between* $\frac{1}{100}$ and $\frac{1}{1000}$ its logarithm must be between -2 and -3. To obtain the logarithm, therefore, we must either *subtract* a certain fractional part from -2, or *add* a fractional part to -3; that is, we must either annex a *negative decimal* to -2, or a *positive one* to -3.

Thus the logarithm
of .008 is either $-2-.09691$, or $-3+.90309$.*

The latter is generally most convenient in practice, and is more commonly written $\bar{3}.90309$. The line over the index

* That these two expressions are of the same value will be evident, if we subtract the same quantity, +.90309 from each. The remainders will be equal, and therefore the quantities from which the subtraction is made must be equal. See note B.

denotes, that *that* is negative, while the *decimal* part of the logarithm is positive.

The logarithm { of 0.3, is $\overline{1}.47712$, of 0.06, is $\overline{2}.77815$, of 0.009, is $\overline{3}.95424$.

And universally,

11. *The negative index of a logarithm shows how far the first significant figure of the natural number. is removed from the place of units, on the right;* in the same manner as a *positive* index shows how far the first figure of the natural number is removed from the place of units, on the left. (Art. 8.) Thus in the examples in the last article,

The decimal 3 is in the *first* place from that of units,
6 is in the *second* place,
9 is in the *third* place;

And the indices of the logarithms are $\overline{1}$, $\overline{2}$, and $\overline{3}$.

12. It is often more convenient, however, to make the *index* of the logarithm positive, as well as the decimal part. This is done by adding 10 to the index.

Thus, for -1, 9 is written; for -2, 8, &c.
Because $-1+10=9$, $-2+10=8$, &c.

With this alteration,

The logarithm { $\overline{1}.90309$, $\overline{2}.90309$, $\overline{3}.90309$ } becomes { 9.90309, 8.90309, 7.90309, &c.

This is making the index of the logarithm 10 too great. But with proper caution, it will lead to no error in practice.

13. The *sum* of the logarithms of two numbers, is the logarithm of the *product* of those numbers; and the *difference* of the logarithms of two numbers, is the logarithm of the *quotient* of one of the numbers divided by the other. (Art. 2.) In Briggs' system, the logarithm of 10 is 1. (Art. 3.) If therefore any number be multiplied or divided by 10, its logarithm will be increased or diminished by 1: and as this is an integer, it will only change the *index* of the logarithm, without affecting the decimal part.

Thus the logarithm of	4730 is 3.67486,
And the logarithm of	10 is 1.
The logarithm of the product	47300 is 4.67486
And the logarithm of the quotient	473 is 2.67486

Here the *index* only is altered, while the decimal part remains the same. We have then this important property,

14. *The* DECIMAL PART *of the logarithm of any number is the same, as that of the number multiplied or divided by* 10, 100, 1000, &c.

Thus the log. of 45670,	is 4.65963,	
4567,	3.65963,	
456.7,	2.65963,	
45.67,	1.65963,	
4.567,	0.65963,	
.4567,	$\bar{1}$.65963,	or 9.65963,
.04567,	$\bar{2}$.65963,	" 8.65963,
.004567,	$\bar{3}$.65963,	" 7.65963.

This property, which is peculiar to Briggs' system, is of great use in abridging the logarithmic tables. For when we have the logarithm of any number, we have only to change the index, to obtain the logarithm of every other number. whether integral, fractional, or mixed, consisting of the same significant figures. The decimal part of the logarithm of a fraction found in this way, is always *positive*. For it is the same as the decimal part of the logarithm of a whole number.

15. In a series of fractions *continually decreasing*, the negative indices of the logarithms *continually increase*. Thus

In the series	1,	.1,	.01,	.001,	.0001,	.00001,	&c.
The logarithms are	0,	−1,	−2,	−3,	−4,	−5,	&c.

If the progression be continued, till the fraction is reduced to 0, the negative logarithm will become greater than any assignable quantity. The logarithm of 0, therefore, is *infinite and negative*. (Alg. 447.)

16. It is evident also, that all *negative* logarithms belong to fractions which are between 1 and 0; while *positive* loga-

rithms belong to natural numbers which are greater than 1. As the whole range of numbers, both positive and negative, is thus exhausted in supplying the logarithms of integral and fractional positive quantities; there can be no other numbers to furnish logarithms for *negative* quantities. On this account the logarithm of a negative quantity is, by some writers, considered as *impossible*. But as there is no difference in the multiplication, division, involution &c. of positive and negative quantities, except in applying the *signs*; they may be considered as all positive, while these operations are performing by means of logarithms; and the proper signs may be *afterwards* affixed.

17. *If a series of numbers be in* GEOMETRICAL *progression, their logarithms will be in* ARITHMETICAL *progression.* For, in a geometrical series ascending, the quantities increase by a common *multiplier*; (Alg. 436.) that is, each succeeding term is the *product* of the preceding term into the ratio. But the *logarithm* of this product is the *sum* of the logarithms of the preceding term and the ratio; that is, the logarithms increase by a common *addition*, and are, therefore, in arithmetical progression. (Alg. 422.) In a geometrical progression *descending*, the terms decrease by a common *divisor* and their logarithms, by a common *difference*.

Thus the numbers 1, 10, 100, 1000, 10000, &c. are in geometrical progression.

And their logarithms 0, 1, 2, 3, 4, &c are in arithmetical progression.

Universally, if in any geometrical series,

a=the least term, r=the ratio,

L=its logarithm, l=its logarithm;

Then the logarithm of ar is $L+l$, (Art. 1.)

of ar^2 is $L+2l$,

of ar^3 is $L+3l$, &c.

Here, the quantities a, ar^2, ar^3, ar^4, &c. are in geometrical progression. (Alg. 436.)

And their logarithms L, $L+l$, $L+2l$, $L+3l$, &c. are in arithmetical progression. (Alg. 423.)

* See note C

THE LOGARITHMIC CURVE.

19. The relations of logarithms, and their corresponding numbers, may be represented by the abscissas and ordinates of a curve. Let the line AC (Fig. 1.) be taken for unity. Let AF be divided into portions, each equal to AC, by the points 1, 2, 3, &c. Let the line a represent the *radix* of a given system of logarithms, suppose it to be 1.3 ; and let a^2, a^3, &c. correspond, in length, with the different powers of a. Then the distances from A to 1, 2, 3, &c. will represent the *logarithms* of a, a^2, a^3, &c. (Art. 2.) The line CH is called the *logarithmic curve*, because its *abscissas* are proportioned to the logarithms of numbers represented by its *ordinates*. (Alg. 527.)

20. As the abscissas are the distances from AC, on the line AF, it is evident, that the abscissa of the point C is 0, which is the logarithm of $1=AC$. (Art. 2.) The distance from A to 1 is the logarithm of the ordinate a, which is the *radix* of the system. For Briggs' logarithms, this ought to be ten times AC. The distance from A to 2 is the logarithm of the ordinate a^2 ; from A to 3 is the logarithm of a^3, &c.

21. The logarithms of numbers less than a unit are *negative*. (Art. 9.) These may be represented by portions of the line AN, on the *opposite side* of AC. (Alg. 507.) The ordinates a^{-1}, a^{-2}, a^{-3}, &c. are less than AC, which is taken for unity; and the abscissas, which are the distances from A to -1, -2, -3, &c. are negative.

22. If the curve be continued ever so far, it will never meet the axis AN, For, as the ordinates are in geometrical progression decreasing, each is a certain portion of the preceding one. They will be diminished more and more, the farther they are carried, but can never be reduced absolutely to nothing. The axis AN is, therefore, an *asymptote* of the curve. (Alg. 545.) As the ordinate decreases, the abscissa increases; so that, when one becomes infinitely small, the other becomes infinitely great. This corresponds with what has been stated, (Art. 15.) that the logarithm of 0 is *infinite* and *negative*.

23. To find the *equation* of this curve,

Let a=the *radix* of the system,
x=any one of the abscissas,
y=the corresponding ordinate.

Then, by the nature of the curve, (Art. 19.) the *ordinate* to any point, is that power of a whose exponent is equal to the *abscissa* of the same point; that is (Alg. 528.)

$$y=a^x.*$$

* For other properties of the logarithmic curve, see Fluxions.

SECTION II.

DIRECTIONS FOR TAKING LOGARITHMS AND THEIR NUMBERS FROM THE TABLES.*

Art. 24. THE purpose which logarithms are intended to answer, is to enable us to perform arithmetical operations with *greater expedition*, than by the common methods. Before any one can avail himself of this advantage, he must become so familiar with the tables, that he can readily find the logarithm of any number ; and, on the other hand, the number to which any logarithm belongs.

In the common tables, the *indices* to the logarithms of the first 100 numbers, are inserted. But, for all other numbers, the *decimal* part only of the logarithm is given ; while the index is left to be supplied, according to the principles in arts. 8 and 11.

25. *To find the logarithm of any number between* 1 *and* 100 ;

Look for the proposed number, on the left ; and against it, in the next column, will be the logarithm, with its index. Thus

The log. of 18 is 1.25527. The log. of 73 is 1.86332.

26. *To find the logarithm of any number between* 100 *and* 1000 ; or of any number consisting of not more than three significant figures, with ciphers annexed.

In the smaller tables, the three first figures of each number, are generally placed in the left hand column ; and the fourth figure is placed at the head of the other columns.

Any number, therefore, between 100 and 1000, may be found on the left hand ; and directly opposite, in the next column, is the decimal part of its logarithm. To this the *index* must be prefixed, according to the rule in art. 8.

* The best English Tables are Hutton's in 8vo. and Taylor's in 4to. In these, the logarithms are carried to seven places of decimals, and proportional parts are placed in the margin. The smaller tables are numerous; and, when accurately printed, are sufficient for common calculations.

The log. of 458 is	2.66087,	The log. of 935 is	2.97081,
of 796	2.90091,	of 386	2.58659.

If there are *ciphers* annexed to the significant figures, the logarithm may be found in a similar manner. For, by art. 14, the *decimal* part of the logarithm of any number is the same, as that of the number multiplied into 10, 100, &c. All the difference will be in the *index*; and this may be supplied by the same general rule.

The log. of 4580 is	3.66087,	The log. of 326000 is	5.51322,
of 79600	4.90091,	of 8010000	6.90363.

27. *To find the logarithm of any number consisting of* FOUR *figures, either with, or without, ciphers annexed.*

Look for the three first figures, on the left hand, and for the fourth figure, at the head of one of the columns. The logarithm will be found, opposite the three first figures, and in the column which, at the head, is marked with the fourth figure.*

The log. of 6234 is	3.79477,	The log. of 783400 is	5.89398,
of 5231	3.71858,	of 6281000	6.79803.

28. *To find the logarithm of a number containing* MORE *than* FOUR *significant figures.*

By turning to the tables, it will be seen, that if the *differences* between several numbers be small, in comparison with the numbers themselves; the differences of the *logarithms* will be nearly proportioned to the differences of the *numbers*. Thus

The log. of 1000 is	3.00000,
of 1001	3.00043,
of 1002	3.00087,
of 1003	3.00130,
of 1004	3.00173, &c.

Here the differences in the numbers are, 1, 2, 3, 4, &c. and the corresponding differences in the logarithms are 43, 87, 130, 173, &c.

Now 43 is nearly half of 87, one third of 130, one fourth of 173, &c.

Upon this principle, we may find the logarithm of a number which is between two other numbers whose logarithms

* In Taylor's, Hutton's and other tables, *four* figures are placed in the left hand column, and the *fifth* at the top of the page.

are given by the tables. Thus the logarithm of 21716 is not to be found, in those tables which give the numbers to four places of figures only.

But by the table, the log. of 21720 is 4.33686
and the log. of 21710 is 4.33666

The difference of the two numbers is 10; and that of the logarithms 20.

Also, the difference between 21710, and the proposed number 21716 is 6.

If, then, a difference of 10 in the numbers
make a difference of 20 in the logarithms:
A difference of 6 in the numbers, will
make a difference of 12 in the logarithms.

That is, 10 : 20 :: 6 : 12.

If, therefore, 12 be added to 4.33666, the log. of 21710;
The sum will be 4.33678, the log. of 21716.
We have, then, this

RULE.

To find the logarithm of a number consisting of more than four figures;

Take out the logarithm of two numbers, one greater, and the other less, than the number proposed: Find the difference of the two numbers, and the difference of their logarithms: Take also the difference between the least of the two numbers, and the proposed number. Then say,

As the difference of the two numbers,
To the difference of their logarithms;
So is the difference between the least of the two numbers, and the proposed number,
To the proportional part to be added to the least of the two logarithms.

It will generally be expedient to make the *four first figures*, in the least of the two numbers, the same as in the proposed number, substituting ciphers, for the remaining figures; and to make the greater number the same as the less, with the addition of a unit to the last significant figure. Thus

For 36843, take 36840, and 36850,
For 792674, 792600, 792700,
For 6537825, 6537000, 6538000, &c.

The first term of the proportion will then be 10, or 100, or 1000, &c.

Ex. 1. Required the logarithm of 362572.

The logarithm of 362600 is 5.55943
of 362500 5.55931

The differences are 100, and 12.

Then 100 : 12 :: 72 : 8.64, or 9 nearly.
And the log. 5.55931+9=5.55940, the log. required.

Ex. 2. The log. of 78264 is 4.89356
3. The log. of 143542 is 5.15698
4. The log. of 1129535 is 6.05290.

By a little practice, such a facility, in abridging these calculations, may be acquired, that the logarithms may be taken out, in a very short time. When great accuracy is not required, it will be easy to make an allowance sufficiently near, without formally stating a proportion. In the larger tables, the proportional parts which are to be added to the logarithms, are already prepared, and placed in the margin.

29. *To find the logarithm of a* DECIMAL FRACTION.

The logarithm of a decimal is the same as that of a whole number, excepting the *index*. (Art. 14.) To find then the logarithm of a decimal, take out that of a whole number consisting of the same figures; *observing to make the negative index equal to the distance of the first significant figure of the fraction from the place of units.* (Art. 11.)

The log. of 0.07643, is $\bar{2}.88326$, or 8.88326, (Art. 12.)
of 0.00259, $\bar{3}.41330$, or 7.41330,
of 0.0006273, $\bar{4}.79732$, or 6.79732.

30. *To find the logarithm of a* MIXED *decimal number.*

Find the logarithm, in the same manner as if *all* the figures were integers; and then prefix the index which belongs to the *integral* part, according to art. 8.

The logarithm of 26.34 is 1.42062.

The index here is 1, because 1 is the index of the logarithm of every number greater than 10, and less than 100. (Art. 7.)

The log. of 2.36 is 0.37291, The log. of 364.2 is 2.56134,
of 27.8 1.44404, of 69.42 1.84148.

31. *To find the logarithm of a* VULGAR FRACTION.

From the nature of a vulgar fraction, the numerator may be considered as a *dividend*, and the denominator as a *divisor*; in other words, the value of the fraction is equal to the quotient, of the numerator divided by the denominator. (Alg. 135.) But in logarithms, division is performed by *subtraction*; that is, the *difference* of the logarithms of two numbers, is the logarithm of the *quotient* of those numbers. (Art. 1.) To find then the logarithm of a vulgar fraction, *subtract the logarithm of the denominator* from *that of the numerator*. The difference will be the logarithm of the fraction. Or the logarithm may be found, by first reducing the vulgar fraction to a *decimal*. If the numerator is less than the denominator, the index of the logarithm must be *negative*, because the value of the fraction is less than a unit. (Art. 9.)

Required the logarithm of $\frac{34}{87}$.

The log. of the numerator is 1.53148
of the denominator 1.93952

of the fraction $\bar{1}.59196$, or 9.59196.
The logarithm of $\frac{362}{7854}$ is $\bar{2}.66362$, or 8.66362.
of $\frac{7}{6329}$ $\bar{3}.04376$, or 7.04376.

32. If the logarithm of a *mixed number* is required, reduce it to an improper fraction, and then proceed as before.

The logarithm of $3\frac{7}{9}=\frac{34}{9}$ is 0. 57724.

---

33. *To find the* NATURAL NUMBER *belonging to any logarithm.*

In computing by logarithms, it is necessary, in the first place, to take from the tables the logarithms of the numbers which enter into the calculation; and, on the other hand, at the close of the operation, to find the number belonging to the logarithm obtained in the result. This is evidently done, by *reversing* the methods in the preceding articles.

Where great accuracy is not required, look in the tables for the logarithm which is *nearest* to the given one ; and directly opposite, on the left hand, will be found the *three first* figures, and at the top, over the logarithm, the *fourth* figure, of the number required. This number, by pointing off decimals, or by adding ciphers, if necessary, must be made to correspond with the *index* of the given logarithm, according to arts. 8 and 11.

The natural number belonging

to 3.86493 is 7327, to $\bar{1}.62672$ is 42.24,

to 2.90141 796.9, to $\bar{2}.89115$ 0.07783.

In the last example, the index requires that the first significant figure should be in the *second* place from units, and therefore a cipher must be prefixed. In other instances, it is necessary to annex ciphers on *the right*, so as to make the number of figures exceed the index by 1.

The natural number belonging

to 6.71567 is 5196000, to $\bar{3}.65677$ is 0.004537,

to 4.67062 46840, to $\bar{4}.59802$ 0.0003963.

34. When great accuracy is required, and the given logarithm is not exactly, or very nearly, found in the tables, it will be necessary to reverse the rule in art. 28.

Take from the tables two logarithms, one the next greater, the other the next less than the given logarithm. Find the difference of the two logarithms, and the difference of their natural numbers ; also the difference between the least of the two logarithms, and the given logarithm. Then say,

As the difference of the two logarithms,
To the difference of their numbers ;
So is the difference between the given
logarithm and the least of the other two,
To the proportional part to be added to
the least of the two numbers.

Required the number belonging to the logarithm 2.67325.

Next great log.	2.67330. Its numb.	471.3. Given log.	2.67325.
Next less	2.67321. Its numb.	471.2. Next less	2.67321.
Differences	9	0.1	4

Then 9 : 0.1 :: 4 : 0.044, which is to be added to the number 471.2

The number required is 471.244.

The natural number belonging
to 4.37627 is 23783.45, to 1.73698 is 54.57357,
to 3.69479 4952.08, to $\bar{1}.09214$ is 0.123635.

---

35. *Correction of the tables.* The tables of logarithms have been so carefully and so repeatedly calculated, by the ablest computers, that there is no room left to question their general correctness. They are not, however, exempt from the common imperfections of the press. But an errour of this kind is easily corrected, by comparing the logarithm with any two others to whose *sum* or *difference* it ought to be equal. (Art. 1.)

Thus $48=24\times2=16\times3=12\times4=8\times6$. Therefore, the logarithm of 48 is equal to the *sum* of the logarithms of 24 and 2, of 16 and 3, &c.

And $3=\frac{6}{2}=\frac{12}{4}=\frac{15}{5}=\frac{18}{6}=\frac{21}{7}$, &c. Therefore, the logarithm of 3 is equal to the *difference* of the logarithms of 6 and 2, of 12 and 4, &c.

SECTION III.

METHODS OF CALCULATING BY LOGARITHMS.

ART. 36. THE arithmetical operations for which logarithms were originally contrived, and on which their great utility depends, are chiefly multiplication, division, involution, evolution, and finding the term required in single and compound proportion. The principle on which all these calculations are conducted, is this;

If the logarithms of two numbers be added, the SUM *will be the logarithm of the* PRODUCT *of the numbers;* and

If the logarithm of one number be subtracted from that of another, the DIFFERENCE *will be the logarithm of the* QUOTIENT *of one of the numbers divided by the other.*

In proof of this, we have only to call to mind, that logarithms are the EXPONENTS *of a series of powers and roots.* (Arts. 2, 5.) And it has been shown, that powers and roots are *multiplied* by *adding* their exponents; and *divided,* by *subtracting* their exponents. (Alg. 233, 237, 280, 286.)

MULTIPLICATION BY LOGARITHMS.

37. ADD THE LOGARITHMS OF THE FACTORS: THE SUM WILL BE THE LOGARITHM OF THE PRODUCT.

In making the addition, 1 is to be carried, for every 10, from the decimal part of the logarithm, to the index. (Art. 7.)

	Numbers.	Logarithms.		Numbers.	Logarithms.
Mult.	36.2 (Art. 30.)	1.55871.	Mult.	640	2.80618
Into	7.84	0.89432.	Into	2.316	0.36474
Prod.	283.8	2.45303	Prod.	1482	3.17092

The logarithms of the two factors are taken from the tables. The product is obtained, by finding, in the tables, the natural number belonging to the sum. (Art. 33.)

Mult. 89.24	1.95056	Mult. 134.	2.12710
Into 3.687	0.56667	Into 25.6	1.40824
Prod. 329.	2.51723	Prod. 3430	3.53534

38. When any or all of the indices of the logarithms are *negative*, they are to be added according to the rules for the addition of positive and negative quantities in algebra. But it must be kept in mind, that the decimal part of the logarithm is *positive*. (Art. 10.) Therefore, that which is carried from the decimal part to the index, must be considered positive also.

Mult. 62.84	1.79824	Mult. 0.0294	$\bar{2}.46835$
Into 0.682	$\bar{1}.83378$	Into 0.8372	$\bar{1}.92283$
Prod. 42.86	1.63202	Prod. 0.0246	$\bar{2}.39118$

In each of these examples, +1 is to be carried from the decimal part of the logarithm. This added to −1, the lower index, makes it 0; so that there is nothing to be added to the upper index.

If any perplexity is occasioned, by the addition of positive and negative quantities, it may be avoided, by borrowing 10 to the index. (Art. 12.)

Mult. 62.84	1.79824	Mult. 0.0294	8.46835
Into 0.682	9.83378	Into 0.8372	9.92283
Prod. 42.86	1.63202	Prod. 0.0246	8.39118

Here 10 is added to the negative indices, and afterwards rejected from the index of the sum of the logarithms.

Multiply	26.83	1.42862		1.42862
Into	0.00069	$\bar{4}.83885$	or	6.83885
Product	0.0185	$\bar{2}.26747$		8.26747

Here +1 carried to −4 makes it −3, which added to the upper index +1, gives −2 for the index of the sum.

Multiply	.00845	$\bar{3}$.92686	or	7.92686
Into	1068.	3.02857		3 02857
Product	9.0246	0.95543		0.95543

The product of 0.0362 into 25.38 is 0.9188
of 0.00467 into 348.1 is 1.625
of 0.0861 into 0.00843 is 0.0007258

39. *Any number of factors* may be multiplied together by adding their logarithms. If there are several *positive*, and several *negative* indices, these are to be reduced to one, as in algebra, by taking the difference between the sum of those which are negative, and the sum of those which are positive, increased by what is carried from the decimal part of the logarithms. (Alg. 78.)

Multiply	6832	3.83455		3.83455
Into	0.00863	$\bar{3}$.93601	or	7.93601
And	0.651	$\bar{1}$.81358		9.81358
And	0.0231	$\bar{2}$.36361	or	8.36361
And	62.87	1.79844		1.79844
Prod.	55.74	1.74619		[illegible].74619

Ex. 2. The prod. of 36.4×7.82×68.91×0.3846 is 7544

3. The prod. of 0.00629×2.647×0.082×278.8×0.00063 is 0.0002398.

40. *Negative* quantities are multiplied, by means of logarithms, in the same manner as those which are positive. (Art. 16.) But, after the operation is ended, the proper sign must be applied to the natural number expressing the product, according to the rules for the multiplication of positive and negative quantities in algebra. The negative index of a *log-*

arithm, must not be confounded with the sign which denotes that the *natural number* is negative. That which the index of the logarithm is intended to show, is not whether the natural number is *positive or negative,* but whether it is *greater or less than a unit.* (Art. 16.)

Mult.	+36.42	1.56134	Mult.	—2.681	0.42830
Into	—67.31	1.82808	Into	+37.24	1.57101
Prod.	—2451	3.38942	Prod.	—99.84	1.99931

In these examples, the logarithms are taken from the tables, and added, in the same manner, as if both factors were positive. But after the product is found, the negative sign is prefixed to it, because + is multiplied into —. (Alg. 105.)

Mult.	0.263	$\bar{1}.41996$	Mult.	0.065	$\bar{2}.81291$
Into	0.00894	$\bar{3}.95134$	Into	0.693	$\bar{1}.84073$
Prod.	0.002351	$\bar{3}.37130$	Prod.	0.04504	$\bar{2}$.[illegible]5364

Here, the indices of the logarithms are negative, but the product is positive, because the factors are both positive.

Mult.	—62.59	1.79650	Mult.	—68.3	1.83442
Into	—0.00863	$\bar{3}.93601$	Into	—0.0096	$\bar{3}.98227$
Prod.	+0.5402	$\bar{1}.73251$	Prod.	+0.6557	$\bar{1}.81669$

Division by Logarithms.

41. **From the logarithm of the DIVIDEND, SUBTRACT the logarithm of the DIVISOR; the DIFFERENCE will be the logarithm of the QUOTIENT.** (Art. 36.)

	Numbers.	Logarithms.		Numbers.	Logarithms.
Divide	6238	3.79505	Divide	896.3	2.95245
By	2982	3.47451	By	9.847	0.89330
Quot.	2.092	0.32054	Quot.	91.02	1.95915

42. The *decimal* part of the logarithm may be subtracted as in common arithmetic. But for the *indices*, when either of them is negative, or the lower one is greater than the upper one, it will be necessary to make use of the general rule for subtraction in algebra; that is, to change the signs of the subtrahend, and then proceed as in addition. (Alg. 82.) When 1 is carried from the decimal part, this is to be considered affirmative, and applied to the index, before the sign is changed.

Divide	0.8697	$\bar{1}$.93937	or	9.93937
By	98.65	1.99410		1.99410
Quot.	0.008816	$\bar{3}$.94527		7.94527

In this example, the upper logarithm being less than the lower one, it is necessary to borrow 10, as in other cases of subtraction; and therefore to carry 1 to the lower index, which then becomes +2. This changed to —2, and added to —1 above it, makes the index of the difference of the logarithms —3.

Divide	29.76	1.47363		1.47363
By	6254	3.79616		3.79616
Quot.	0.00476	$\bar{3}$.67747	or	7.67747

Here, 1 carried to the lower index, makes it +4. This changed to —4, and added to 1 above it, gives —3 for the index of the difference of the logarithms.

Divide 6.832	0.83455	Divide 0.00634	$\bar{3}.80209$
By .0362	$\bar{2}.55871$	By 62.18	1.79365
Quot. 188.73	2.27584	Quot. 0.000102	$\bar{4}.00844$

The quotient of 0.0985 divided by 0.007241, is 13.6.

The quotient of 0.0621 divided by 3.68, is 0.01687.

43. To divide *negative* quantities, proceed in the same manner as if they were positive, (Art. 40) and prefix to the quotient, the sign which is required by the rules for division in algebra.

Divide +3642	3.56134	Divide —0.657	$\bar{1}.81757$
By —23.68	1.37438	By +0.0793	$\bar{2}.89927$
Quot. —153.8	2.18696	Quot. —8.285	0.91830

In these examples, the sign of the divisor being different from that of the dividend, the sign of the quotient must be negative. (Alg. 123.)

Divide —0.364	$\bar{1}.56110$	Divide —68.5	1.83569
By —2.56	0.40824	By +0.094	$\bar{2}.97313$
Quot. +0.1422	$\bar{1}.15286$	Quot. —728.7	2.86256

Involution by Logarithms.

44. Involving a quantity is multiplying it into itself. By means of logarithms, multiplication is performed by addition If, then, the logarithm of any quantity be *added to itself*, the

logarithm of a *power* of that quantity will be obtained. But adding a logarithm, or any other quantity, to itself, is *multiplication*. The involution of quantities, by means of logarithms, is therefore performed, by multiplying the logarithms.

Thus the logarithm

of 100 is 2

of 100×100, that is, of $\overline{100}^2$ is $2+2 = 2 \times 2$.

of $100 \times 100 \times 100$, $\overline{100}^3$ is $2+2+2 = 2 \times 3$.

of $100 \times 100 \times 100 \times 100$, $\overline{100}^4$ is $2+2+2+2 = 2 \times 4$.

On the same principle, the logarithm of $\overline{100}^n$ is $2 \times n$.
And the logarithm of x^n, is $(\log. x) \times n$. Hence,

45. To involve a quantity by logarithms. MULTIPLY THE LOGARITHM OF THE QUANTITY, BY THE INDEX OF THE POWER REQUIRED.

The reason of the rule is also evident, from the consideration, that logarithms are the exponents of powers and roots, and a power or root is involved, by *multiplying* its index into the index of the power required. (Alg. 220, 288.)

Ex. 1. What is the cube of 6.296?

Root 6.296,	its log.	0.79906
	Index of the power	3
Power 249.6		2 39718

2. Required the 4th power of 21.32

Root 21.32	log.	1.32879
	Index	4
Power 206614		5.31516

3. Required the 6th power of 1.689

Root 1.689	log.	0.22763
	Index	6
Power 23.215		1.36578

4. Required the 144th power of 1.003

Root 1.003	log.	0.00130
	Index	144
Power 1.539		0.18720

46. It must be observed, as in the case of multiplication, (Art. 38.) that what is carried from the *decimal* part of the logarithm is *positive*, whether the index itself is positive or negative. Or, if 10 be added to a negative index, to render it positive, (Art. 12.) this will be multiplied, as well as the other figures, so that the logarithm of the square, will be 20 too great ; of the cube, 30 too great, &c.

Ex. 1. Required the cube of 0.0649

Root 0.0649	log.	$\bar{2}.81224$	or 8.81224
	Index	3	3
Power 0.0002733		$\bar{4}.43672$	6.43672

2. Required the 4th power of 0.1234

Root 0.1234	log.	$\bar{1}.09132$	or 9.09132
	Index	4	4
Power 0.0002319		$\bar{4}.36528$	6.36528

3 Required the 6th power of 0.9977

Root 0.9977	log.	$\bar{1}.99900$	or 9.99900
	Index	6	6
Power 0.9863		$\bar{1}.99400$	9.99400

4. Required the cube of 0.08762.

Root 0.08762	log.	$\bar{2}.94260$	or 8.94260
	Index	3	3
Power 0.0006727		$\bar{4}.82780$	6.82780

5. The 7th power of 0.9061 is 0.5015.
6. The 5th power of 0.9344 is 0.7123.

Evolution by Logarithms.

47. Evolution is the opposite of involution. Therefore as quantities are involved, by the *multiplication* of logarithms, roots are extracted by the *division* of logarithms; that is,

To extract the root of a quantity by logarithms, DIVIDE THE LOGARITHM OF THE QUANTITY, BY THE NUMBER EXPRESSING THE ROOT REQUIRED.

The reason of the rule is evident also, from the fact, that logarithms are the exponents of powers and roots, and evolution is performed, by dividing the exponent, by the number expressing the root required. (Alg. 257.)

1. Required the square root of 648.3.

	Numbers.	Logarithms.
Power	648.3	2)2.81178
Root	25.46	1.40589

2. Required the cube root of 897.1.

Power	897.1	3)2.95284
Root	9.645	0.98428

In the first of these examples, the logarithm of the given number is divided by 2; in the other, by 3.

3. Required the 10th root of 6948.

Power	6948	10)3.84186
Root	2.422	0.38418

4. Required the 100th root of 983.

Power	983	100)2.99255
Root	1.071	0.02992

The division is performed here, as in other cases of decimals, by removing the decimal point to the left.

5. What is the ten thousandth root of 49680000 ?

Power	49680000	10000)7.69618
Root	1.00179	0.00077

We have, here, an example of the great rapidity with which arithmetical operations are performed by logarithms.

48. If the index of the logarithm is *negative*, and is *not divisible* by the given divisor, without a remainder, a difficulty will occur, unless the index be altered.

Suppose the cube root of 0.0000892 is required. The logarithm of this is $\overline{5}.95036$. If we divide the index by 3, the quotient will be—1, with—2 remainder. This remainder, if it were positive, might, as in other cases of division, be prefixed to the next figure. But the remainder is *negative*, while the decimal part of the logarithm is positive ; so that, when the former is prefixed to the latter, it will make neither +2.9 nor—2.9, but—2+.9. This embarrassing intermixture of positives and negatives may be avoided, by adding to the index another negative number, to make it exactly divisible by the divisor. Thus, if to the index—5 there be added—1, the sum—6 will be divisible by 3. But this addition of a negative number must be *compensated*, by the addition of an equal positive number, which may be prefixed to the decimal part of the logarithm. The division may then be continued, without difficulty, through the whole.

Thus, if the logarithm $\overline{5}.95036$ be altered to $\overline{6}+1.95036$ it may be divided by 3, and the quotient will be $\overline{2}.65012$. We have then this rule,

49. *Add to the index*, if necessary, *such a negative number as will make it exactly divisible by the divisor, and prefix an equal positive number to the decimal part of the logarithm.*

1. Required the 5th root of 0.009642

Power	0.009642	log. $\overline{3}.98417$
		or $\overline{5}+2.98417$
Root	0.3952	$\overline{1}.59683$

2. Required the 7th root of 0 0004935

Power	0.0004935	log. $\overline{4}.69329$
		or 7)$\overline{7}+3.69329$
Root	0.337	$\overline{1}.52761$

50. If, for the sake of performing the division conveniently, the negative index be rendered *positive*, it will be expedient to borrow as many tens, as there are units in the number denoting the root.

What is the fourth root of 0.03698 ?

Power	0.03698	$4)\overline{2}.56797$	or $4)38.56797$
Root	0.4385	$\overline{1}.64199$	9.64199

Here the index, by borrowing, is made 40 too great, that is, +38 instead of —2. When, therefore, it is divided by 4, it is still 10 too great, +9 instead of —1.

What is the 5th root of 0.008926 ?

Power	0.008926	$5)\overline{3}.95066$	or $5)47.95066$
Root	0.38916	$\overline{1}.59013$	9.59013

51. A *power of a root* may be found by first *multiplying* the logarithm of the given quantity into the index of the power, (Art. 45.) and then *dividing* the product by the number expressing the root. (Art. 47.)

1. What is the value of $(53)^{\frac{6}{7}}$, that is, the 6th power of the 7th root of 53 ?

Given number 53	log. 1.72428
Multiplying by	6
Dividing by	7)10.34568
Power required 30.06	1.47795

2. What is the 8th power of the 9th root of 654 ?

Proportion by Logarithms.

52. In a proportion, when three terms are given, the fourth is found, in common arithmetic, by multiplying together the second and third, and dividing by the first. But when logarithms are used, *addition* takes the place of multiplication, and *subtraction*, of division.

To find then, by logarithms, the fourth term in a proportion, ADD THE LOGARITHMS OF THE SECOND AND THIRD TERMS, AND *from the sum* SUBTRACT THE LOGARITHM OF

THE FIRST TERM. The remainder will be the logarithm of the term required.

Ex. 1. Find a fourth proportional to 7964, 378, and 27960.

	Numbers.	Logarithms.
Second term	378	2.57749
Third term	27960	4.44654
		7.02403
First term	7964	3.90113
Fourth term	1327	3.12290

2. Find a 4th proportional to 768, 381 and 9780.

Second term	381	2.58092
Third term	9780	3.99034
		6.57126
First term	768	2.88536
Fourth term	4852	3.68590

Arithmetical Complement.

53. When one number is to be subtracted from another, it is often convenient, first to subtract it from 10, then to *add the difference* to the other number, and afterwards to reject the 10.

Thus, instead of $a-b$, we may put $10-b+a-10$.

In the first of these expressions, b is subtracted from a. In the other, b is subtracted from 10, the difference is added to a, and 10 is afterwards taken from the sum. The two expressions are equivalent, because they consist of the same terms, with the addition, in one of them, of $10-10=0$. The alteration is, in fact, nothing more than borrowing 10, for the sake of convenience, and then rejecting it in the result.

Instead of 10, we may borrow, as occasion requires, 100, 1000, &c.

Thus $a-b=100-b+a-100=1000-b+a-1000$, &c.

54. The DIFFERENCE *between a given number and* 10, *or* 100, *or* 1000, *&c. is called the* ARITHMETICAL COMPLEMENT *of that number.*

The arithmetical complement of a number consisting of *one* integral figure, either with or without decimals, is found, by subtracting the number from 10. If there are *two* integral figures, they are subtracted from 100 ; if *three*, from 1000, &c.

Thus the arithmetical compl't of 3.46 is 10—3.46=6.54
of 34.6 is 100—34.6=65.4
of 346. is 1000—346.=654. &c.

According to the rule for subtraction in arithmetic, any number is subtracted from 10, 100, 1000, &c. by beginning on the right hand, and taking each figure from 10, after *increasing* all except the first, by *carrying* 1.

Thus, if from	10.00000
We subtract	7.63125

The difference, or arith'l comp't is 2.36875, which is obtained, by taking 5 from 10, 3 from 10, 2 from 10, 4 from 10, 7 from 10, and 8 from 10. But, instead of taking each figure, *increased by* 1, from 10 ; we may take it *without being increased*, from 9.

Thus 2 from 9 is the same as 3 from 10,
3 from 9, the same as 4 from 10, &c. Hence,

55. *To obtain the* ARITHMETICAL COMPLEMENT *of a number, subtract the right hand significant figure from* 10, *and each of the other figures from* 9. If, however, there are *ciphers* on the right hand of all the significant figures, they are to be set down without alteration.

In taking the arithmetical complement of a logarithm, if the index is *negative*, it must be *added* to 9 ; for adding a negative quantity is the same as subtracting a positive one. (Alg. 81.) The difference between—3 and+9, is not 6, but 12.

The arithmetical complement

of 6.24897 is	3.75103	of $\bar{2}$.70649 is	11.29351
of 2.98643	7.01357	of 3.64200	6.35800
of 0.62430	9.37570	of 9.35001	0.64999

56. The principal use of the arithmetical complement, is in working proportions by logarithms; where some of the terms are to be *added*, and one or more to be *subtracted*. In the Rule of Three or simple proportion, two terms are to be added, and from the sum, the first term is to be subtracted. But if, instead of the logarithm of the first term, we substitute its arithmetical complement, this may be *added* to the sum of the other two, or more simply, all three may be added together, by one operation. After the index is diminished by 10, the result will be the same as by the common method. For subtracting a number is the same, as adding its arithmetical complement, and then rejecting 10, 100, or 1000, from the sum. (Art. 53.)

It will generally be expedient, to place the terms in the same order, in which they are arranged in the statement of the proportion.

1.	As 6273 *a. c.*	6.20252	2.	As 253 *a. c*	7.59688
	Is to 769.4	2.88615		Is to 672.5	2.82769
	So is 37.61	1.57530		So is 497	2.69636
	To 4.613	0.66397		To 1321.1	3.12093
3.	As 46.34 *a. c.*	8.33404	4	As 9.85 *a. c.*	9.00656
	Is to 892.1	2 95041		Is to 643	2.80821
	So is 7.633	0.88298		So is 76.3	1.88252
	To 147	2.16743		To 4981	3.69729

Compound Proportion

57. In compound, as in single proportion, the term required may be found by logarithms, if we substitute addition for multiplication, and subtraction for division.

Ex. 1. If the interest of $365, for 3 years and 9 months, be $82.13; what will be the interest of $8940, for 2 years and 6 months?

In common arithmetic, the statement of the question is made in this manner,

* See Webber's Arithmetic.

$\left.\begin{matrix}365 \text{ dollars} \\ 3.75 \text{ years}\end{matrix}\right\}$: 82.13 dollars : : $\left\{\begin{matrix}8940 \text{ dollars} \\ 2.5 \text{ years}\end{matrix}\right\}$:

And the method of calculation is, to *divide* the *product* of the third, fourth, and fifth terms, by the *product* of the two first.* This, if logarithms are used, will be to *subtract* the *sum* of the logarithms of the two first terms, from the *sum* of the logarithms of the other three.

Two first terms	365	log.	2.56229
	3.75		0.57403
Sum of the logarithms			3.13632
Third term	82.13		1.91450
Fourth and fifth terms	8940		3.95134
	2.5		0.39794
Sum of the logs. of the 3d, 4th, and 5th			6.26378
Do.	1st and 2d		3.13632
Term required	1341		3.12746

58. The calculation will be more simple, if, instead of *subtracting* the logarithms of the two first terms, we *add* their *arithmetical complements*. But it must be observed, that *each* arithmetical complement increases the index of the logarithm by 10. If the arithmetical complement be introduced into *two* of the terms, the index of the sum of the logarithms will be 20 too great; if it be in *three* terms, the index will be 30 too great, &c.

Two first terms	365	a. c.	7.43771
	3.75	a. c.	9.42597
Third term	82 13		1.91450
Fourth and fifth terms	8940		3.95134
	2.5		0.39794
Term required	1341		23.12746

The result is the same as before, except that the index of the logarithm is 20 too great.

Ex. 2. If the wages of 53 men for 42 days be 2200 dollars; what will be the wages of 87 men for 34 days?

53 men } : 2200 : : { 87 men
42 days } { 34 days

Two first terms	{ 53	*a. c.*	8.27572
	{ 42	*a. c.*	8.37675
Third term	2200		3.34242
Fourth and fifth terms	{ 87		1.93952
	{ 34		1.53148
Term required 2923.5			3.46589

59. In the same manner, if the product of *any number* of quantities, is to be divided, by the product of several others; we may add together the logarithms of the quantities to be divided, and the arithmetical complements of the logarithms of the divisors.

Ex. If 29.67×346.2 be divided by $69.24 \times 7.862 \times 497$; what will be the quotient?

Numbers to be divided	{ 29.67		1.47232
	{ 346.2		2.53933
Divisors	{ 69.24	*a. c.*	8.15964
	{ 7.862	*a. c.*	9.10447
	{ 497	*a. c.*	7.30364
Quotient	0.03797		8.5794

In this way, the calculations in *Conjoined Proportion* may be expeditiously performed.

Compound Interest.

60. In calculating compound interest, the amount for the first year, is made the principal for the second year; the amount for the second year, the principal for the third year, &c. Now the amount at the end of each year, must be proportioned to the principal at the beginning of the year. If

the principal for the first year be 1 dollar, and if the amount of 1 dollar for 1 year $=a$; then, (Alg. 377.)

$1 : a ::$ $\begin{cases} a : a^3 = \text{the amount for the 2d year, or the principal for the 3d;} \\ a^2 : a^3 = \text{the amount for the 3d year, or the principal for the 4th;} \\ a^3 : a^4 = \text{the amount for the 4th year, or the principal for the 5th.} \end{cases}$

That is, the amount of 1 dollar for any number of years is obtained, by finding the amount for 1 year, and involving this to a power whose index is equal to the number of years. And the amount of any other principal, for the given time, is found, by multiplying the amount of 1 dollar, into the number of dollars, or the fractional part of a dollar.

If logarithms are used, the *multiplication* required here may be performed by *addition;* and the *involution,* by *multiplication.* (Art. 45.) Hence,

61. To calculate Compound Interest, *Find the amount of 1 dollar for 1 year; multiply its logarithm by the number of years; and to the product, add the logarithm of the principal.* The sum will be the logarithm of the *amount* for the given time. From the amount subtract the principal, and the remainder will be the *interest.*

If the interest becomes due *half yearly* or *quarterly;* find the amount of one dollar, for the half year or quarter, and multiply the logarithm, by the number of half years or quarters in the given time.

If P$=$the principal,
$a$$=$the amount of 1 dollar for 1 year,
$n$$=$any number of years, and
A$=$the amount of the given principal for n years; then

$$A = a^n \times P.$$

Taking the logarithms of both sides of the equation, and reducing it, so as to give the value of each of the four quantities, in terms of the others, we have

1. $\text{Log. } A = n \times \text{log. } a + \text{log. } P.$
2. $\text{Log. } P = \text{log } A - n \times \text{log. } a.$
3. $\text{Log. } a = \dfrac{\text{log. } A - \text{log. } P.}{n}$
4. $n = \dfrac{\text{log. } A - \text{log. } P.}{\text{log. } a.}$

Any three of these quantities being given, the fourth may be found.

Ex. 1. What is the amount of 20 dollars, at 6 per cent compound interest, for 100 years?

Amount of 1 dollar for 1 year	1.06	log. 0.0253059
Multiplying by		100
		2.53059
Given principal	20	1.30103
Amount required	$6786	3.83162

2. What is the amount of 1 cent, at 6 per cent compound interest, in 500 years?

Amount of 1 dollar for 1 year	1.06	log. 0.0253059
Multiplying by		500
		12.65295
Given principal	0.01	−2.00000
Amount	$44,973,000,000	10.65295

More exact answers may be obtained, by using logarithms of a greater number of decimal places.

3. What is the amount of 1000 dollars, at 6 per cent compound interest, for 10 years? Ans. 1790.80.

4. What principal, at 4 per cent. interest, will amount to 1643 dollars in 21 years? Ans. 721.

5. What principal, at 6 per cent. will amount to 202 dollars in 4 years? Ans. 160.

6. At what rate of interest, will 400 dollars amount to $569\frac{1}{3}$, in 9 years? Ans. 4 per cent.

7. In how many years will 500 dollars amount to 900 at 5 per cent. compound interest? Ans. 12 years.

8. In what time will 10,000 dollars amount to 16,288, at 5 per cent. compound interest? Ans. 10 years.

9. At what rate of interest, will 11,106 dollars amount to 20,000 in 15 years? Ans. 4 per cent.

10. What principal, at 6 per cent. compound interest, will amount to 3188 dollars in 8 years? Ans. $2000.

11. What will be the amount of 1200 dollars, at 6 per cent. compound interest, in 10 years, if the interest is converted into principal every *half-year?*

Ans. 2167.3 dollars.

12. In what time will a sum of money double, at 6 per cent. compound interest? Ans. 11.9 years.

13. What is the amount of 5000 dollars, at 6 per cent. compound interest, for $28\frac{1}{4}$ years? Ans. 25,942 dollars.

Increase of Population.

61. *b.* The natural increase of population in a country, may be calculated in the same manner as compound interest; on the supposition, that the yearly rate of increase is regularly proportioned to the actual number of inhabitants. From the population at the beginning of the year, the *rate* of increase being given, may be computed the whole increase during the year. This *added* to the number at the beginning, will give the amount, on which the increase of the *second* year is to be calculated, in the same manner as the first year's interest on a sum of money, added to the sum it-

self, gives the amount on which the interest for the second year is to be calculated.

If P=the population at the beginning of the year,
a=1+the fraction which expresses the rate of increase,
n=any number of years ; and
A=the amount of the population at the end of n years;
then, as in the preceding article,

$$A=a^n \times P, \text{ and}$$

1. $\text{Log. } A=n \times \text{log. } a+\text{log. } P.$
2. $\text{Log. } P=\text{log. } A-n \times \text{log. } a.$
3. $\text{Log. } a=\dfrac{\text{log. } A-\text{log. } P.}{n}$
4. $n=\dfrac{\text{log. } A-\text{log. } P.}{\text{log. } a}$

Ex. 1. The population of the United States in 1820 was 9,625,000. Supposing the yearly rate of increase to be $\frac{1}{34}$th part of the whole, what will be the population in 1830?

Here P=9,625,000. $n=10$. $a=1+\frac{1}{34}=\frac{35}{34}$.

And log. A=10×log. $\frac{35}{34}$+log. (9,625,000,)
Therefore, A=12,860,000, the population in 1830.

2. If the number of inhabitants in a country be five millions, at the beginning of a century; and if the yearly rate of increase be $\frac{1}{30}$; what will be the number, at the end of the century? Ans. 132,730,000.

3. If the population of a country, at the end of a century, is found to be 45,860,000; and if the yearly rate of increase has been $\frac{1}{120}$; what was the population, at the commencement of the century? Ans. 20 millions.

4. The population of the United States in 1810 was 7,240,000; in 1820, 9,625,000. What was the annual rate of increase between these two periods, supposing the increase each year to be proportioned to the population at the beginning of the year?

Here $\log. a = \frac{\log. 9,625,000 - \log. 7.240,000}{10}$

Therefore, $a=1.029$; and $\frac{29}{1000}$, or 2.9 per cent. is the rate of increase.

5. In how many years, will the population of a country advance from two millions to five millions; supposing the yearly rate of increase to be $\frac{7}{360}$? Ans. $47\frac{1}{2}$ years.

6. If the population of a country, at a given time, be seven millions; and if the yearly rate of increase be $\frac{1}{25}$th; what will be the population at the end of 35 years?

7. The population of the United States in 1800 was 5,306,000. What was it in 1780, supposing the yearly rate of increase to be $\frac{1}{28}$?

8. In what time, will the population of a country advance from four millions to seven millions, if the ratio of increase be $\frac{3}{100}$?

9. What must be the rate of increase, that the population of a place may change from nine thousand to fifteen thousand, in 12 years?

If the population of a country is not affected by immigration or emigration, the rate of increase will be equal to the difference between the ratio of the *births*, and the ratio of the *deaths*, when compared with the whole population.

Ex. 10. If the population of a country, at any given time, be ten millions; and the ratio of the annual number of births to the whole population be $\frac{1}{24}$, and the ratio of deaths $\frac{1}{45}$, what will be the number of inhabitants, at the end of 60 years?

Here the yearly rate of increase $=\frac{1}{24}-\frac{1}{45}=\frac{7}{360}$.
And the population, at the end of 60 years $=31,750,000$.

The rate of increase or decrease from *immigration* or *emigration*, will be equal to the difference between the ratio of immigration and the ratio of emigration; and if this difference be added to, or subtracted from, the difference between the ratio of the births and that of the deaths, the whole rate of increase will be obtained.

Ex. 11. If in a country, the ratio of births be $\frac{1}{30}$,
the ratio of deaths $\frac{1}{40}$,
the ratio of immigration $\frac{1}{50}$,
the ratio of emigration $\frac{1}{60}$;
and if the population this year be 10 millions, what will it be 20 years hence?

The rate of the natural increase $=\frac{1}{30}-\frac{1}{40}=\frac{1}{120}$;
That of increase from immigration$=\frac{1}{50}-\frac{1}{60}=\frac{1}{300}$;
The sum of the two is $\frac{7}{600}$;
And the population at the end of 20 years, is 12,611,000.

12. If the ratio of the births be $\frac{1}{20}$,
of the deaths, $\frac{1}{30}$,
of immigration, $\frac{1}{40}$,
of emigration, $\frac{1}{50}$,
in what time will three millions increase to four and a half millions?

If the period in which the population will *double* be given; the numbers for several successive periods, will evidently be in a geometrical progression, of which the ratio is 2; and as the number of periods will be one less than the number of terms;

If P=the first term,
A=the last term,
n=the number of periods;
Then will $A=P\times2^n$, (Alg. 439.)
Or log. A=log. P.=log. $P+n\times$log. 2.

Ex. 1. If the descendants of a single pair double once in 25 years, what will be their number, at the end of one thousand years?

The number of periods here is 40.
And $A=2\times2^{40}=2,199,200,000,000$.

2. If the descendants of Noah, beginning with his three sons and their wives, doubled once in 20 years for 300 years; what was their number, at the end of this time?

Ans. 196,608.

3. The population of the United States in 1820 being 9,625,000; what must it be in the year 2020, supposing it to double once in 25 years? Ans. 2,464,000,000.

4. Supposing the descendants of the first human pair to double once in 50 years, for 1650 years, to the time of the deluge, what was the population of the world, at that time?

Exponential Equations.

62. An Exponential equation is one in which the letter expressing the unknown quantity is an *exponent*.

Thus $a^x=b$, and $x^x=bc$, are exponential equations. These are most easily solved by logarithms. As the two members of an equation are equal, their logarithms must also be equal. If the logarithm of each side be taken, the equation may then be reduced, by the rules given in algebra.

Ex. What is the value of x in the equation $3^x=243$?

Taking the logarithms of both sides log. (3^x)=log. 243

But the logarithm of a *power* is equal to the logarithm of the root, multiplied into the index of the power. (Art. 45.)

Therefore (log.3)$\times x$=log. 243; and dividing by log. 3,

$$x=\frac{\text{log. } 243}{\text{log. } 3}=\frac{2.38561}{0.47712}=5. \quad \text{So that } 3^5=243.$$

63. The preceding is an exponential equation of the simplest form. Other cases, after the logarithm of each side is taken, may be solved by *Trial and Errour*, in the same manner as affected equations. (Alg. 503.) For this purpose, make two suppositions of the value of the unknown quantity, and find their errours; then say,

As the difference of the errours, to the difference of the assumed numbers ;
So is the least errour, to the correction required in the corresponding assumed number.

Ex. 1. Find the value of x in the equation $x^x=256$
Taking the logarithms of both sides $(\log. x)\times x=\log. 256$
Let x be supposed equal to 3.5, or 3 6.

By the first supposition.	By the second supposition.
$x=3.5$, and $\log. x=0.54407$	$x=3.6$, and $\log. x=0.55630$
Multiplying by 3.5	Multiplying by 3.6
$(\log. x)\times x=1.90424$	$(\log. x)\times x=2.00268$
$\log. 256=2.40824$	$\log. 256=2.40824$
Errour -0.50400	Errour -0.40556
Difference of the errours	0.09844

Then 0.09844 : 0.1 :: 0.40556 : 0.4119, the correction. This added to 3.6, the second assumed number, makes the value of $x=4.0119$.

To correct this farther, suppose $x=4.011$, or 4.012.

By the first supposition.	By the second supposition.
$x=4.011$, and $\log. x=0.60325$	$x=4.012$, and $\log. x=0.60336$
Multiplying by 4.011	Multiplying by 4.012
$(\log. x)\times x=2.41963$	$(\log. x)\times x=2.42068$
$\log. 256=2.40824$	$\log. 256=2.40824$
Errour $+0.01139$	Errour $+0.01244$
Difference of the errours	0.00105

Then 0.00105 : 0.001 :: 0.01139 : 0.011 very nearly.

Subtracting this correction from the first assumed number 4.011, we have the value of $x=4$, which satisfies the conditions of the proposed equation ; for $4^4=256$.

2. Reduce the equation $4x^x=100x^3$. Ans. $x=5$.

3. Reduce the equation $x^x=9x$.

64. The exponent of a power may be itself a power, as in the equation

$$a^{m^x}=b.$$

where x is the exponent of the power m^x, which is the exponent of the power a^{m^x}.

Ex. 4. Find the value of x, in the equation $9^{3^x}=1000$.
$3^x\times(\text{log. }9)=\text{log. }1000$. Therefore $3^x=\frac{\text{log. }1000}{\text{log. }9}=3.14$
Then as $3 \;=3.14$. $x\,(\text{log. }3)=\text{log. }(3.\ 14.)$

Therefore $x=\frac{\text{log. }(3.\ 14)}{\text{log. }3.}=\frac{.4969296}{.4771213}=1.04$.

In cases like this, where the factors, divisors, &c. are logarithms, the calculation may be facilitated, by taking the *logarithms of the logarithms.* Thus the value of the fraction $\frac{.4969296}{.4771213}$ is most easily found, by subtracting the logarithm of the logarithm which constitutes the denominator, from the logarithm of that which forms the numerator.

Find the value of x, in the equation $\frac{ba^x+d}{c}=m$.

Ans. $x=\frac{\text{log. }(cm-d)-\text{log. }b.}{\text{log. }a.}$

SECTION IV.

DIFFERENT SYSTEMS OF LOGARITHMS, AND COMPUTATION OF THE TABLES.

65. FOR the common purposes of numerical computation, Briggs' system of logarithms has a decided advantage over every other. But the theory of logarithms is an important instrument of investigation, in the higher departments of mathematical science. In its numerous applications, there is frequent occasion to compare the common system with others ; especially with that which was adopted, by the celebrated inventor of logarithms, Lord Napier. In conducting these investigations, it is often expedient to express the logarithm of a number, in the form of a *series.*

If $a^x=N$, then x is the logarithm of N. (Art. 2.)

To find the value of x, in a series, let the quantities a and N be put into the form of a binomial, by making $a=1+b$, and $N=1+n$. Then $(1+b)^x=(1+n)$, and extracting the root y of both sides, we have

$$(1+b)^{\frac{x}{y}}=(1+n)^{\frac{1}{y}}$$

By the binomial theorem

$$(1+b)^{\frac{x}{y}}=1+\frac{x}{y}(b)+\frac{x}{y}(\frac{x}{y}-1)\left(\frac{b^2}{2}\right)+\frac{x}{y}(\frac{x}{y}-1)(\frac{x}{y}-2)\left(\frac{b^3}{2.3}\right)$$
$$+\ \&c.$$

$$(1+n)^{\frac{1}{y}}=1+\frac{1}{y}(n)+\frac{1}{y}(\frac{1}{y}-1)\left(\frac{n^2}{2}\right)+\frac{1}{y}(\frac{1}{y}-1)(\frac{1}{y}-2)\left(\frac{n^3}{2.3}\right)$$
$$+\ \&c.$$

As these expressions will be the same, whatever be the value of y, let y be taken indefinitely great ; then $\frac{x}{y}$ and $\frac{1}{y}$ being indefinitely small, in comparison with the numbers -1, -2, &c. with which they are connected, may be cancelled (Alg. 456.) leaving

$$1+\frac{x}{y}b-\frac{x}{y}\left(\frac{b^2}{2}\right)+\frac{x}{y}\left(\frac{b^3}{3}\right)-\frac{x}{y}\left(\frac{b^4}{4}\right)\&c.=1+\frac{1}{y}n-\frac{1}{y}\left(\frac{n^2}{2}\right)$$
$$+\frac{1}{y}\left(\frac{n^3}{3}\right)-\frac{1}{y}\left(\frac{n^4}{4}\right)\&c.$$

Rejecting 1 from each side of the equation, multiplying by y, and dividing by the compound factor into which x is multiplied, we have

$$x=\text{Log. N}=\frac{n-\frac{1}{2}n^2+\frac{1}{3}n^3-\frac{1}{4}n^4+\text{\&c.}}{b-\frac{1}{2}b^2+\frac{1}{3}b^3-\frac{1}{4}b^4+\text{\&c.}} \quad \text{A}$$

Or, as $n = \text{N}-1$, and $b=a-1$,

$$\text{Log. N}=\frac{(\text{N}-1)-\frac{1}{2}(\text{N}-1)^2+\frac{1}{3}(\text{N}-1)^3-\frac{1}{4}(\text{N}-1)^4+\text{\&c.}}{(a-1)-\frac{1}{2}(a-1)^2+\frac{1}{3}(a-1)^3-\frac{1}{4}(a-1)^4+\text{\&c.}}$$

Which is a general expression, for the logarithm of any number N, in any system in which the base is a. The numerator is expressed in terms of N only; and the denominator in terms of a only: So that, whatever be the number, the denominator will remain the same, unless the base is changed. The reciprocal of this constant denominator, viz.

$$\frac{1}{(a-1)-\frac{1}{2}(a-1)^2+\frac{1}{3}(a-1)^3-\frac{1}{4}(a-1)^4+\text{\&c.}}$$

is called the *Modulus* of the system of which a is the base. If this be denoted by M, then

$$\text{Log. N}=\text{M}\times\left((\text{N}-1)-\tfrac{1}{2}(\text{N}-1)^2+\tfrac{1}{3}(\text{N}-1)^3-\tfrac{1}{4}(\text{N}-1)^4+\text{\&c.}\right)$$

66. The foundation of Napier's system of Logarithms is laid, by making the modulus equal to *unity*. From this condition the *base* is determined. Taking the equation marked A. and making the denominator equal to 1, we have

$$x=n-\tfrac{1}{2}n^2+\tfrac{1}{3}n^3-\tfrac{1}{4}n^4+\tfrac{1}{5}n^5-\text{\&c.}$$

By reverting this equation*

$$n=x+\frac{x^2}{2}+\frac{x^3}{2.3}+\frac{x^4}{2.3.4}+\frac{x^5}{2.3.4.5}+\text{\&c.}$$

Or, as by the notation, $n+1=\text{N}=a^x$,

$$a^x=1+x+\frac{x^2}{2}+\frac{x^3}{2.3}+\frac{x^4}{2.3.4}+\frac{x^5}{2.3.4.5}+\text{\&c.}$$

If then x be taken equal to 1, we have

$$a=1+1+\frac{1}{2.3}+\frac{1}{2.3.4}+\frac{1}{2.3.4.5}\text{\&c.}$$

Adding the first fifteen terms, we have

2.7182818284

Which is the base of Napier's system, correct to ten places of decimals.

* See note D.

Napier's logarithms are also called *hyperbolic* logarithms, from certain relations which they have to the spaces between the asymptotes and the curve of an hyperbola; although these relations are not, in fact, peculiar to Napier's system.

67. The logarithms of *different* systems are compared with each other, by means of the modulus. As in the series

$$\frac{(N-1)-\frac{1}{2}(N-1)^2+\frac{1}{3}(N-1)^3-\frac{1}{4}(N-1)^4+\&c.}{(a-1)-\frac{1}{2}(a-1)^2+\frac{1}{3}(a-1)^3-\frac{1}{4}(a-1)^4+\&c.}$$

which expresses the logarithm of N, the *denominator* only is affected by a change of the base a; and as the value of fractions, whose numerators are given, are reciprocally as their denominators. (Alg. 360. cor. 2.)

The logarithm of a given number, in one system,
Is to the logarithm of the same number in another system;
As the modulus of one system,
To the modulus of the other.

So that, if the modulus of each of the systems be given, and the logarithm of any number be calculated in one of the systems; the logarithm of the same number in the other system may be calculated by a simple proportion. Thus if M be the modulus in Brigg's system, and M′ the modulus in Napier's; l the logarithm of a number in the former, and l' the logarithm of the same number in the latter; then,

$$M : M' :: l : l',$$
Or, as $M'=1$
$$M : 1 :: l : l'$$

Therefore, $l=l'\times M$, that is, the common logarithm of a number, is equal to Napier's logarithm of the same, multiplied into the modulus of the common system.

To find this modulus, let a be the base of Brigg's system, and e the base of Napier's; and let $l.a$ denote the common logarithm of a, and $l'.a$ denote Napier's logarithm of a.

Then $M : 1 :: l.a : l'a$. Therefore $M=\frac{l.a}{l'.a}$

But in the common system, $a=10$, and $l.a=1$.

So that, $M=\frac{1}{l'.10}$, that is, the modulus of Brigg's system, is equal to 1 divided by Napier's logarithm of 10.

Again $M : 1 :: l.e : l'.e$

But as e denotes Napier's base, $l'.e=1$.

So that $M=l.e$, that is, the modulus of the common system, is equal to the common logarithm of Napier's base.

Therefore, either of the expressions, $l.e$, or $\frac{1}{l'.a}$, may be used, to convert the logarithms of one of the systems into those of the other.

The ratio of the logarithms of two numbers to each other, is the same in one system as in another. If N and n be the two numbers;

Then, $\quad l.N : l'.N :: M : M'$

$\quad l.n :: l'.n :: M : M'$

Therefore, $\quad l.N : l.n :: l'.N : l'.n$

Computation of Logarithms.

68. The logarithms of most numbers can be calculated by approximation only, by finding the sum of a sufficient number of terms, in the series which expresses the value of the logarithms. According to art. 65.

$$\text{Log. } N=M\times((N-1)-\tfrac{1}{2}(N-1)^2+\tfrac{1}{3}(N-1)^3, \text{ \&c.})$$

Or, putting as before, $n=N-1$,

$$\text{Log. } (1+n)=M(n-\tfrac{1}{2}n^2+\tfrac{1}{3}n^3-\tfrac{1}{4}n^4+\tfrac{1}{5}n^5-\text{\&c.})$$

But this series will not converge, when n is a whole number, greater than unity. To convert it into another which will converge, let $(1-n)$ be expanded in the same manner as $(1+n)$, (Art. 65.) The formula will be the same, except that the odd powers of n will be negative instead of positive. We shall then have,

$$\text{Log. } (1+n)=M(n-\tfrac{1}{2}n^2+\tfrac{1}{3}n^3-\tfrac{1}{4}n^4+\tfrac{1}{5}n^5-\text{\&c.})$$
$$\text{Log. } (1-n)=M(-n-\tfrac{1}{2}n^2+\tfrac{1}{3}n^3-\tfrac{1}{4}n^4+\tfrac{1}{5}n^5-\text{\&c.})$$

Subtracting the one from the other the even powers of n disappear, and we have

$$M(2n+\tfrac{2}{3}n^3+\tfrac{2}{5}n^5+\tfrac{2}{7}n^7+\text{\&c.})$$

or

$$2M(n+\tfrac{1}{3}n^3+\tfrac{1}{5}n^5+\tfrac{1}{7}n^7+\text{\&c.})$$

But this, which is the *difference* of the logarithms of $(1+n)$ and $(1-n)$ is the logarithm of the *quotient* of the one divided by the other. (Art. 36.)

That is, $\text{Log.}\ \frac{1+n}{1-n}=2\text{M}\ (n+\frac{1}{3}n^3+\frac{1}{5}n^5+\frac{1}{7}n^7+\&\text{c.})$

Now put $n=\frac{1}{z-1}$

$$\text{Then,}\quad \frac{1+n}{1-n}=\frac{1+\frac{1}{z-1}}{1-\frac{1}{z-1}}=\frac{\frac{z}{z-1}}{\frac{z-2}{z-1}}=\frac{z}{z-2}$$

Therefore, substituting $\frac{z}{z-2}$ for $\frac{1+n}{1-n}$, and $\frac{1}{z-1}$ for n, we have

$$\text{Log.}\ \frac{z}{z-2}=2\text{M}\left(\frac{1}{(z-1)}+\frac{1}{3(z-1)^3}+\frac{1}{5(z-1)^5}+\&\text{c.}\right)$$

Or, (Art. 36,)

$$\text{Log.}\ z-\text{log.}\ (z-2)=2\text{M}\left(\frac{1}{(z-1)}+\frac{1}{3(z-1)^3}+\frac{1}{5(z-1)^5}+\&\text{c.}\right)$$

Therefore,

$$\text{Log.}\ z=\text{log.}\ (z-2)+2\text{M}\left(\frac{1}{(z-1)}+\frac{1}{3(z-1)^3}+\frac{1}{5(z-1)^5}+\&\text{c.}\right)$$

This series may be applied to the computation of any number greater than 2.

To find the logarithm of 2, let. $z=4$,

Then $(z-1)=3$, and the preceding series, after transposing log. $(z-2)$ becomes

$$\text{Log.}\ 4-\text{log.}\ 2=2\text{M}\left(\frac{1}{3}+\frac{1}{3.3^3}+\frac{1}{5.3^5}+\frac{1}{7.3^7}\ \&\text{c.}\right)$$

But as 4 is the square of 2 log. $4=2\,\text{log.}\ 2$. (Alg. 44.) So that log. $4-\text{log.}\ 2=\text{log.}\ 2$. We have then

$$\text{Log. } 2=2M\left(\frac{1}{3}+\frac{1}{3.3^3}+\frac{1}{5.3^5}+\frac{1}{7.3^7}+\frac{1}{9.3^9}+\text{\&c.}\right)$$

When the logarithms of the *prime* numbers are computed, the logarithms of all other numbers may be found, by simply adding the logarithms of the factors of which the numbers are composed. (Art. 36.)

69. In Napier's system, where M=1, the logarithms may be computed, as in the following table.

Napier's or Hyperbolic Logarithms.

$\text{Log. } 2=2\left(\frac{1}{3}+\frac{1}{3.3^3}+\frac{1}{5.3^5}+\frac{1}{7.3^7}\text{\&c.}\right)$	=0.693147
$\text{Log. } 3=2\left(\frac{1}{2}+\frac{1}{3.2^3}+\frac{1}{5.2^5}+\frac{1}{7.2^7}\text{ \&c.}\right)$	=1.098612
Log. 4=2 log. 2	1.386294
$\text{Log. } 5=\text{log. } 3+2\left(\frac{1}{4}+\frac{1}{3.4^3}+\frac{1}{5.4^5}+\frac{1}{7.4^7}\text{\&c.}\right)$	=1.609438
Log. 6=log. 3+log. 2.	=1.791759
$\text{Log. } 7=\text{log. } 5+2\left(\frac{1}{6}+\frac{1}{3.6^3}+\frac{1}{5.6^5}+\frac{1}{7.6^7}\text{ \&c.}\right)$	=1.955900
Log. 8=log. 4+log. 2.	=2.079441
Log. 9=2 log. 3.	=2.197224
Log. 10=log. 5+log. 2.	=2.302585
&c. &c.	&c.

70. To compute the logarithms of the common system, it will be necessary to find the value of the *modulus*. This is equal to 1 divided by Napiers' logarithm of 10 (Art. 67.) that is,

$$\frac{1}{2.302585}=.43429448.$$

This number substituted for M, or twice the number viz. .86858896 substituted for 2 M, in the series in art. 68. will enable us to calculate the common logarithm of any number.

Common or Brigg's Logarithms.

$\text{Log. } 2 = .86858896\left(\frac{1}{3} + \frac{1}{3.3^3} + \frac{1}{5.3^5} + \frac{1}{7.3^7} \text{ \&c.}\right) = 0.301030$

$\text{Log. } 3 = 86858896\left(\frac{1}{2} + \frac{1}{3 \cdot 2^3} + \frac{1}{5.2^5} + \frac{1}{7.2^7} \text{ \&c.}\right) = 0.477121$

Log. 4=2 log. 2. =0.602060

Log. 5=log. 10−log. 2=1 −log. 2. =0.698970

Log. 6=log. 3.+log. 2. =0.778151

$\text{Log. } 7 = 86858896\left(\frac{1}{6} + \frac{1}{3.6^3} + \frac{1}{5.6^5} + \frac{1}{7.6^7} \text{ \&c.}\right)$

+log. 5 =0.855098

Log. 8=3 log. 2. =0.903090

Log. 9=2 log. 3. =0.954243

Log. 10. =1.000000

&c. &c.

TRIGONOMETRY.

SECTION I.

SINES, TANGENTS, SECANTS, &c.

ART. 71. TRIGONOMETRY *treats of the relations of the sides and angles of* TRIANGLES. Its first object is, to determine the length of the sides, and the quantity of the angles. In addition to this, from its principles are derived many interesting methods of investigation in the higher branches of analysis, particularly in physical astronomy. Scarcely any department of mathematics is more important, or more extensive in its applications. By trigonometry, the mariner traces his path on the ocean; the geographer determines the latitude and longitude of places, the dimensions and positions of countries, the altitude of mountains, the courses of rivers, &c. and the astronomer calculates the distances and magnitudes of the heavenly bodies, predicts the eclipses of the sun and moon, and measures the progress of light from the stars.

72. Trigonometry is either *plane* or *spherical*. The former treats of triangles bounded by *right lines*; the latter, of triangles bounded by *arcs of circles*.

Divisions of the Circle.

73. In a triangle there are two classes of quantities which are the subjects of inquiry, the *sides* and the *angles*. For the purpose of measuring the latter, a *circle* is introduced.

The periphery of every circle, whether great or small, is supposed to be divided into 360 equal parts called *degrees*, each degree into 60 *minutes*, each minute into 60 *seconds*, each second into 60 *thirds*, &c. marked with the characters °, ′, ″, ‴, &c. Thus 32° 24′ 13″ 22‴ is 32 degrees 24 minutes 13 seconds 22 thirds.*

A degree, then, is not a magnitude of a given *length*; but a certain *portion* of the whole circumference of any circle. It is evident, that the 360th part of a large circle is greater, than the same part of a small one. On the other hand, the *number* of degrees, in a small circle, is the same as in a large one.

The fourth part of a circle is called a *quadrant*, and contains 90 degrees.

74. To *measure* an angle, a circle is so described that its centre shall be the angular point, and its periphery shall cut the two lines which include the angle. The *arc* between the two lines is considered a *measure of the angle*, because, by Euc. 33. 6, angles at the centre of a given circle, have the same ratio to each other, as the arcs on which they stand. Thus the arc AB, (Fig. 2.) is a measure of the angle ACB.

It is immaterial what is the size of the circle, provided it cuts the lines which include the angle. Thus the angle ACD (Fig 4.) is measured by either of the arcs AG, *ag*. For ACD is to ACH, as AG to AH, or as *ag* to *ah*. (Euc. 33. 6.)

75. In the circle ADGH (Fig. 2.) let the two diameters AG and DH be perpendicular to each other. The angles ACD, DCG, GCH, and HCA will be right angles; and the periphery of the circle will be divided into four equal parts, each containing 90 degrees. As a right angle is subtended by an arc of 90°, the angle itself is said to contain 90°. Hence, in two right angles, there are 180°, in four right angles 360°; and in any other angle, as many degrees, as in the arc by which it is subtended.

76 The sum of the three angles of any triangle being equal to two right angles, (Euc. 32. 1.) is equal to 180°. Hence, there can never be more than one obtuse angle in a triangle. For the sum of two obtuse angles is more than 180°.

77. *The* COMPLEMENT *of an arc or an angle, is the difference between the arc or angle and* 90 *degrees.*

The complement of the arc AB (Fig. 2.) is DB; and the complement of the angle ACB is DCB. The complement of the arc BDG is also DB.

* See note E.

The complement of 10° is 80°,	of 60° is 30°,
of 20° is 70°,	of 120 is 30°,
of 50° is 40°,	of 170 is 80°, &c.

Hence an acute angle and its complement are always equal to 90°. The angles ACB and DCB are together equal to a right angle. The two acute angles of a right angled triangle are equal to 90° : therefore each is the complement of the other.

78. *The* SUPPLEMENT *of an arc or an angle is the difference between the arc or angle and* 180 *degrees.*

The supplement of the arc BDG (Fig. 2.) is AB ; and the supplement of the angle BCG is BCA.

The supplement of 10° is 170°,	of 120° is 60°.
of 80° is 100°,	of 150° is 30°, &c.

Hence an angle and its supplement are always equal to 180°. The angles BCA and BCG are together equal to two right angles.

79. Cor. As the three angles of a plane triangle are equal to two right angles, that is, to 180° (Euc. 32 1.) the sum of any two of them is the supplement of the other. So that the third angle may be found, by subtracting the sum of the other two from 180°. Or the sum of any two may be found, by subtracting the third from 180°.

80. A straight line drawn from the centre of a circle to any part of the periphery, is called a *radius* of the circle. In many calculations, it is convenient to consider the radius, whatever be its length, as *a unit.* (Alg. 510.) To this must be referred the numbers expressing the lengths of other lines. Thus 20 will be twenty times the radius, and 0.75, three fourths of the radius.

Definitions of Sines, Tangents, Secants, &c.

81. To facilitate the calculations in trigonometry, there are drawn, within and about the circle, a number of straight lines, called *Sines, Tangents, Secants, &c.* With these the learner should make himself perfectly familiar.

82. *The* SINE *of an arc is a straight line drawn from one end of the arc, perpendicular to a diameter which passes through the other end.*

Thus BG (Fig. 3.) is the sine of the arc AG. For BG is a line drawn from the end G of the arc, perpendicular to the diameter AM which passes through the other end A of the arc.

Cor. The sine is *half the chord* of *double the arc.* The sine BG is half PG, which is the chord of the arc PAG, double the arc AG.

83. *The* VERSED SINE *of an arc is that part of the diameter which is between the sine and the arc.*

Thus BA is the versed sine of the arc AG.

84. *The* TANGENT *of an arc, is a straight line drawn perpendicular from the extremity of the diameter which passes through one end of the arc, and extended till it meets a line drawn from the centre through the other end.*

Thus AD (Fig. 3.) is the tangent of the arc AG.

85. *The* SECANT *of an arc, is a straight line drawn from the centre, through one end of the arc, and extended to the tangent which is drawn from the other end.*

Thus CD (Fig. 3.) is the secant of the arc AG.

86. In Trigonometry, the terms *tangent* and *secant* have a more limited meaning, than in Geometry. In both, indeed, the tangent *touches* the circle, and the secant *cuts* it. But in Geometry, these lines are of no determinate length; whereas, in Trigonometry, they extend from the diameter to the point in which they intersect each other.

87. The lines just defined are sines, tangents and secants of *arcs*. BG (Fig. 3.) is the sine of the arc AG. But this arc subtends the *angle* GCA. BG is then the sine of the arc which subtends the angle GCA. This is more concisely expressed, by saying that BG is the sine of the angle GCA. And universally, the sine, tangent, and secant of an *arc*, are said to be the sine, tangent, and secant of the *angle* which stands at the centre of the circle, and is subtended by the arc. Whenever, therefore, the sine, tangent, or secant of an angle is spoken of; we are to suppose a circle to be drawn whose centre is the angular point; and that the lines mentioned belong to that arc of the periphery which subtends the angle.

88. The *sine*, and *tangent* of an acute angle, are *opposite* to the angle. But the *secant* is one of the lines which *include* the angle. Thus the sine BG, and the tangent AD (Fig. 3.) are opposite to the angle DCA. But the secant CD is one of the lines which include the angle.

89. *The sine complement or* COSINE *of an angle, is the sine of the* COMPLEMENT *of that angle.* Thus, if the diameter HO (Fig. 3.) be perpendicular to MA, the angle HCG is the complement of ACG; (Art. 77.) and LG, or its equal CB, is the sine of HCG. (Art. 82.) It is, therefore, the *cosine* of GCA. On the other hand GB is the sine of GCA, and the cosine of GCH.

So also the *cotangent* of an angle is the tangent of the *complement* of the angle. Thus HF is the cotangent of GCA. And the *cosecant* of an angle is the secant of the *complement* of the angle. Thus CF is the cosecant of GCA.

Hence, as in a right angled triangle, one of the acute angles is the complement of the other; (Art. 77.) the sine, tangent, and secant of one of these angles, are the cosine, cotangent, and cosecant of the other.

90. The sine, tangent, and secant of the *supplement* of an angle, are each equal to the sine, tangent, and secant of the angle itself. It will be seen, by applying the definition (Art. 82.) to the figure, that the sine of the obtuse angle GCM is BG, which is also the sine of the acute angle GCA. It should be observed, however, that the sine of an acute angle is *opposite* to it; while the sine of an obtuse angle *falls without* the angle, and is opposite to its supplement. Thus BG, the sine of the angle MCG, is not opposite to MCG, but to its supplement ACG.

The *tangent* of the obtuse angle MCG is MT, or its equal AD, which is also the tangent of ACG. And the *secant* of MCG is CD, which is also the secant of ACG.

91. But the *versed sine* of an angle *is not the same*, as that of its *supplement.* The versed sine of an *acute* angle is equal to the *difference* between the cosine and radius. But the versed sine of an *obtuse* angle is equal to the *sum* of the cosine and radius. Thus the versed sine of ACG is AB=AC −BC. (Art. 83.) But the versed sine of MCG is MB=MC +BC.

Relations of Sines, Tangents, Secants, &c. to each other.

92. The relations of the sine, tangent, secant, cosine, &c. to each other, are easily derived from the proportions of the sides of similar triangles. (Euc. 4. 6.) In the quadrant ACH, (Fig. 3.) these lines form three similar triangles, viz. ACD, BCG or LCG, and HCF. For, in each of these, there is one

right angle, because the sines and tangents are, by definition, perpendicular to AC; as the cosine and cotangent are to CH. The lines CH, BG, and AD are parallel, because CA makes a right angle with each. (Euc. 27. 1.) For the same reason, CA, LG, and HF are parallel. The alternate angles GCL, BGC, and the opposite angle CDA are equal; (Euc. 29. 1.) as are also the angles GCB, LGC, and HFC. The triangles ACD, BCG, and HCF are therefore similar.

It should also be observed, that the line BC, between the sine and the centre of the circle, is parallel and equal to the cosine; and that LC, between the cosine and centre, is parallel and equal to the sine; (Euc. 34. 1.) so that one may be taken for the other, in any calculation.

93. From these similar triangles, are derived the following proportions; in which R is put for radius,

sin for sine,	*cos* for cosine,
tan for tangent,	*cot* for cotangent,
sec for secant,	*cosec* for cosecant.

By comparing the triangles CBG and CAD,

1. AC : BC::AD : BG, that is, R : *cos*::*tan* : *sin*.
2. CG : CD::BG : AD R : *sec*::*sin* : *tan*.
3. CB : CA::CG : CD *cos* : R::R : *sec*.

Therefore $R^2 = cos \times sec$.

By comparing the triangles CLG and CHF,

4. CH : CL::HF : LG, that is, R : *sin*::*cot* : *cos*.
5. CG : CF::LG : HF R : *cosec*::*cos* : *cot*.
6. CL : CH::CG : CF *sin* : R::R : *cosec*.

Therefore $R^2 = sin \times cosec$.

By comparing the triangles CAD and CHF,

7. CH : AD::CF : CD, that is R : *tan*::*cosec* : *sec*.
8. CA : HF::CD : CF R : *cot*::*sec* : *cosec*.
7. AD : AC::CH : HF *tan* : R::R : *cot*.

Therefore $R^2 = tan \times cot$.

It will not be necessary for the learner to commit these proportions to memory. But he ought to make himself so familiar with the manner of stating them from the figure, as to be able to explain them, whenever they are referred to.

94. Other relations of the sine, tangent, &c. may be derived from the proposition, that the square of the hypothenuse is equal to the sum of the squares of the perpendicular sides. (Euc. 47. 1.)

In the right angled triangles CBG, CAD, and CHF, (Fig. 3.)

1. $\overline{CG}^2 = \overline{CB}^2 + \overline{BG}^2$, that is $R^2 = cos^2 + sin^2$,*
2. $\overline{CD}^2 = \overline{CA}^2 + \overline{AD}^2$ $\qquad sec^2 = R^2 + tan^2$,
3. $\overline{CF}^2 = \overline{CH}^2 + \overline{HF}^2$ $\qquad cosec^2 = R^2 + cot^2$.

And, extracting the root of both sides, (Alg. 296.)

$$R = \sqrt{cos^2 + sin^2} = \sqrt{sec^2 - tan^2} = \sqrt{cosec^2 - cot^2}$$

Hence, if $R = 1$, (Alg. 510.)

$$Sine = \sqrt{1 - cos^2} \qquad Sec = \sqrt{1 + tan^2}$$
$$Cos = \sqrt{1 - sin^2} \qquad Cosec = \sqrt{1 + cot^2}$$

95. *The sine of* 90°, *The chord of* 60°, *And the tangent of* 45° } *are*, in any circle, *each equal to the radius*, and therefore equal to each other.

Demonstration.

1 In the quadrant ACH, (Fig. 5.) the arc AH is 90°. The sine of this, according to the definition, (Art. 82.) is CH, the radius of the circle.

2. Let AS be an arc of 60°. Then the angle ACS, being measured by this arc, will also contain 60°; (Art. 75.) and the triangle ACS will be equilateral. For the sum of the three angles is equal to 180°. (Art. 76.) From this, taking the angle ACS, which is 60°, the sum of the remaining two is 120°. But these two are *equal*, because they are subtended by the equal sides, CA and CS, both radii of the circle. Each, therefore, is equal to *half* 120°, that is to 60°.

* Sine2 is here put for the square of the sine, cos2 for the square of the cosine &c.

All the angles being equal, the sides are equal, and therefore AS, the chord of 60°, is equal to CS the radius.

3. Let AR be an arc of 45°. AD will be its tangent, and the angle ACD subtended by the arc, will contain 45°. The angle CAD is a right angle, because the tangent is, by definition, perpendicular to the radius AC. (Art. 84.) Subtracting ACD, which is 45°, from 90°, (Art. 77.) the other acute angle ADC will be 45° also. Therefore the two legs of the triangle ACD are equal, because they are subtended by equal angles; (Euc. 6. 1.) that is, AD the tangent of 45°, is equal to AC the radius.

Cor. The *cotangent* of 45° is also equal to radius. For the complement of 45° is itself 45°. Thus HD, the cotangent of ACD, (Fig. 5. is equal to AC the radius.

96. The sine of 30° is equal to *half radius*. For the sine of 30° is equal to half the chord of 60°. (Art. 82. cor.) But by the preceding article, the chord of 60° is equal to radius. Its half, therefore, which is the sine of 30° is equal to half radius.

Cor. 1. The *cosine* of 60° is equal to half radius. For the cosine of 60° is the sine of 30° (Art. 89.)

Cor. 2. The cosine of $30°=\frac{1}{2}\sqrt{3}$. For

$$Cos^2 30°=R-sin^2 30=1-\tfrac{1}{4}=\tfrac{3}{4}.$$

Therefore,

$$Cos\ 30°=\sqrt{\tfrac{3}{4}}=\tfrac{1}{2}\sqrt{3}.$$

96. *b*. The sine of $45°=\frac{1}{\sqrt{2}}$. For

$$R^2=1=sin^2\ 45°+cos^2\ 45°=2\ sin^2\ 45°$$

Therefore, $Sin\ 45°=\sqrt{\tfrac{1}{2}}=\frac{1}{\sqrt{2}}$

97. The *chord* of any arc is a *mean proportional*, between the *diameter* of the circle, and the *versed sine* of the arc.

Let ADB (Fig. 6.) be an arc, of which AB is the chord, BF the sine, and AF the versed sine. The angle ABH is a right angle, (Euc. 31. 3.) and the triangles ABH and ABF are similar. (Euc. 8. 6.) Therefore,

$$AH : AB :: AB : AF.$$

That is, the diameter is to the chord, as the chord to the versea sine.

In Fig. 6th, let the arc AD$=a$, and ADB$=2a$. Draw BF perpendicular to AH. This will divide the right angled triangle ABH into two similar triangles. (Euc. 8. 6.) The angles ACD and AHB are equal. (Euc. 20. 3.) Therefore the four triangles ACG, AHB, FHB, and FAB are similar; and the line BH is twice CG, because BH : CG : : HA : CA.

The sides of the four triangles are

AG$=\sin a$,	CG$=\cos a$,	HF$=$*vers. sup.* $2a$,
AB$=2\sin a$,	BH$=2\cos a$,	AC$=$the radius,
BF$=\sin 2a$,	AF$=$*vers* $2a$,	AH$=$the diameter.

A variety of proportions may be stated, between the homologous sides of these triangles: For instance,

By comparing the triangles ACG and ABF,

AC : AG : : AB : AF, that is, R : *sin* a : : 2*sin* a : *vers* $2a$
AC : CG : : AB : BF R : *cos* a : : 2*sin* a : *sin* $2a$
AG : CG : : AF : BF *Sin* a : *cos* a : : *vers* $2a$: *sin* $2a$

Therefore,

$$R\times \text{vers } 2a=2\sin^2 a$$
$$R\times \sin 2a \;=2\sin a\times \cos a$$
$$\text{Sin } a\times \sin 2a=\text{vers } 2a\times \cos a$$

By comparing the triangles ACG and BFH,

AC : CG : : BH : HF, that is, R : *cos* a : : 2*cos* a : *vers. sup.* $2a$
AG : CG : : BF : HF *Sin* a : *cos* a : : *sin* $2a$: *vers. sup.* $2a$

Therefore,

$$R\times \text{vers. sup. } 2a=2\cos^2 a$$
$$\text{Sin } a\times \text{vers. sup. } 2a=\cos a\times \sin 2a$$
&c. &c.

That is, the product of radius into the versed sine of the supplement of twice a given arc, is equal to twice the square of the cosine of the arc.

And the product of the sine of an arc, into the versed sine of the supplement of twice the arc, is equal to the product of the cosine of the arc, into the sine of twice the arc, &c. &c.

SECTION II.

THE TRIGONOMETRICAL TABLES.

Art. 98. TO facilitate the operations in trigonometry, the sine, tangent, secant, &c. have been calculated for every degree and minute, and in some instances, for every second, of a quadrant, and arranged in tables. These constitute what is called the *Trigonometrical Canon.** It is not necessary to extend these tables beyond 90°; because the sines, tangents, and secants, are of the same magnitude, in one of the quadrants of a circle, as in the others. Thus the sine of 30° is equal to that of 150°. (Art. 90.)

99. And in any instance, if we have occasion for the sine, tangent, or secant of an *obtuse angle*, we may obtain it, by looking for its equal, the sine, tangent, or secant of the *supplementary acute angle*.

100. The tables are calculated for a circle whose radius is supposed to be a *unit*. It may be an inch, a yard, a mile, or any other denomination of length. But the *sines*, *tangents*, &c. must always be understood to be of the same denomination as the radius.

101. All the *sines*, except that of 90°, are *less than the radius*, (Art. 82, and Fig. 3.) and are expressed in the tables by decimals.

Thus the sine of 20° is 0.34202, of 60° is 0.86603,
of 40° is 0.64279, of 89° is 0.99985, &c.

When the tables are intended to be very exact, the decimal is carried to a greater number of places.

The *tangents* of all angles less than 45° are also less than radius. (Art. 95.) But the tangents of angles greater than 45°, are *greater* than radius, and are expressed by a whole number and a decimal. It is evident that all the *secants* also

* For the *construction* of the Canon, see Section VIII.

must be greater than radius, as they extend from the centre, to a point without the circle.

102. The numbers in the table here spoken of, are called *natural* sines, tangents, &c. They express the lengths of the several lines which have been defined in arts. 82, 83, &c. By means of them, the angles and sides of triangles may be accurately determined. But the calculations must be made by the tedious processes of multiplication and division. To avoid this inconvenience, another set of tables has been provided, in which are inserted the *logarithms* of the natural sines, tangents, &c. By the use of these, addition and subtraction are made to perform the office of multiplication and division. On this account, the tables of logarithmic, or as they are sometimes called, *artificial* sines, tangents, &c. are much more valuable, for practical purposes, than the *natural* sines, &c. Still it must be remembered, that the former are derived from the latter. The artificial sine of an angle, is the logarithm of the natural sine of that angle. The artificial tangent is the logarithm of the natural tangent, &c.

103. One circumstance, however, is to be attended to, in comparing the two sets of tables. The radius to which the *natural* sines, &c. are calculated, is *unity*. (Art. 100.) The secants, and a part of the tangents are, therefore, *greater* than a unit; while the sines, and another part of the tangents, are *less* than a unit. When the logarithms of these are taken, some of the indices will be *positive*, and others *negative;* (Art. 9.) and the throwing of them together in the same table, if it does not lead to error, will at least be attended with inconvenience. To remedy this, 10 is added to each of the indices. (Art. 12.) They are then all positive. Thus the natural sine of 20° is 0.34202. The logarithm of this is $\bar{1}.53405$. But the index, by the addition of 10, becomes $10-1=9$. The logarithmic sine in the tables is therefore 9.53405.*

Directions for taking Sines, Cosines, &c. from the tables.

104. The *cosine*, *cotangent*, and *cosecant* of an angle, are the sine, tangent, and secant of the *complement* of the angle. (Art. 89.) As the complement of an angle is the difference between the angle and 90°, and as 45 is the half of 90; if any given angle within the quadrant is greater than 45°, its

*Or the tables may be supposed to be calculated to the radius 10000000000, whose logarithm is 10.

complement is less ; and, on the other hand, if the angle is *less* than 45°, its complement is greater. Hence, every cosine, cotangent, and cosecant of an angle greater than 45°, has its equal, among the sines, tangents, and secants of angles less than 45°, and *v. v.*

Now, to bring the trigonometrical tables within a small compass, the same column is made to answer for the *sines* of a number of angles *above* 45°, and for the *cosines* of an equal number *below* 45°.

Thus 9.23967 is the log. *sine* of 10°, and the *cosine* of 80°,
9.53405 the *sine* of 20°, and the *cosine* of 70°, &c.

The tangents and secants are arranged in a similar manner. Hence,

105. *To find the Sine, Cosine, Tangent, &c. of any number of degrees and minutes.*

If the given angle is *less* than 45°, look for the degrees at the *top* of the table, and the minutes on the *left* ; then, opposite to the minutes, and under the word sine at the head of the column, will be found the sine ; under the word tangent, will be found the tangent ; &c.

The log. sin	of 43° 25′ is	9.83715	The tan	of 17° 20′ is	9.49430
	of 17° 20′	9.47411		of 8° 46′	9.18812
The cos	of 17° 20′	9.97982	The cot	of 17° 20′	10.50570
	of 8° 46′	9.99490		of 8° 46′	10.81188

The first figure is the index ; and the other figures are the decimal part of the logarithm.

106. If the given angle is between 45° and 90° ; look for the degrees at the *bottom* of the table, and the minutes on the *right* ; then, opposite to the minutes, and *over* the word sine at the foot of the column, will be found the sine ; over the word tangent, will be found the tangent, &c.

Particular care must be taken, when the angle is less than 45°, to look for the title of the column, at the *top*, and for the minutes, on the *left* ; but when the angle is between 45° and 90°, to look for the title of the column, at the *bottom* and for the minutes, on the *right*.

The log. sine	of 81° 21′ is	9.99503
The cosine	of 72° 10′	9.48607
The tangent	of 54° 40′	10.14941
The cotangent	of 63° 22′	9.70026

107. If the given angle is *greater* than 90°, look for the sine, tangent, &c. of its *supplement*. (Art. 98, 99.)

The log. sine	of 96° 44′ is	9.99699
The cosine	of 171° 16′	9.99494
The tangent	of 130° 26′	10.06952
The cotangent	of 156° 22′	10.35894

108. *To find the sine, cosine, tangent, &c. of any number of degrees, minutes, and* SECONDS.

In the common tables, the sine, tangent, &c. are given only to every *minute* of a degree.* But they may be found to *seconds*, by taking *proportional parts* of the difference of the numbers as they stand in the tables. For, within a single minute, the variations in the sine, tangent, &c. are nearly proportional to the variations in the angle. Hence,

To find the sine, tangent, &c. to seconds: Take out the number corresponding to the given degree and minute; and also that corresponding to the next greater minute, and find their difference. Then state this proportion;

As 60, to the given number of seconds;

So is the difference found, to the correction for the seconds.

This correction, in the case of sines, tangents, and secants, is to be *added* to the number answering to the given degree and minute; but for cosines, cotangents, and cosecants, the correction is to be *subtracted*

For, as the sines *increase*, the cosines *decrease*.

Ex. 1. What is the logarithmic sine of 14° 43′ 10″ ?

The sine of 14° 43′ is	9.40490
of 14° 44′	9.40538
Difference	48

Here it is evident, that the sine of the required angle is greater than that of 14° 43′, but less than that of 14° 44′. And as the difference corresponding to a whole minute or

* In the very valuable tables of Michael Taylor, the sines and tangents are given to *every second*.

60″ is 48; the difference for 10″ must be a proportional part of 48. That is,

60″ : 10″ :: 48 : 8 the correction to be *added* to the sine of 14° 43′.

Therefore the sine of 14° 43′ 10″ is 9.40498

2. What is the logarithmic cosine of 32° 16′ 45″?

The cosine of	32° 16′ is	9.92715
of	32° 17′	9.92707
Difference		8

Then 60″ : 45″ :: 8 : 6 the correction to be *subtracted* from the cosine of 32° 16′.

Therefore the cosine of 32° 16′ 45″ is 9.92709

The tangent of	24° 15′ 18″	is	9.65376
The cotangent of	31° 50′ 5″	is	10.20700
The sine of	58° 14′ 32″	is	9.92956
The cosine of	55° 10′ 26″	is	9.75670

If the given number of seconds be any even part of 60, as $\frac{1}{2}$, $\frac{1}{3}$, $\frac{1}{4}$, &c. the correction may be found, by taking a like part of the difference of the numbers in the tables, without stating a proportion in form.

109. *To find the degrees and minutes belonging to any given sine, tangent, &c.*

This is reversing the method of finding the sine, tangent, &c. (Art. 105, 6, 7.)

Look in the column of the same name, for the sine, tangent, &c. which is *nearest* to the given one; and if the title be at the *head* of the column, take the degrees at the *top* of the table, and the minutes on the *left*; but if the title be at the *foot* of the column, take the degrees at the *bottom*, and the minutes on the *right*.

Ex. 1. What is the number of degrees and minutes belonging to the logartihmic sine 9.62863?

The nearest sine in the tables is 9.62865. The title of sine is at the head of the column in which these numbers are found. The degrees at the top of the page are 25, and the minutes on the left are 10. The angle required is, therefore 25° 10′.

The angle belonging to

the sine	9.87993	is 49° 20′	the cos	9.97351	is 19° 48
the tan	9.97955	43° 39′	the cotan	9.75791	60° 12′
the sec	10.65396	77° 11′	the cosec	10.49066	18° 51′

110. *To find the degrees, minutes, and* SECONDS *belonging to any given sine, tangent, &c.*

This is reversing the method of finding the sine, tangent, &c. to seconds. (Art. 108.)

First find the difference between the sine, tangent, &c. next greater than the given one, and that which is next less; then the difference between this less number and the given one; Then

As the difference first found, is to the other difference;

So are 60 seconds, to the number of seconds, which, in the case of sines, tangents, and secants, are to be *added* to the degrees and minutes belonging to the least of the two numbers taken from the tables; but for cosines, cotangents, and cosecants, are to be *subtracted.*

Ex. 1. What are the degrees, minutes, and seconds, belonging to the logarithmic sine 9.40498?

Sine next greater	14° 44′	9.40538	Given sine	9.40498
Next less	14° 43′	9.40490	Next less	9.40490
Difference		48	Difference	8

Then 48 : 8 :: 60″ : 10″, which added to 14° 43′ gives 14° 43′ 10″ for the answer.

2 What is the angle belonging to the cosine 9.09773?

Cosine next greater	82° 48′	9.09807	Given cosine	9.09773
Next less	82° 49′	9.09707	Next less	9.09707
Difference		100	Difference	66

Then 100 : 66::60″ : 40″, which subtracted from 82° 49′, gives 82° 48′ 20″ for the answer.

It must be observed here, as in all other cases, that of the two angles, the less has the greater cosine.

The angle belonging to

the sin 9.20621 is	9° 15′ 6″	the tan 10.43434 is	69° 48′ 16″
the cos 9.98157	16° 34′ 30″	the cot 10.33554	24° 47′ 16″

Method of Supplying the Secants and Cosecants.

111. In some trigonometrical tables, the secants and cosecants are not inserted. But they may be easily obtained from the sines and cosines. For, by art. 93, proportion 3d,

$$cos \times sec = R^2$$

That is, the product of the cosine and secant, is equal to the square of radius. But, in logarithms, addition takes the place of multiplication; and, in the tables of logarithmic sines, tangents, &c. the radius is 10. (Art. 103.) Therefore, in these tables,

$$\cos + \sec = 20. \qquad \text{Or } \sec = 20 - \cos.$$

Again, by art. 93, proportion 6,

$$\sin \times \text{cosec} = R^2.$$

Therefore, in the tables,

$$\sin + \text{cosec} = 20. \qquad \text{Or cosec} = 20 - \sin. \quad \text{Hence,}$$

112. To obtain the *secant*, subtract the cosine from 20; and to obtain the *cosecant*, subtract the sine from 20.

These subtractions are most easily performed, by taking the right hand figure from 10, and the others from 9, as in finding the arithmetical complement of a logarithm; (Art. 55.) observing however, to add 10 to the index of the secant or cosecant. In fact, the secant is the arithmetical complement of the cosine, with 10 added to the index.

For the secant	=20—cos
And the ar. comp. of cos	=10—cos. (Art. 54.)

So also the cosecant is the arithmetical complement of the sine, with 10 added to the index. The tables of secants and cosecants are, therefore, of use in furnishing the artihmetical complement of the sine and cosine, in the following simple manner:

113. For the arithmetical complement of the *sine*, subtract 10 from the index of the cosecant; and for the arithmetical complement of the *cosine*, subtract 10 from the index of the secant.

By this, we may save the trouble of taking each of the figures from 9.

SECTION III.

SOLUTIONS OF RIGHT-ANGLED TRIANGLES.

ART. 114. IN a triangle, there are *six parts*, three sides, and three angles. In every trigonometrical calculation, it is necessary that some of these should be known, to enable us to find the others. *The number of parts which must be given, is* THREE, *one of which must be a* SIDE.

If only two parts be given, they will be either two sides, a side and an angle, or two angles; neither of which will limit the triangle to a particular form and size.

If *two sides* only be given, they may make any angle with each other; and may, therefore, be the sides of a thousand different triangles. Thus the two lines a and b (Fig. 7.) may belong either to the triangle ABC, or ABC′, or ABC″. So that it will be impossible, from knowing two of the sides of a triangle, to determine the other parts.

Or, if a *side and an angle* only be given, the triangle will be indeterminate. Thus, if the side AB (Fig. 8.) and the angle at A be given; they may be parts either of the triangle ABC, or ABC′, or ABC.″

Lastly, if two *angles*, or even if *all* the angles be given, they will not determine the length of the sides. For the triangles ABC, A′B′C′, A″B″C″. (Fig. 9.) and a hundred others which might be drawn, with sides parallel to these, will all have the same angles. So that one of the parts given must always be a side. If this and any other two parts, either sides or angles, be known, the other three may be found, as will be shown, in this and the following section.

115. Triangles are either *right angled* or *oblique angled*. The calculations of the former are the most simple, and those which we have the most frequent occasion to make. A great portion of the problems in the mensuration of heights and distances, in surveying, navigation and astronomy, are solved by rectangular trigonometry. Any triangle whatever may be divided into two right angled triangles, by drawing a perpendicular from one of the angles to the opposite side.

116. One of the six parts in a right angled triangle, is always given, viz, the right angle. This is a *constant* quantity; while the other angles and the sides are variable. It is also to be observed, that, if one of the *acute* angles is given, the other is known of course. For one is the complement of the other. (Art. 76, 77.) So that, *in a right angled triangle, subtracting one of the acute angles from 90° gives the other.* There remain, then, only *four* parts, one of the acute angles, and the three sides to be sought by calculation. If any *two* of these be given, with the right angle, the others may be found.

117. To illustrate the method of calculation, let a case be supposed in which a right angled triangle CAD (Fig. 10.) has one of its sides equal to the radius to which the trigonometrical tables are adapted.

In the first place, let the *base* of the triangle be equal to the tabular radius. Then, if a circle be described, with this radius, about the angle C as a centre, DA will be the *tangent*, and DC the *secant* of that angle. (Art. 84, 85.) So that the radius, the tangent, and the secant of the angle at C, constitute the three sides of the triangle. The *tangent*, taken from the tables of natural sines, tangents, &c. will be the length of the *perpendicular;* and the *secant* will be the length of the *hypothenuse.* If the tables used be logarithmic, they will give the *logarithms* of the lengths of the two sides.

In the same manner, *any* right angled triangle whatever, whose base is equal to the radius of the tables, will have its other two sides found among the tangents and secants. Thus, if the quadrant AH (Fig. 11.) be divided into portions of 15° each; then, in the triangle

CAD, AD will be the tan, and CD the sec of 15°,
In CAD′, AD′ will be the tan, and CD′ the sec of 30°,
In CAD″, AD″ will be the tan, and CD″ the sec of 45°,
[&c.

118. In the next place, let the *hypothenuse* of a right angled triangle CBF (Fig. 12.) be equal to the radius of the tables. Then, if a circle be described, with the given radius, and about the angle C as a centre; BF will be the *sine*, and BC the *cosine* of that angle. (Art. 82. 89.) Therefore the sine of the angle at C, taken from the tables, will be the length

of the *perpendicular*, and the cosine will be the length of the *base*.

And any right angled triangle whatever, whose hopothenuse is equal to the tabular radius, will have its other two sides found among the sines and cosines. Thus if the quadrant AH (Fig. 13.) be divided into portions of 15° each, in the points F, F,′ F″, &c.; then, in the triangle,

CBF, FB will be the sin, and CB the cos, of 15°,
In CB′F′, FB′ will be the sin, and CB′ the cos, of 30°,
In CB″F″, F″B″ will be the sin, and CB″ the cos, of 45°,
[&c.

119. By merely *turning to the tables*, then, we may find the parts of any right angled triangle which has one of its sides equal to the radius of the tables. But for determining the parts of triangles which have *not* any of their sides equal to the tabular radius, the following proportion is used:

As the radius of one circle,
To the radius of any other;
So is a sine, tangent, or secant, in one,
To the sine, tangent, or secant, of the same number of degrees, in the other.

In the two concentric circles AHM, *ahm*, (Fig. 4,) the arcs AG and *ag* contain the same number of degrees. (Art. 74.) The sines of these arcs are BG and *bg*, the tangents AD and *ad*, and the secants CD and *cd*. The four triangles, CAD, CBG, C*ad*, and C*bg*, are similar. For each of them, from the nature of sines and tangents, contains one right angle; the angle at C is common to them all; and the other acute angle in each is the complement of that at C. (Art. 77.) We have, then, the following proportions. (Euc. 4. 6.)

1. CG : C*g*::BG : *bg*.

That is, one radius is to the other, as one *sine* to the other.

2. CA : C*a*::DA : *da*.

That is, one radius is to the other, as one *tangent* to the other.

3. CA : C*a*::CD : C*d*.

That is, one radius is to the other, as one *secant* to the other.

Cor. BG : *bg*::DA : *da*::CD : C*d*.

That is, as the sine in one circle, to the sine in the other; so is the tangent in one, to the tangent in the other; and so is the secant in one, to the secant in the other.

This is a general principle, which may be applied to most trigonometrical calculations. If one of the sides of the proposed triangle be made radius, each of the other sides will be the sine, tangent, or secant, of an arc described by this radius. Proportions are then stated, between these lines, and the *tabular* radius, sine, tangent, &c.

120. A line is said to be *made radius*, when a circle is described, or supposed to be described, whose semidiameter is equal to the line, and whose centre is at one end of it.

121. In any right angled triangle, *if the* HYPOTHENUSE *be made radius, one of the legs will be a* SINE *of its opposite angle, and the other leg a* COSINE *of the same angle.*

Thus, if to the triangle ABC (Fig. 14.) a circle be applied, whose radius is AC, and whose centre is A, then BC will be the *sine*, and BA the *cosine*, of the angle at A. (Art. 82, 89.)

If, while the same line is radius, the other end C be made the centre, then BA will be the *sine*, and BC the *cosine*, of the angle at C.

122. *If either of the* LEGS *be made radius, the other leg will be a* TANGENT *of its opposite angle, and the hypothenuse will be a* SECANT *of the same angle;* that is, of the angle between the secant and the radius.

Thus, if the *base* AB (Fig. 15.) be made radius, the centre being at A, BC will be the *tangent*, and AC the *secant*, of the angle at A. (Art. 84, 85.)

But, if the *perpendicular* BC (Fig. 16.) be made radius, with the centre at C, then AB will be the *tangent*, and AC the *secant*, of the angle at C.

123. As the side which is the sine, tangent, or secant of one of the acute angles, is the cosine, cotangent, or cosecant of the other; (Art. 89.) the *perpendicular* BC (Fig. 14.) is the *sine* of the angle A, and the *cosine* of the angle C; while the *base* AB is the *sine* of the angle C, and the *cosine* of the angle A.

If the base is made radius, as in Fig. 15, the *perpendicular* BC is the *tangent* of the angle A, and the *cotangent* of the angle C; while the *hypothenuse* is the *secant* of the angle A, and the *cosecant* of the angle C.

If the perpendicular is made radius, as in Fig. 16, the *base* AB is the *tangent* of the angle C, and the *cotangent* of the angle A; while the *hypothenuse* is the *secant* of the angle C, and the *cosecant* of the angle A.

124. Whenever a right angled triangle is proposed, whose sides or angles are required; a *similar* triangle may be formed, from the sines, tangents, &c. of the *tables*. (Art. 117, 118.) The parts required are then found, by stating proportions between the similar sides of the two triangles. If the triangle proposed be ABC, (Fig. 17.) another *abc* may be formed, having the same angles with the first, but differing from it in the length of its sides, so as to correspond with the numbers in the tables. If similar sides be made radius in both, the remaining similar sides will be lines of *the same name;* that is, if the perpendicular in one of the triangles be a *sine*, the perpendicular in the other will be a sine; if the base in one be a *cosine*, the base in the other will be a cosine, &c.

If the *hypothenuse* in each triangle be made radius, as in Fig. 14, the perpendicular *bc* will be the *tabular sine* of the angle at *a;* and the perpendicular BC will be a sine of the equal angle A, in a circle of which AC is radius.

If the *base* in each triangle be made radius, as in Fig. 15, then the perpendicular *bc* will be the *tabular tangent* of the angle at *a;* and BC will be a tangent of the equal angle A, in a circle of which AB is radius, &c.

125. From the relations of the similar sides of these triangles, are derived the two following *theorems*, which are sufficient for calculating the parts of any right angled triangle whatever, when the requisite data are furnished. One is used, when a *side* is to be found; the other, when an *angle* is to be found.

Theorem I.

126. When a *side* is required;

As the tabular sine, tangent, &c. of
the same name with the given side,
To the given side;
So is the tabular sine, tangent, &c. of
same name with the required side,
To the required side.

It will be readily seen, that this is nothing more than a statement, in general terms, of the proportions between the similar sides of two triangles, one proposed for solution, and the other formed from the numbers in the tables.

Thus if the hypothenuse be *given*, and the base or perpendicular be *required;* then, in Fig. 14, where *ac* is the tabular radius, *bc* the tabular sine of *a*, or its equal A, and *ab* the tabular sine of C; (Art. 124.)

ac : AC::*bc* : BC, that is, R : AC::Sin A : BC.
ac : AC::*ab* : AB, R : AC::Sin C : AB.

In Fig. 15, where *ab* is the tabular radius, *ac* the tabular secant of A, and *bc* the tabular tangent of A;

ac : AC::*bc* : BC, that is Sec A : AC::Tan A : BC.
ac : AC::*ab* : AB, Sec A : AC::R : AB.

In Fig. 16, where *bc* is the tabular radius, *ac* the tabular secant of C, and *ab* the tabular tangent of C;

ac : AC::*bc* : BC, that is, Sec C : AC::R : BC.
ac : AC::*ab* : AB, Sec C : AC::Tan C : AB.

Theorem II.

127. When an *angle* is required;

As the given side made radius,
To the tabular radius;
So is another given side,
To the tabular sine, tangent, &c. of the same name.

Thus if the side made radius, and one other side be given, then, in Fig. 14,

AC : *ac*::BC : *bc*, that is, AC : R::BC : Sin A.
AC : *ac*::AB : *ab* AC : R::AB : Sin C.

In Fig. 15,

AB : *ab*::BC : *bc*, that is, AB : R::BC : Tan A.
AB : *ab*::AC : *ac* AB : R::AC : Sec A.

In Fig. 16.

BC : *bc*::AB : *ab*, that is, BC : R::AB : Tan C.
BC : *bc*::AC : *ac* BC : R::AC : Sec C.

It will be observed, that in these theorems, *angles* are not introduced, though they are among the quantities which are either given or required, in the calculation of triangles. But the tabular sines, tangents, &c. may be considered the *representatives* of angles, as one may be found from the other, by merely turning to the tables.

128. In the theorem for finding a *side*, the first term of the proportion is a *tabular number*. But, in the theorem for finding an *angle*, the first term is a side. Hence, in applying the proportions to particular cases, this rule is to be observed,

To find a SIDE, *begin with a tabular number,*
To find an ANGLE, *begin with a side.*

Radius is to be reckoned among the tabular numbers.

129. In the theorem for finding an *angle*, the first term is a *side made radius*. As in every proportion, the three first terms must be given, to enable us to find the fourth, it is evident, that where this theorem is applied, the side made radius must be a *given* one. But, in the theorem for finding a *side*, it is not necessary that either of the terms should be radius. Hence,

130. *To find a* SIDE, ANY *side may be made radius.*
To find an ANGLE, *a* GIVEN *side must be made radius.*

It will generally be expedient, in both cases, to make radius one of the terms in the proportion; because, in the tables of natural sines, tangents, &c. radius is 1, and in the logarithmic tables it is 10. (Art. 103.)

131. The proportions in Trigonometry are of the same nature as other simple proportions. The fourth term is found, therefore, as in the Rule of Three in Arithmetic, by *multiplying together the second and third terms, and dividing their product by the first term.* This is the mode of calculation, when the tables of *natural* sines, tangents, &c. are used. But the operation by logarithms is so much more expeditious, that it has almost entirely superseded the other method. In logarithmic calculations, addition takes the place of multiplication; and subtraction the place of division.

The logarithms expressing the lengths of the *sides* of a triangle, are to be taken from the tables of common logarithms. The logarithms of the *sines*, *tangents*, &c. are found in the tables of artificial sines, &c. The calculation is then made *by adding the second and third terms, and subtracting the first.* (Art. 52.)

132. The logarithmic radius 10, or, as it is written in the tables, 10.00000, is so easily added and subtracted, that the three terms of which it is one, may be considered as, in effect, reduced to two. Thus, if the tabular radius is in the *first* term, we have only to add the other two terms, and then take 10 from the index ; for this is subtracting the first term. If radius occurs in the *second* term, the first is to be subtracted from the third, after its index is increased by 10. In the same manner, if radius is in the *third* term, the first is to be subtracted from the second.

133. Every species of right angled triangles may be be solved upon the principle, that the sides of similar triangles are proportional, according to the two theorems mentioned above. There will be some advantages, however, in giving the examples in distinct classes.

There must be given, in a right angled triangle, *two* of the parts, besides the right angle. (Art. 116.) These may be ;

1. The hypothenuse and an angle ; or
2. The hypothenuse and a leg ; or
3. A leg and an angle ; or
4. The two legs.

Case I.

134. Given { The hypothenuse, And an angle : } to find { The base and Perpendicular.

Ex. 1. If the hypothenuse AC (Fig. 17.*) be 45 miles, and the angle at A 32° 20′, what is the length of the base AB, and the perpendicular BC ?

In this case, as *sides* only are required, *any* side may be made radius. (Art. 130.)

If the hypothenuse be made radius, as in Fig. 14, BC will be the sine of A, and AB the sine of C, or the cosine of A. (Art. 121.) And if *abc* be a similar triangle, whose hypothenuse is equal to the *tabular* radius, *bc* will be the tabular sine of A, and *ab* the tabular sine of C. (Art. 124.)

* The parts which are *given* are distinguished by a mark across the line, or at the opening of the angle, and the parts *required*, by a cipher.

To find the *perpendicular*, then, by Theorem I, we have this proportion ;

$$ac : AC :: bc : BC.$$
$$\text{Or} \quad R : AC :: \text{Sin } A : BC.$$

Whenever the terms Radius, Sine, Tangent, &c. occur in a proportion like this, the *tabular* Radius, &c. is to be understood, as in arts. 126, 127.

The numerical calculation, to find the length of BC, may be made, either by *natural* sines, or by *logarithms*. See art. 131.

By natural Sines.

$$1 : 45 :: 0.53484 : 24.068 = BC.$$

Computation by Logarithms.

As Radius		10.00000
To the hypothenuse	45	1.65321
So is the Sine of A	32° 20′	9.72823
To the perpendicular	24.068	1.38144

Here, the logarithms of the second and third terms are added, and from the sum, the first term 10 is subtracted. (Art. 132.) The remainder is the logarithm of 24.068=BC.

Subtracting the angle at A from 90°, we have the angle at C=57° 40′. (Art. 116.) Then, to find the *base* AB ;

$$ac : AC :: ab : AB$$
$$\text{Or} \quad R : AC :: \text{Sin } C : AB = 38.023.$$

Both the sides required are now found, by making the hypothenuse radius. The results here obtained may be verified, by making either of the other sides radius.

If the *base* be made radius, as in Fig. 15, the perpendicular will be the *tangent*, and the hypothenuse the *secant* of the angle at A. (Art. 122.) Then,

$$\text{Sec } A : AC :: R : AB$$
$$R : AB :: \text{Tan } A : BC$$

By making the arithmetical calculations, in these two proportions, the values of AB and BC will be found the same as before.

If the *perpendicular* be made radius, as in Fig. 16, AB will be the *tangent*, and AC the *secant* of the angle at C. Then,

$$\text{Sec C} : \text{AC} :: \text{R} : \text{BC}$$
$$\text{R} : \text{BC} :: \text{Tan C} : \text{AB}$$

Ex. 2. If the hypothenuse of a right angled triangle be 250 rods, and the angle at the base 46° 30′; what is the length of the base and perpendicular?

Ans. The base is 172.1 rods, and the perpendic. 181.35.

Case II.

135. Given { The hypothenuse, And one leg } to find { The angles and The other leg.

Ex. 1. If the hypothenuse (Fig. 18.) be 35 leagues, and the base 26; what is the length of the perpendicular, and the quantity of each of the acute angles?

To find the angles it is necessary that one of the *given* sides be made radius. (Art. 130.)

If the *hypothenuse* be radius, the base and perpendicular will be sines of their opposite angles. Then,

$$\text{AC} : \text{R} :: \text{AB} : \text{Sin C} = 47° \ 58'\tfrac{1}{2}$$

And to find the *perpendicular* by Theorem I;

$$\text{R} : \text{AC} :: \text{Sin A} : \text{BC} = 23.43$$

If the *base* be radius, the perpendicular will be *tangent*, and the hypothenuse *secant* of the angle at A. Then,

$$\text{AB} : \text{R} :: \text{AC} : \text{Sec A}$$
$$\text{R} : \text{AB} :: \text{Tan A} : \text{BC}$$

In this example, where the hypothenuse and base are given, the angles can not be found by making the *perpendicular* radius. For to find an angle, a *given* side must be made radius. (Art. 130.)

136. **Ex. 2.** If the hypothenuse (Fig.19.) be 54 miles, and the perpendicular 48 miles, what are the angles, and the base ?

Making the *hypothenuse* radius.

AC : R : : BC : Sin A
R : AC : : Sin C : AB

The numerical calculation will give A=62° 44′, and AB =24.74.

Making the *perpendicular radius.*

BC : R : : AC : Sec C
R : BC : : Tan C : AB

The angles can not be found by making the *base* radius, when its length is not given.

Case III.

137. Given { The angles, And one leg } to find { The hypothenuse, And the other leg.

Ex. 1. If the base (Fig. 20,) be 60, and the angle at the base 47° 12′, what is the length of the hypothenuse and the perpendicular ?

In this case, as *sides* only are required, *any* side may be radius.

Making the *hypothenuse* radius.

Sin C : AB : : R : AC=88.31
R : AC : : Sin A : BC=64.8

Making the *base* radius.

R : AB : : Sec A : AC
R : AB : : Tan A : BC

Making the *perpendicular* radius.

Tan C : AB : : R : BC
R : BC : : Sec C : AC

138. Ex. 2. If the perpendicular (Fig. 21.) be 74, and the angle C 61° 27′, what is the length of the base and the hypothenuse ?

Making the *hypothenuse* radius.

Sin A : BC : : R : AC
R : AC : : Sin C : AB

Making the *base* radius.

Tan A : BC : : R : AB
R : AB : : Sec A : AC

Making the *perpendicular* radius.

R : BC : : Sec C : AC
R : BC : : Tan C : AB

The hypothenuse is 154.83 and the base 136.

Case IV.

139. Given { The base, and Perpendicular } to find { The hypothenuse, And the angles.

Ex. 1. If the base (Fig. 22, be 284, and the perpendicular 192, what are the angles, and the hypothenuse ?

In this case, one of the legs must be made radius, to find an angle; because the hypothenuse is not given.

Making the *base* radius.

AB : R : : BC : Tan A=34° 4′
R : AB : : Sec A : AC=342.84

Making the *perpendicular* radius.

BC : R : : AB : Tan C
R : BC : : Sec C : AC

Ex. 2. If the base be 640, and the perpendicular 480, what are the angles and hypothenuse ?

Ans. The hypothenuse is 800, and the angle at the base 36° 52′ 12″.

Examples for practice.

1. Given the hypothenuse 68, and the angle at the base 39° 17′; to find the base and perpendicular.
2. Given the hypothenuse 850, and the base 594, to find the angles, and the perpendicular.
3. Given the hypothenuse 78 and perpendicular 57, to find the base, and the angles.
4. Given the base 723, and the angle at the base 64° 18′, to find the hypothenuse and perpendicular.
5. Given the perpendicular 632, and the angle at the base 81° 36′, to find the hypothenuse and the base.
6. Given the base 32, and the perpendicular 24, to find the hypothenuse, and the angles.

140. The preceding solutions are all effected, by means of the tabular sines, tangents, and secants. But, when any *two sides* of a right angled triangle are given, the third side may be found, without the aid of the trigonometrical tables, by the proposition, that *the square of the hypothenuse is equal to the sum of the squares of the two perpendicular sides* (Euc. 47. 1.)

If the legs be given, extracting the square root of the *sum* of their squares, will give the hypothenuse. Or, if the hypothenuse and one leg be given, extracting the square root of the *difference* of the squares, will give the other leg.

Let h=the hypothenuse, p=the perpendicular, b=the base } of a right angled triangle.

Then $h^2=b^2+p^2$, or (Alg. 296.) $h=\sqrt{b^2+p^2}$

By trans. $b^2=h^2-p^2$, or $b=\sqrt{h^2-p^2}$

And $p^2=h^2-b^2$, or $p=\sqrt{h^2-b^2}$

Ex. 1. If the base is 32, and the perpendicular 24, what is the hypothenuse? Ans. 40.

2. If the hypothenuse is 100, and the base 80, what is the perpendicular? Ans. 60.

3. If the hypothenuse is 300, and the perpendicular 220, what is the base?

Ans. $\overline{300}^2-\overline{220}^2=4160$, the root of which is 204 nearly.

141. It is generally most convenient to find the difference of the squares by *logarithms*. But this is not to be done by *subtraction*. For subtraction, in logarithms, performs the office of *division*. (Art. 41.) If we subtract the logarithm of b^2 from the logarithm of h^2, we shall have the logarithm, not of the *difference* of the squares, but of their *quotient*. There is, however, an indirect, though very simple method, by which the difference of the squares may be obtained by logarithms. It depends on the principle, that the *difference of the squares of two quantities is equal to the product of the sum and difference of the quantities*. (Alg. 235.) Thus

$$h^2-b^2=(h+b)\times(h-b)$$

as will be seen at once, by performing the multiplication. The two factors may be multiplied by *adding* their logarithms. Hence,

142. *To obtain the difference of the squares of two quantities, add the logarithm of the sum of the quantities, to the logarithm of their difference.* After the logarithm of the difference of the squares is found; the *square root* of this difference is obtained, by dividing the logarithm by 2. (Art. 47.)

Ex. 1. If the hypothenuse be 75 inches, and the base 45, what is the length of the perpendicular?

Sum of the given sides	120	log.	2.07918
Difference of do.	30		1.47712
	Dividing by		2)3.55630
Side required	60		1.77815

2. If the hypothenuse is 135, and the perpendicular 108, what is the length of the base? Ans. 81.

SECTION IV.

SOLUTIONS OF OBLIQUE ANGLED TRIANGLES.

ART. 143. THE sides and angles of oblique angled triangles may be calculated by the following theorems.

THEOREM I.

In any plane triangle, THE SINES OF THE ANGLES ARE AS THEIR OPPOSITE SIDES.

Let the angles be denoted by the letters A, B, C, and their opposite sides by a, b, c, as in Fig. 23 and 24. From one of the angles, let the line p be drawn perpendicular to the opposite side. This will fall either within or without the triangle.

1. Let it fall *within* as in Fig. 23. Then, in the right angled triangles ACD and BCD, according to art. 126,

$$R : b :: \text{Sin A} : p$$
$$R : a :: \text{Sin B} : p$$

Here, the two *extremes* are the same in both proportions. The other four terms are, therefore, *reciprocally* proportional: (Alg. 387.*) that is,

$$a : b :: \text{Sin A} : \text{Sin B}.$$

2. Let the perpendicular p fall *without* the triangle, as in Fig. 24. Then, in the right angled triangles ACD and BCD;

$$R : b :: \text{Sin A} : p$$
$$R : a :: \text{Sin B} : p$$

Therefore as before,

$$a : b :: \text{Sin A} : \text{Sin B}.$$

* Euclid 23. 5.

Sin A is here put both for the sine of DAC, and for that of BAC. For, as one of these angles is the *supplement* of the other, they have the same sine. (Art. 90.)

The sines which are mentioned here, and which are used in calculation, are *tabular* sines. But the proportion will be the same, if the sines be adapted to any other radius. (Art. 119.)

Theorem II.

144. In a plane triangle,

As the sum of any two of the sides,
To their difference;
So is the tangent of half the sum of the opposite angles;
To the tangent of half their difference.

Thus the sum of AB and AC (Fig. 25) is to their difference; as the tangent of half the sum of the angles ACB and ABC, to the tangent of half their difference.

Demonstration.

Extend CA to G, making AG equal to AB; then CG is *the sum of the two sides* AB and AC. On AB, set off AD equal to AC; then BD is *the difference of the sides* AB and AC.

The sum of the two angles ACB and ABC, is equal to the sum of ACD and ADC; because each of these sums is the supplement of CAD. (Art. 79.) But, as AC=AD by construction, the angle ADC=ACD. (Euc. 5. 1.) Therefore ACD is *half the sum* of ACB and ABC. As AB=AG, the angle AGB=ABG or DBE. Also GCE or ACD=ADC=BDE. (Euc. 15. 1.) Therefore, in the triangles GCE and DBE, the two remaining angles DEB and CEG are equal; (Art. 79.) So that CE is perpendicular to BG. (Euc. Def. 10. 1.) If then CE is made radius, GE is the tangent of GCE, (Art. 84.) that is, *the tangent of half the sum of the angles opposite to* AB and AC.

If from the greater of the two angles ACB and ABC, there be taken ACD their half sum; the remaining angle ECB will be their half difference. (Alg. 341.) The tangent of this angle, CE being radius, is EB, that is, *the tangent of half the difference of the angles opposite to* AB and AC. We have then,

CG=the sum of the sides AB and AC;
DB=their difference;
GE=the tangent of half the sum of the opposite angles;
EB=the tangent of half their difference.

But, by similar triangles,

CG : DB::GE : EB. Q. E. D.

THEOREM III.

145. If upon the longest side of a triangle, a perpendicular be drawn from the opposite angle;

AS THE LONGEST SIDE,
TO THE SUM OF THE TWO OTHERS;
SO IS THE DIFFERENCE OF THE LATTER,
TO THE DIFFERENCE OF THE SEGMENTS MADE BY THE PERPENDICULAR.

In the triangle ABC, (Fig. 26.) if a perpendicular be drawn from C upon AB;

$$AB : CB+CA :: CB-CA : BP-PA.*$$

Demonstration.

Describe a circle on the centre C, and with the radius BC. Through A and C, draw the diameter LD, and extend BA to H. Then by Euc. 35, 3,

$$AB\times AH=AL\times AD$$

Therefore,

$$AB : AD :: AL : AH$$

But $AD=CD+CA=CB+CA$
And $AL=CL-CA=CB-CA$
And $AH=HP-PA=BP-PA$ (Euc. 3. 3.)

If then, for the three last terms in the proportion, we substitute their equals, we have,

$$AB : CB+CA :: CB-CA : PB-PA$$

146. It is to be observed, that the greater segment is next the greater side. If BC is greater than AC, (Fig. 26.) PB is

* See note F.

greater than AP. With the radius AC, describe the arc AN. The segment NP=AP. (Euc. 3. 3.) But BP is greater than NP.

147. The two segments are to each other, as the tangents of the opposite angles, or the cotangents of the adjacent angles. For, in the right angled triangles ACP and BCP, (Fig. 26.) if CP be made radius, (Art. 126.)

R : PC::Tan ACP : AP
R : PC::Tan BCP : BP

Therefore, by equality of ratios, (Alg. 384.*)
Tan ACP : AP::Tan BCP : BP

That is, the segments are as the tangents of the opposite angles. And the tangents of these are the *cotangents* of the angles A and B. (Art. 89.)

Cor. The greater segment is opposite to the greater angle. And of the angles at the base, the less is next the greater side. If BP is greater than AP, the angle BCP is greater than ACP; and B is less than A. (Art. 77.)

---

148. To enable us to find the sides and angles of an oblique angled triangle, *three* of them must be *given*. (Art. 114.)

These may be, either

1. Two angles and a side, or
2. Two sides and an angle *opposite* one of them, or
3. Two sides and the *included* angle, or
4. The three sides.

The two first of these cases are solved by theorem I, (Art. 143.) the third by theorem II, (Art. 144.) and the fourth by theorem III, (Art. 145.)

149. In making the calculations, it must be kept in mind, that the greater side is always opposite to the greater angle, (Euc. 18, 19. 1.) that there can be only one *obtuse* angle in a triangle, (Art. 76.) and therefore, that the angles opposite to the two least sides must be *acute*.

* Euc. 11, 5

Case I.

150. Given,

Two angles, and A side, } to find { The remaining angle, and The other two sides.

The third angle is found, by merely subtracting the sum of the two which are given from 180°. (Art. 79)

The sides are found, by stating, according to theorem I, the following proportion;

> As the sine of the angle opposite the *given* side,
> To the length of the given side;
> So is the sine of the angle opposite the *required* side,
> To the length of the required side.

As a *side* is to be found, it is necessary to begin with a *tabular number*.

Ex. 1. In the triangle ABC (Fig. 27.) the side b is given 32 rods, the angle A 56° 20′, and the angle C 49° 10′, to find the angle B, and the sides a and c.

The sum of the two given angles 56° 20′+49° 10′=105° 30′; which subtracted from 180°, leaves 74° 30′ the angle B.

Then,

$$\text{Sin B} : b :: \begin{cases} \text{Sin A} : a \\ \text{Sin C} : c \end{cases}$$

Calculation by logarithms.

As the Sine of B	74° 30′	a. c.	0.01609
To the side b	32		1.50515
So is the Sine of A	56° 20′		9.92027
To the side a	27.64		1.44151

As the Sine of B	74° 30′	a. c.	0.01609
To the side b	32		1.50515
So is the Sine of C	49° 10′		9.87887
To the side c	25.13		1.40011

The *arithmetical complement* used in the first term here, may be found, in the usual way, or by taking out the *cosecant* of the given angle, and rejecting 10 from the index. (Art. 113.)

Ex. 2. Given the side b 71. the angle A 107° 6′ and the angle C 27° 40 ; to find the angle B, and the sides a and c. The angle B is 45° 14′. Then

$$\text{Sin B} : b :: \begin{cases} \text{Sin A} : a=95.58 \\ \text{Sin C} : c=46.43 \end{cases}$$

When one of the given angles is *obtuse*, as in this example, the sine of its *supplement* is to be taken from the tables. (Art. 99.)

Case II.

151. Given

Two sides, and An opposite angle,	to find	The remaining side, and The other two angles.

One of the required angles is found, by beginning with a side, and, according to theorem I, stating the proportion,

> As the side opposite the given angle,
> To the sine of that angle ;
> So is the side opposite the required angle,
> To the sine of that angle.

The third angle is found, by subtracting the sum of the other two from 180° ; and the remaining side is found, by the proportion in the preceding article.

152. In this second case, if the side opposite to the given angle be shorter than the other given side, the solution will be *ambiguous*. Two different triangles may be formed, each of which will satisfy the conditions of the problem.

Let the side b, (Fig. 28.) the angle A, and the length of the side opposite this angle be given. With the latter for radius, (if it be shorter than b,) describe an arc, cutting the line AH in the points B and B′. The lines BC and B′C will be equal. So that, with the same data, there may be formed two different triangles, ABC and AB′C.

There will be the same ambiguity in the numerical calculation. The answer found by the proportion will be the *sine* of an angle. But this may be the sine, either of the *acute* angle AB′C, or of the *obtuse* angle ABC. For, BC being equal to B′C, the angle CB′B is equal to CBB′. Therefore ABC, which is the supplement of CBB′ is also the supplement of CB′B. But the sine of an angle is the same, as the sine of its supplement. (Art. 90.) The result of the calculation will, therefore be ambiguous. In practice however, there will generally be some circumstances which will determine whether the angle required is acute or obtuse.

If the side opposite the given angle be *longer* than the other given side, the angle which is subtended by the latter, will necessarily be acute. For there can be but one obtuse angle in a triangle, and this is always subtended by the longest side. (Art. 149.)

If the *given* angle be obtuse, the other two will, of course, be acute. There can, therefore, be no ambiguity in the solution.

Ex. 1. Given the angle A, (Fig. 28.) 35° 20′, the opposite side a 50, and the side b 79; to find the remaining side, and the other two angles.

To find the angle opposite to b, (Art. 151.)

$$a : \text{Sin A} :: b : \text{Sin B}$$

The calculation here gives the acute angle AB′C 54°3′ 50″. and the obtuse angle ABC 125° 56′ 10″. If the latter be added to the angle at A 35° 20′, the sum will be 161° 16′ 10″, the supplement of which 18° 43′ 50″ is the angle ACB. Then in the triangle ABC, to find the side c=AB,

$$\text{Sin A} : a :: \text{Sin C} : c = 27.76$$

If the *acute* angle AB′C 54° 3′ 50″ be added to the angle at A 35° 20′, the sum will be 89° 23′ 50″, the supplement of which 90° 36′ 10″ is the angle ACB′. Then, in the triangle AB′C,

$$\text{Sin A} : \text{CB}' :: \text{Sin C} : \text{AB}' = 86.45$$

Ex. 2. Given the angle at A 63° 35′ (Fig. 29.) the side b 64, and the side a 72; to find the side c, and the angles B′ and C.

a : Sin A : : b : Sin B$=52°\ 45'\ 25''$
Sin A : a : : Sin C : $c=72.05$

The sum of the angles A and B is 116° 20′ 25″, the supplement of which 63° 39′ 35″ is the angle C.

In this example the solution is *not ambiguous*, because the side opposite the given angle is longer than the other given side.

Ex. 3. In a triangle of which the angles are A, B, and C, and the opposite sides a, b, and c, as before; if the angle A be 121° 40′, the opposite side a 68 rods, and the side b 47 rods; what are the angles B and C, and what is the length of the side c? Ans. B is 36° 2′ 4″, C 22° 17′ 56″, and c 30.3.

In this example also, the solution is not ambiguous, because the *given* angle is obtuse.

Case III.

153. Given

Two sides, and The included angle,	to find	The remaining side, and The other two angles.

In this case, the angles are found by theorem II. (Art. 144.) The required side may be found by theorem I.

In making the solutions, it will be necessary to observe, that by subtracting the given angle from 180°, the *sum* of the other two angles is found; (Art. 79.) and, that *adding half the difference of two quantities to their half sum gives the greater quantity, and subtracting the half difference from the half sum gives the less.* (Alg. 341.) The latter proposition may be geometrically demonstrated thus;

Let AE (Fig. 32.) be the greater of two magnitudes, and BE the less. Bisect AB in D, and make AC equal to BE. Then

AB is the *sum* of the two magnitudes;
CE their *difference;*
DA or DB *half* their *sum;*
DE or DC *half* their *difference;*
But DA+DE=AE the *greater* magnitude;
And DB−DE=BE the *less.*

Ex. 1. In the triangle ABC (Fig. 30.) the angle A is given 26° 14′ the side b 39, and the side c 53; to find the angles B and C, and the side a.

The *sum* of the sides b and c is	$53+39=92$,
And their *difference*	$53-39=14$.

The sum of the angles B and $C=180°-26°\ 14'=153°\ 46'$
And *half the sum* of B and C is 76° 53′

Then, by theorem II,

$$(b+c):(b-c)::\text{Tan}\ \tfrac{1}{2}(B+C):\text{Tan}\ \tfrac{1}{2}(B-C)$$

To and from the half sum	76° 53′
Adding and subtracting the half difference	33 8 50
We have the greater angle	110 1 50
And the less angle	43 44 10

As the greater of the two given sides is c, the greater angle is C, and the less angle B. (Art. 149.)

To find the side a, by theorem I,

$$\text{Sin B} : b :: \text{Sin A} : a = 24.94$$

Ex. 2. Given the angle A 101° 30′ the side b 76, and the side c 109; to find the angles B and C, and the side a.

B is 30° 57½′, C 47° 32½′, and a 144.8.

Case IV.

154. Given the three sides, to find the angles.

In this case, the solutions may be made, by drawing a perpendicular to the longest side, from the opposite angle. This will divide the given triangle into two *right angled* triangles. The two segments may be found by theorem III. (Art. 145.) There will then be given, in each of the right angled triangles, the hypothenuse and one of the legs, from which the angles may be determined, by rectangular trigonometry. (Art. 135.)

Ex. 1. In the triangle ABC (Fig. 31.) the side AB is 39, AC 35, and BC 27. What are the angles?

Let a perpendicular be drawn from C, dividing the longest side AB into the two segments AP and BP. Then by theorem III,

$$AB : AC+BC :: AC-BC : AP-BP$$

As the longest side	39 *a. c.*	8.40894
To the sum of the two others	62	1.79239
So is the difference of the latter	8	0.90309
To the difference of the segments	12.72	1.10442

The greater of the two segments is AP, because it is next the side AC, which is greater than BC. (Art. 146.)

To and from half the sum of the segments	19.5
Adding and subtracting half their difference, (Art.153.)	6.36
We have the greater segment AP	25.86
And the less BP	13.14

Then, in each of the right angled triangles APC and BPC, we have given the hypothenuse and base ; and by art. 135,

$$AC : R :: AP : \text{Cos } A=42° 21' 57''$$
$$BC : R :: BP : \text{Cos } B=60° 52' 42''$$

And subtracting the sum of the angles A and B from 180°, we have the remaining angle ACB=76° 45′ 21″.

Ex. 2. If the three sides of a triangle are 78, 96, and 104 ; what are the angles ?

Ans. 45° 41′ 48″, 61° 43′ 27″, and 72° 34′ 45″.

Examples for Practice.

1. Given the angle A 54° 30′, the angle B 63° 10′, and the side a 164 rods; to find the angle C, and the sides b and c.
2. Given the angle A 45° 6′ the opposite side a 93, and the side b 108 ; to find the angles B and C, and the side c.
3. Given the angle A 67° 24′, the opposite side a 62, and the side b 46 ; to find the angles B and C, and the side c.
4. Given the angle A 127° 42′, the opposite side a 381, and the side b 184; to find the angles B and C, and the side c.

5. Given the side *b* 58, the side *c* 67, and the included angle A=36°; to find the angles B and C, and the side *a*.
6. Given the three sides, 631, 268, and 546; to find the angles.

155. The three theorems demonstrated in this section, have been here applied to *oblique angled* triangles only. But they are equally applicable to *right angled* triangles.

Thus, in the triangle ABC, (Fig. 17.) according to theorem I, (Art. 143.)

Sin B : AC : : Sin A : BC

This is the same proportion as one stated in art. 134, except that, in the first term here, the *sine of* B is substituted for *radius*. But, as B is a right angle, its sine is *equal* to *radius*. (Art. 95.)

Again, in the triangle ABC, (Fig. 21.) by the same theorem;

Sin A : BC : : Sin C : AB

This is also one of the proportions in rectangular trigonometry, when the hypothenuse is made radius.

The other two theorems might be applied to the solution of right angled triangles. But, when one of the angles is *known* to be a right angle, the methods explained in the preceding section, are much more simple in practice.*

* For the application of Trigonometry to the Mensuration of Heights and Distances, see Navigation and Surveying.

SECTION V.

GEOMETRICAL CONSTRUCTION OF TRIANGLES, BY THE PLANE SCALE.

ART. 156. TO facilitate the construction of geometrical figures, a number of graduated lines are put upon the common two feet scale; one side of which is called the *Plane Scale*, and the other side, *Gunter's Scale*. The most important of these are the scales of *equal parts*, and the line of *chords*. In forming a given triangle, or any other right lined figure, the parts which must be made to agree with the conditions proposed, are the *lines*, and the *angles*. For the former, a scale of equal parts is used; for the latter, a line of chords.

157. The line on the upper side of the plane scale, is divided into *inches* and *tenths* of an inch. Beneath this, on the left hand, are two *diagonal* scales of equal parts,* divided into inches and half inches, by perpendicular lines. On the larger scale, one of the inches is divided into tenths, by lines which pass *obliquely* across, so as to intersect the parallel lines which run from right to left. The use of the oblique lines is to measure *hundredths* of an inch, by inclining more and more to the right, as they cross each of the parallels.

To take off, for instance, an extent of 3 inches, 4 tenths, and 6 hundredths;

Place one foot of the compasses at the intersection of the perpendicular line marked 3 with the parallel line marked 6, and the other foot at the intersection of the latter with the oblique line marked 4,

The other diagonal scale is of the same nature. The divisions are smaller, and are numbered from left to right.

158. In geometrical constructions, what is often required, is to make a figure, not *equal* to a given one, but only *similar*. Now figures are similar which have equal angles, and the

* These lines are not represented in the plate, as the learner is supposed to have the scale before him.

sides about the equal angles *proportioned.* (Euc. Def. 1. 6.) Thus a land surveyor, in plotting a field, makes the several lines in his plan to have the same proportion to each other, as the sides of the field. For this purpose, a scale of equal parts may be used, of any dimensions whatever. If the sides of the field are 2, 5, 7, and 10 *rods*, and the lines in the plan are 2, 5, 7, and 10 *inches*, and if the angles are the same in each, the figures are similar. One is a copy of the other, upon a smaller scale.

So any two right lined figures are similar, if the angles are the same in both, and if the number of smaller parts in each side of one, is equal to the number of larger parts in the corresponding sides of the other. The several divisions on the scale of equal parts may, therefore, be considered as representing any measures of length, as feet, rods, miles, &c. All that is necessary is, that the scale be not changed, in the construction of the same figure; and that the several divisions and subdivisions be properly proportioned to each other. If the larger divisions, on the diagonal scale, are units, the smaller ones are tenths and hundredths. If the larger are tens, the smaller are units and tenths.

159. In laying down an *angle*, of a given number of degrees, it is necessary to *measure* it. Now the proper measure of an angle is an arc of a circle. (Art. 74.) And the measure of an arc, where the radius is given, is its *chord*. For the chord is the distance, in a straight line, from one end of the arc to the other. Thus the chord AB (Fig. 33.) is a measure of the arc ADB, and of the angle ACB.

To form the *line of chords*, a circle is described, and the lengths of its chords determined for every degree of the quadrant. These measures are put on the plane scale, on the line marked CHO.

160. The chord of 60° is equal to *radius*. (Art. 95.) In laying down or measuring an angle, therefore, an arc must be drawn, with a radius which is equal to the extent from 0 to 60 on the line of chords. There are generally on the scale, two lines of chords. Either of these may be used; but the angle must be measured by the same line from which the radius is taken.

161. To *make an angle*, then, of a given number of degrees; From one end of a straight line as a centre, and with a radius equal to the chord of 60° on the line of chords, describe an arc of a circle cutting the straight line. From the

point of intersection, extend the chord of the given number of degrees, applying the other extremity to the arc; and through the place of meeting, draw the other line from the angular point.

If the given angle is *obtuse*, take from the scale the chord of *half* the number of degrees, and apply it *twice* to the arc. Or make use of the chords of any two arcs whose *sum* is equal to the given number of degrees.

A *right angle* may be constructed, by drawing a perpendicular without using the line of chords.

Ex. 1. To make an angle of 32 degrees. (Fig. 33.) With the point C, in the line CH, for a centre, and with the chord of 60° for radius, describe the arc ADF. Extend the chord of 32° from A to B; and through B, draw the line BC. Then is ACB an angle of 32 degrees.

2. To make an angle of 140 degrees. (Fig. 34.) On the line CH, with the chord of 60°, describe the arc ADF; and extend the chord of 70° from A to D, and from D to B. The arc $ADB=70° \times 2=140.°$

On the other hand,

162. To *measure an angle;* On the angular point as a centre, and with the chord of 60° for radius, describe an arc to cut the two lines which include the angle. The distance between the points of intersection, applied to the line of chords, will give the measure of the angle in degrees. If the angle be *obtuse,* divide the arc into two parts.

Ex. 1. To measure the angle ACB. (Fig. 33.) Describe the arc ADF cutting the lines CH and CB. The distance AB will extend 32° on the line of chords.

2. To measure the angle ACB. (Fig. 34.) Divide the arc ADB into two parts, either equal or unequal, and measure each part. by applying its chord to the scale. The sum of the two will be 140°.

163. Besides the lines of chords, and of equal parts, on the plane scale; there are also lines of natural *sines*, *tangents*, and *secants*, marked Sin. Tan. and Sec. of *semitangents*, marked S. T. of *longitude*, marked Lon. or M. L. of *rhumbs*, marked Rhu. or Rum. &c. These are not necessary in trigonometrical constructions. Some of them are used in Navigation; and some of them, in the projections of the Sphere.

164. In Navigation, the quadrant, instead of being graduated in the usual manner, is divided into *eight* portions, called

Rhumbs. The *Rhumb line,* on the scale, is a line of chords, divided into rhumbs and quarter-rhumbs, instead of degrees.

165. The line of *Longitude* is intended to show the number of geographical miles in a degree of longitude, at different distances from the equator. It is placed over the line of chords, with the numbers in an inverted order: so that the figure above shows the length of a degree of longitude, in any latitude denoted by the figure below.* Thus at the equator, where the latitude is 0, a degree of longitude is 60 geographical miles. In latitude 40, it is 46 miles; in latitude 60, 30 miles, &c.

166. The graduation on the line of *secants* begins where the line of sines ends. For the greatest sine is only equal to radius; but the secant of the least arc is greater than radius.

167. The *semitangents* are the tangents of *half* the given arcs. Thus the semitangent of 20° is the tangent of 10°. The line of semitangents is used in one of the projections of the sphere.

---

168. In the construction of *triangles,* the sides and angles which are *given,* are laid down according to the directions in arts. 158, 161. The parts *required* are then measured, according to arts. 158, 162. The following problems correspond with the four cases of oblique angled triangles; (Art. 148.) but are equally adapted to right angled triangles.

169. Prob. I. *The angles and one side* of a triangle being given; to find, by construction, the other two sides.

Draw the given side. From the ends of it, lay off two of the given angles. Extend the other sides till they intersect; and then measure their lengths on a scale of equal parts.

Ex. 1. Given the side b 32 rods, (Fig. 27.) the angle A 56° 20′, and the angle C 49° 10′; to construct the triangle, and find the lengths of the sides a and c.

Their lengths will be 25 and 27½.

2. In a right angled triangle, (Fig. 17.) given the hypothenuse 90, and the angle A 32° 20′, to find the base and perpendicular.

The length of AB will be 76, and of BC 48.

* Sometimes the line of longitude is placed *under* the line of chords.

3. Given the side AC 68, the angle A 124°, and the angle C 37°: to construct the triangle.

170 PROB. II. *Two sides and an opposite angle* being given, to find the remaining side, and the other two angles.

Draw one of the given sides; from one end of it, lay off the given angle; and extend a line indefinitely for the required side. From the other end of the first side, with the remaining given side for radius. describe an arc cutting the indefinite line. The point of intersection will be the end of the required side.

If the side opposite the given angle be less than the other given side, the case will be *ambiguous*. (Art. 152.)

Ex. 1. Given the angle A 63° 35′ (Fig. 29.) the side b 32, and the side a 36.

The side AB will be 36 nearly, the angle B 52° 45½′ and C 63° 39½′

2. Given the angle A (Fig. 28.) 35° 20′; the opposite side a 25, and the side b 35.

Draw the side b 35, make the angle A 35° 20′, and extend AH indefinitely. From C with radius 25, describe an arc cutting AH in B and B′. Draw CB and CB′, and two triangles will be formed, ABC and AB′C, each corresponding with the conditions of the problem.

3. Given the angle A 116°, the opposite side a 38, and the side b 26; to construct the triangle.

171. PROB. III. *Two sides and the included angle* being given; to find the other side and angles.

Draw one of the given sides. From one end of it lay off the given angle, and draw the other given side. Then connect the extremities of this and the first line.

Ex. 1. Given the angle A (Fig. 30.) 26° 14′, the side b 78, and the side c 106; to find B, C, and a.

The side a will be 50, the angle B 43° 44′, and C 110° 2.′

2. Given A 86°, b 65, and c 83; to find B, C, and a.

172. PROB. IV. The *three sides* being given; to find the angles.

Draw one of the sides, and from one end of it, with an extent equal to the second side, describe an arc. From the other end, with an extent equal to the third side, describe a second arc cutting the first; and from the point of intersection draw the two sides. (Euc. 22. 1.)

Ex. 1. Given AB (Fig. 31.) 78, AC 70, and BC, 54; to find the angles.

The angles will be A 42° 22′, B 60° 52$\frac{2}{3}$′, and C 76° 45$\frac{1}{3}$′.

2. Given, the three sides 68, 39, and 46; to find the angles.

173. Any right lined figure whatever, whose sides and angles are given, may be constructed, by laying down the sides from a scale of equal parts, and the angles from a line of chords.

Ex. Given the sides AB (Fig. 35.) =20, BC=22, CD=30, DE=12; and the angles B=102°, C=130°, D=108°, to construct the figure.

Draw the side AB=20, make the angle B=102°, draw BC=22, make C=130°, draw CD=30, make D=108°, draw DE=12, and connect E and A.

The last line EA may be measured on the scale of equal parts; and the angles E and A, by a line of chords.

SECTION VI.

DESCRIPTION AND USE OF GUNTER'S SCALE.

ART. 174. AN expeditious method of solving the problems in trigonometry, and making other logarithmic calculations, in a mechanical way, has been contrived by Mr. Edmund Gunter. The logarithms of numbers, of sines, tangents, &c. are represented by *lines*. By means of these, multiplication, division, the rule of three, involution, evolution, &c. may be performed much more rapidly, than in the usual method by figures.

The logarithmic lines are generally placed on one side only of the scale in common use. They are,

A line of artificial *Sines* divided into *Rhumbs*, and marked		S. R.
A line of artificial *Tangents*,	do.	T. R.
A line of the logarithms of *numbers*,		Num.
A line of artificial *Sines*, to every *degree*,		SIN.
A line of artificial *Tangents*,	do.	TAN.
A line of *Versed Sines*,		V. S.

To these are added a line of *equal parts*, and a line of *Meridional Parts*, which are not logarithmic. The latter is used in Navigation.

The Line of Numbers.

175. Portions of the line of *Numbers*, are intended to represent the *logarithms* of the natural series of numbers 2, 3, 4, 5, &c.

The logarithms of 10, 100, 1000, &c. are 1, 2, 3, &c. (Art. 3.)

If then, the log. of 10 be represented by a line of 1 foot;
the log. of 100 will be repres'd by one of 2 feet;
the log. of 1000 by one of 3 feet;
the lengths of the several lines being proportional to the corresponding logarithms in the tables. *Portions* of a foot will represent the logarithms of numbers between 1 and 10;

and portions of a line 2 feet long, the logarithms of numbers between 1 and 100.

On Gunter's scale, the line of the logarithms of numbers begins at a brass pin on the left, and the divisions are numbered 1, 2, 3, &c. to another pin near the middle. From this the numbers are repeated, 2, 3, 4, &c. which may be read 20, 30, 40, &c. The logarithms of numbers between 1 and 10 are represented by portions of the first half of the line; and the logarithms of numbers between 10 and 100, by portions greater than half the line, and less than the whole.

176. The logarithm of 1, which is 0, is denoted, not by any extent of line, but by a *point* under 1, at the commencement of the scale. The distances from this point to different parts of the line, represent other logarithms, of which the *figures* placed over the several divisions are the *natural numbers*. For the intervening logarithms, the intervals between the figures, are divided into tenths, and sometimes into smaller portions. On the right hand half of the scale, as the divisions which are numbered are *tens*, the subdivisions are units.

Ex. 1. To take from the scale the logarithm of 3.6; set one foot of the compasses under 1 at the beginning of the scale, and extend the other to the 6th division after the first figure 3.

2. For the logarithm of 47; extend from 1 at the beginning, to the 7th subdivision after the second figure 4.*

177. It will be observed, that the divisions and subdivisions *decrease*, from left to right; as in the tables of *logarithms*, the differences decrease. The difference between the logarithms of 10 and 100 is no greater, than the difference between the logarithms of 1 and 10.

178. The line of numbers, as it has been here explained, furnishes the logarithms of all numbers between 1 and 100. And if the indices of the logarithms be neglected, the same scale may answer for all numbers whatever. For the *decimal* part of the logarithm of any number is the same, as that of the number multiplied or divided by 10, 100, &c. (Art. 14.) In logarithmic calculations, the use of the indices is to determine the distance of the several figures of the natural numbers from the place of units. (Art. 11.) But in those cases in which the logarithmic line is commonly used, it will

* If the compasses will not reach the distance required; first open them so as to take off *half*, or any part of the distance, and then the remaining part.

not generally be difficult to determine the local value of the figures in the result.

179. We may, therefore, consider the *point* under 1 at the left hand, as representing the logarithm of 1, or 10, or 100; or $\frac{1}{10}$, or $\frac{1}{100}$, &c. for the decimal part of the logarithm of each of these is 0. But if the first 1 is reckoned 10, all the succeeding numbers must also be increased in a tenfold ratio; so as to read, on the first half of the line, 20, 30, 40, &c. and on the other half, 200, 300, &c.

The whole extent of the logarithmic line,

is from 1	to 100,		or from 0.1	to 10,	
or from 10	to 1000,		or from 0.01	to 1,	
or from 100	to 10000, &c.		or from 0.001	to 0.1, &c.	

Different values may, on different occasions, be assigned to the several numbers and subdivisions marked on this line. But for any one calculation, the value must remain the same.

Ex. Take from the scale 365.

As this number is between 10 and 1000, let the 1 at the beginning of the scale, be reckoned 10. Then, from this point to the second 3 is 300; to the 6th dividing stroke is 60; and half way from this to the next stroke is 5.

180. Multiplication, division, &c. are performed by the line of numbers, on the same principle, as by common logarithms. Thus,

To *multiply* by this line, *add* the logarithms of the two factors; (Art. 37.) that is, take off, with the compasses, that length of line which represents the logarithm of *one* of the factors, and apply this so as to extend forward from the end of that which represents the logarithm of the *other* factor. The sum of the two will reach to the end of the line representing the logarithm of the product.

Ex. Multiply 9 into 8. The extent from 1 to 8, added to that from 1 to 9, will be equal to the extent from 1 to 72 the product.

181. To *divide* by the logarithmic line, *subtract* the logarithm of the divisor from that of the dividend; (Art. 41.) that is, take off the logarithm of the divisor, and this extent set back from the end of the logarithm of the dividend, will reach to the logarithm of the quotient.

Ex. Divide 42 by 7. The extent from 1 to 7, set back from 42, will reach to 6 the quotient.

182. *Involution* is performed in logarithms, by multiplying the logarithm of the quantity into the index of the power;

(Art. 45.) that is, by *repeating* the logarithms as many times as there are units in the index. To involve a quantity on the scale, then, take in the compasses the linear logarithm, and *double it*, *treble it*, &c. according to the index of the proposed power.

Ex. 1. Required the square of 9. Extend the compasses from 1 to 9. *Twice* this extent will reach to 81 the square.

2. Required the cube of 4. The extent from 1 to 4, repeated *three times*, will reach to 64 the cube of 4.

183. On the other hand, to perform *evolution* on the scale; take *half*, *one third*, &c. of the logarithm of the quantity, according to the index of the proposed root.

Ex. 1. Required the square root of 49. *Half* the extent from 1 to 49, will reach from 1 to 7 the root.

2. Required the cube root of 27. *One third* the distance from 1 to 27, will extend from 1 to 3 the root.

184. The *Rule of Three* may be performed on the scale, in the same manner as in logarithms, by adding the two middle terms, and from the sum, subtracting the first term. (*Art.* 52.) But it is more convenient in practice to *begin* by subtracting the first term from one of the others. If four quantities are proportional, the quotient of the first divided by the second, is equal to the quotient of the third divided by the fourth. (Alg. 364.)

Thus if $a : b :: c : d$, then $\frac{a}{b}=\frac{c}{d}$, and $\frac{a}{c}=\frac{b}{d}$. (Alg. 380.)

But in logarithms, *subtraction* takes the place of division; so that,

log. a – log. b = log. c – log. d. Or log. a – log. c = log. b – log. d.

Hence,

185. On the scale, *the difference between the first and second terms of a proportion, is equal to the difference between the third and fourth.* Or, the difference between the first and third terms, is equal to the difference between the second and fourth.

The difference between the two terms is taken, by extending the compasses from one to the other. If the second term be greater than the first; the fourth must be greater than the third; if less, less. (Alg. 395.*) Therefore if the compasses extend *forward* from *left to right*, that is, from a

* Euc. 14. 5.

less number to a greater, from the first term to the second; they must also extend forward from the third to the fourth. But if they extend *backward*, from the first term to the second; they must extend the same way, from the third to the fourth.

Ex. 1. In the proportion 3 : 8::12 : 32, the extent from 3 to 8, will reach from 12 to 32; Or, the extent from 3 to 12, will reach from 8 to 32.

2. If 54 yards of cloth cost 48 dollars, what will 18 yards cost? 54 : 48::18 : 16

The extent from 54 to 48, will reach *backwards* from 18 to 16.

3. If 63 gallons of wine cost 81 dollars, what will 35 gallons cost? 63 : 81::35 : 45

The extent from 63 to 81, will reach from 35 to 45.

The Line of Sines.

186. The line on Gunter's scale marked SIN. is a line of logarithmic sines, made to correspond with the line of numbers. The whole extent of the line of numbers, (Art. 179.)
is from 1 to 100, whose logs. are 0.00000 and 2.00000,
or from 10 to 1000, whose logs. are 1.00000 and 3.00000,
or from 100 to 10000, whose logs. are 2.00000 4.00000,
the *difference of the indices* of the two extreme logarithms being in each case 2.

Now the logarithmic sine of 0° 34′ 22″ 41‴ is 8.00000
And the sine of 90° (Art. 95.) is 10.00000

Here also the difference of the indices is 2. If then the point directly beneath one extremity of the line of numbers, be marked for the sine of 0° 34′ 22″ 41‴; and the point beneath the other extremity, for the sine of 90°; the interval may furnish the intermediate sines; the divisions on it being made to correspond with the decimal part of the logarithmic sines in the tables.*

* To represent the sines *less* than 34′ 22″ 41‴, the scale must be extended on the left indefinitely. For, as the sine of an arc approaches to 0, its logarithm, which is negative, increases without limit. (Art. 15.)

The first dividing stroke in the line of Sines is generally at 0° 40′, a little farther to the right than the beginning of the line of Numbers. The next division is at 0° 50′; then begins the numbering of the degrees, 1, 2, 3, 4, &c. from left to right.

The line of Tangents.

187. The first 45 degrees on this line are numbered from left to right, nearly in the same manner as on the line of Sines.

The logarithmic tangent of 0° 34′ 22″ 35‴ is 8.00000
And the tangent of 45°, (Art. 95.) is 10.00000

The difference of the indices being 2, 45 degrees will reach to the end of the line. For those above 45° the scale ought to be continued much farther to the right. But as this would be inconvenient, the numbering of the degrees, after reaching 45, is *carried back* from right to left. The same dividing stroke answers for an arc and its *complement*, one above and the other below 45°. For, (Art. 93. Propor. 9.)

$$\tan : R :: R : \cot$$

In logarithms, therefore, (Art. 184.)

$$\tan - R = R - \cot.$$

That is, the *difference* between the tangent and radius, is equal to the difference between radius and the cotangent: in other words, one is as much *greater* than the tangent of 45°, as the other is *less*. In taking, then, the tangent of an arc greater than 45°, we are to suppose the distance between 45 and the division marked with a given number of degrees, to be added to the whole line, in the same manner as if the line were continued out. In working proportions, extending the compasses *back*, from a less number to a greater, must be considered the same as carrying them *forward* in other cases. See art. 185.

Trigonometrical Proportions on the Scale.

188. In working proportions in trigonometry by the scale; *the extent from the first term to the middle term of the same*

name, will reach from the other middle term to the fourth term. (Art. 185.)

In a trigonometrical proportion, two of the terms are the lengths of sides of the given triangle; and the other two are tabular sines, tangents, &c. The former are to be taken from the line of numbers; the latter, from the lines of logarithmic sines and tangents. If one of the terms is a *secant*, the calculation cannot be made on the scale, which has commonly no line of secants. It must be kept in mind that *radius* is equal to the sine of 90°, or to the tangent of 45°. (Art. 95.) Therefore, whenever radius is a term in the proportion, one foot of the compasses must be set on the end of the line of sines or of tangents.

189. The following examples are taken from the proportions which have already been solved by numerical calculation.

Ex. 1. In Case I of right angled triangles, (Art. 134. ex. 1.)

R : 45::Sin 32° 20' : 24

Here the third term is a *sine;* the first term radius is, therefore, to be considered as the sine of 90°. Then the extent from 90° to 32° 20' on the line of sines, will reach from 45 to 24 on the line of numbers. As the compasses are set *back* from 90° to 32° 20'; they must also be set back from 45. (Art. 185.)

2. In the same case, if the base be made radius, (Page 60.)

R : 38::Tan 32° 20' : 24

Here, as the third term is a *tangent*, the first term radius is to be considered the tangent of 45°. Then the extent from 45° to 32° 20' on the line of tangents, will reach from 38 to 24 on the line of numbers.

3. If the perpendicular be made radius, (Page 60.)

R : 24::Tan 57° 40' : 38

The extent from 45° to 57° 40' on the line of tangents, will reach from 24 to 38 on the line of numbers. For the tangent of 57° 40' on the scale look for its *complement* 32° 20'. (Art. 187.) In this example, although the compasses extend *back* from 45° to 57° 40′; yet, as this is from a *less* number to a *greater*, they must extend *forward* on the line of numbers. (Art. 185. 187.)

4. In art. 135, 35 : R : : 26 : Sin 48°

The extent from 35 to 26 will reach from 90° to 48°.

5. In art. 136, R : 48 : : Tan 27°$\frac{1}{4}$: 24$\frac{3}{4}$

The extent from 45° to 27°$\frac{1}{4}$, will reach from 48 to 24$\frac{3}{4}$.

6. In art. 150, ex. 1. Sin 74° 30′ : 32 : : Sin 56° 20′ : 27$\frac{1}{2}$.

For other examples, see the several cases in Sections III. and IV.

190. Though the solutions in trigonometry may be effected by the logarithmic scale, or by geometrical construction, as well as by arithmetical computation; yet the latter method is by far the most accurate. The first is valuable principally for the *expedition* with which the calculations are made by it. The second is of use, in presenting the *form* of the triangle to the eye. But the accuracy which attends arithmetical operations, is not to be expected, in taking lines from a scale with a pair of compasses.*

* See Note G.

SECTION VII.*

THE FIRST PRINCIPLES OF TRIGONOMETRICAL ANALYSIS.

ART. 191. IN the preceding sections, sines, tangents, and secants, have been employed in calculating the sides and angles of triangles. But the use of these lines is not confined to this object. Important assistance is derived from them, in conducting many of the investigations in the higher branches of analysis, particularly in physical astronomy. It does not belong to an elementary treatise of trigonometry, to prosecute these inquiries to any considerable extent. But this is the proper place for *preparing the formulæ*, the applications of which are to be made elsewhere.

Positive and negative SIGNS *in trigonometry.*

192. Before entering on a particular consideration of the algebraic expressions which are produced by combinations of the several trigonometrical lines, it will be necessary to attend to the positive and negative *signs* in the different quarters of the circle. The sines, tangents, &c. in the tables, are calculated for a single quadrant only. But these are made to answer for the whole circle. For they are of the same length in each of the four quadrants. (Art. 90.) Some of them, however, are *positive;* while others are *negative.* In algebraic processes, this distiction must not be neglected.

193. For the purpose of tracing the changes of the signs, in different parts of the circle, let it be supposed that a straight line CT (Fig. 36.) is fixed at one end C, while the other end is carried round, like a rod moving on a pivot; so that the point S shall describe the circle ABDH. If the two diameters AD and BH be perpendicular to each other, they will divide the circle into quadrants.

* Euler's Analysis of Infinites, Hutton's Mathematics, Lacroix's Differential Calculus, Mansfield's Essays, Legendre's, Lacroix's, Playfair's, Cagnoli's, and Woodhouse's Trigonometry.

194. In the *first quadrant* AB, the sine, cosine, tangent, &c. are considered *all positive*. In the *second quadrant* BD, the sine P′S′ continues *positive;* because it is still on the *upper* side of the diameter AD, from which it is measured. But the *cosine*, which is measured from BH, becomes *negative*, as soon as it changes from the *right* to the *left* of this line. (Alg. 507.) In the *third quadrant*, the *sine* becomes negative, by changing from the upper side to the under side of DA. The *cosine* continues negative, being still on the left of BH. In the *fourth quadrant*, the *sine* continues negative. But the *cosine* becomes positive, by passing to the right of BH.

195. The signs of the *tangents* and *secants* may be derived from those of the sines and cosines. The relations of these several lines to each other must be such, that a uniform method of calculation may extend through the different quadants.

In the first quadrant, (Art. 93. Propor. 1.)

$$R : \cos :: \tan : \sin, \text{ that is, } \operatorname{Tan} = \frac{R \times \sin}{\cos}.$$

The sign of the quotient is determined from the signs of the divisor and dividend. (Alg. 123.) The radius is considered as always positive. If then the sine and cosine be both positive, or both negative, the tangent will be positive. But if one of these be positive, while the other is negative, the tangent will be negative.

Now, by the preceding article,

In the 2d quadrant, the sine is positive, and the cosine negative.

The tangent must therefore be *negative*.

In the 3d quadrant, the sine and cosine are both negative.

The tangent must therefore be *positive*.

In the 4th quadrant, the sine is negative, and the cosine positive.

The tangent must therefore be *negative*.

196. By the 9th, 3d, and 6th proportions in art. 93.

$$1. \operatorname{Tan} : R :: R : \cot, \text{ that is, } \operatorname{Cot} = \frac{R^2}{\tan}.$$

Therefore, as radius is uniformly positive, the *cotangent* must have the same sign as the tangent.

2. Cos : R::R : sec, that is, $\text{Sec}=\frac{R^2}{\cos}$.

The *secant*, therefore, must have the same sign as the cosine.

3. Sin : R::R : cosec, that is, $\text{Cosec}=\frac{R^2}{\sin}$.

The *cosecant*, therefore, must have the same sign as the sine.

The *versed sine*, as it is measured from A, in one direction only, is invariably positive.

197. The *tangent* AT (Fig. 36.) increases, as the arc extends from A towards B. See also Fig. 11. Near B the increase is very rapid; and when the difference between the arc and 90°, is *less* than any assignable quantity, the tangent is *greater* than any assignable quantity, and is said to be *infinite*. (Alg. 447.) If the arc is *exactly* 90 degrees, it has, strictly speaking, *no* tangent. For a tangent is a line, drawn perpendicular to the diameter which passes through one end of the arc, and extended till it *meets* a line proceeding from the centre, through the other end. (Art. 84.) But if the arc is 90 degrees, as AB, (Fig. 36.) the angle ACB is a right angle, and therefore AT is *parallel* to CB; so that, if these lines be extended ever so far, they can never meet. Still, as an arc infinitely near to 90° has a tangent infinitely great, it is frequently said, in concise terms, that the tangent of 90° is infinite.

In the second quadrant, the tangent is, at first, infinitely great, and gradually diminishes, till at D it is reduced to nothing. In the third quadrant it increases again, becomes infinite near H, and is reduced to nothing at A.

The *cotangent* is inversely as the tangent. It is therefore nothing at B and H, (Fig. 36.) and infinite near A and D.

198. The *secant* increases with the tangent, through the first quadrant, and becomes infinite near B; it then diminishes, in the second quadrant, till at D it is equal to the radius CD. In the third quadrant it increases again, becomes infinite near H, after which it diminishes, till it becomes equal to radius.

The *cosecant* decreases, as the secant increases, and *v. v.* It is therefore equal to radius at B and H, and infinite near A and D.

199. The *sine* increases through the first quadrant, till at B (Fig. 36.) it is equal to radius. See also Fig. 13. It then diminishes, and is reduced to nothing at D. In the third quadrant. it increases again, becomes equal to radius at H, and is reduced to nothing at A.

The *cosine* decreases through the first quadrant, and is reduced to nothing at B. In the second quadrant, it increases, till it becomes equal to radius at D. It then diminishes again, is reduced to nothing at H, and afterwards increases till it becomes equal to radius at A.

In all these cases, the arc is supposed to *begin* at A, and to extend round in the direction of BDH.

200. The *sine* and *cosine* vary from nothing to radius, which they never exceed. The *secant* and *cosecant* are never less than radius, but may be greater than any given length. The *tangent* and *cotangent* have every value from nothing to infinity. Each of these lines, after reaching its *greatest* limit, begins to *decrease;* and as soon as it arrives at its *least* limit, begins to *increase.* Thus the sine begins to decrease, after becoming equal to radius, which is its greatest limit. But the secant begins to increase after becoming equal to radius, which is its least limit.

201. The substance of several of the preceding articles is comprised in the following tables. The first shows the *signs* of the trigonometrical lines, in each of the quadrants of the circle. The other gives the *values* of these lines, at the extremity of each quadrant.

Quadrant	1st	2d	3d	4th
Sine and cosecant	+	+	−	−
Cosine and secant	+	−	−	+
Tangent and cotangent	+	−	+	−

	0°	90°	180°	270°	360°
Sine	0	r	0	r	0
Cosine	r	0	r	0	r
Tangent	0	∞	0	∞	0
Cotangent	∞	0	∞	0	∞
Secant	r	∞	r	∞	r
Cosecant	∞	r	∞	r	∞

Here r is put for radius, and ∞ for infinite.

202. By comparing these two tables, it will be seen, that each of the trigonometrical lines changes from positive to negative, or from negative to positive, in that part of the circle

in which the line is either *nothing* or *infinite.* Thus the tangent changes from positive to negative, in passing from the first quadrant to the second, through the place where it is infinite. It becomes positive again, in passing from the second quadrant to the third, through the point in which it is nothing.

203. There can be no more than 360 degrees in any circle. But a body may have a number of successive revolutions in the same circle; as the earth moves round the sun, nearly in the same orbit, year after year. In astronomical calculations, it is frequently necessary to add together parts of different revolutions. The sum may be more than 360° But a body which has made more than a complete revolution in a circle, is only brought back to a point which it had passed over before. So the sine, tangent, &c. of an arc greater than 360°, is the same as the sine, tangent, &c. of some arc less than 360°. If an entire circumference, or a number of circumferences be added to any arc, it will terminate in the same point as before. So that, if C be put for a whole circumference, or 360°, and x be any arc whatever;

$$\sin x = \sin(C+x) = \sin(2C+x) = \sin(3C+x), \text{ \&c.}$$
$$\tan x = \tan(C+x) = \tan(2C+x) = \tan(3C+x), \text{ \&c.}$$

204. It is evident also, that, in a number of successive revolutions, in the same circle;

The first quadrant must coincide with the 5th, 9th, 13th, 17th,
The second, with the 6th, 10th, 14th, 18th, &c.
The third, with the 7th, 11th, 15th, 19th, &c.
The fourth, with the 8th, 12th, 16th, 20th, &c.

205. If an arc extending in a certain direction from a given point, be considered *positive;* an arc extending from the same point, in an *opposite* direction, is to be considered *negative.* (Alg. 507.) Thus, if the arc extending from A to S (Fig. 36.) be positive; an arc extending from A to S‴ will be negative. The latter will not terminate in the *same quadrant* as the other; and the sines of the tabular lines must be accommodated to this circumstance. Thus the sine of AS will be positive, while that of AS‴ will be negative. (Art. 194.) When a greater arc is subtracted from a less, if the latter be positive, the *remainder* must be negative. (Alg. 58, 9.)

TRIGONOMETRICAL FORMULÆ.

206. From the view which has here been taken of the changes in the trigonometrical lines, it will be easy to see, in

what parts of the circle each of them increases or decreases. But this does not determine their exact values, except at the extremities of the several quadrants. In the analytical investigations which are carried on by means of these lines, it is necessary to calculate the changes produced in them, by a given increase or diminution of the arcs to which they belong. In this there would be no difficulty, if the sines, tangents, &c. were *proportioned* to their arcs. But this is far from being the case. If an arc is doubled, its sine is *not* exactly doubled. Neither is its tangent or secant. We have to inquire, then, in what manner, the sine, tangent, &c. of one arc may be obtained, from those of other arcs already known.

The problem on which almost the whole of this branch of analysis depends, consists in deriving, from the sines and cosines of two given arcs, expressions for the sine and cosine of their *sum* and *difference*. For, by addition and subtraction, a few arcs may be so combined and varied, as to produce others of almost every dimension. And the expressions for the tangents and secants may be deduced from those of the sines and cosines.

Expressions for the SINE *and* COSINE *of the* SUM *and* DIFFERENCE *of arcs.*

207. Let a=AH, the greater of the given arcs,
And b=HL=HD, the less. (Fig. 37.)

Then $a+b$=AH+HL=AL, the *sum* of the two arcs,
And $a-b$=AH−HD=AD, their *difference*.

Draw the chord DL, and the radius CH, which may be represented by R. As DH is, by construction, equal to HL; DQ is equal to QL, and therefore DL is perpendicular to CH. (Euc. 3. 3.) Draw DO, HN, QP, and LM, each perpendicular to AC; and DS and QB parallel to AC.

From the definitions of the sine and cosine, (Art. 82, 9,) it is evident, that

$$\text{The sine}\begin{cases}\text{of AH, that is, } \sin a=\text{HN},\\ \text{of HL, that is, } \sin b=\text{QL},\\ \text{of AL, i.e.} \sin(a+b)=\text{LM},\\ \text{of AD, i.e.} \sin(a-b)=\text{DO}.\end{cases}$$

The cosine $\begin{cases} \text{of AH, that is, } \cos a = CN, \\ \text{of HL, that is, } \cos b = CQ, \\ \text{of AL, i.e.} \cos(a+b) = CM, \\ \text{of AD, i.e.} \cos(a-b) = CO. \end{cases}$

The triangle CHN is obviously similar to CQP ; and it is also similar to BLQ, because the sides of the one are perpendicular to those of the other, each to each. We have, then,

1. CH : CQ::HN : QP, that is, $R : \cos b :: \sin a : QP$.
2. CH : QL::CN : BL, $R : \sin b :: \cos a : BL$,
3. CH : CQ::CN : CP, $R : \cos b :: \cos a : CP$,
4. CH : QL::HN : QB, $R : \sin b :: \sin a : QB$,

Converting each of these proportions into an equation ;

1. $QP = \frac{\sin a \cos b}{R}$* 3. $CP = \frac{\cos a \cos b}{R}$

2. $BL = \frac{\sin b \cos a}{R}$ 4. $QB = \frac{\sin a \sin b}{R}$

Then adding the first and second,

$$QP + BL = \frac{\sin a \cos b + \sin b \cos a}{R}$$

Subtracting the second from the first.

$$QP - BL = \frac{\sin a \cos b - \sin b \cos a}{R}$$

Subtracting the fourth from the third.

$$CP - QB = \frac{\cos a \cos b - \sin a \sin b}{R}$$

Adding the third and fourth,

$$CP + QB = \frac{\cos a \cos b + \sin a \sin b}{R}$$

* In these formulæ, the sign of multiplication is omitted ; $\sin a \cos b$ being put for $\sin a \times \cos b$, that is the product of the sine of a into the cosine of b.

But it will be seen, from the figure, that

$$QP+BL=BM+BL=LM=\sin(a+b)$$
$$QP-BL=QP-QS=DO=\sin(a-b)$$
$$CP-QB=CP-PM=CM=\cos(a+b)$$
$$CP+QB=CP+SD=CO=\cos(a-b)$$

208. If then, for the first member of each of the four equations above, we substitute its value, we shall have,

$$\text{I.}\quad \sin(a+b)=\frac{\sin a\cos b+\sin b\cos a}{R}$$

$$\text{II.}\quad \sin(a-b)=\frac{\sin a\cos b-\sin b\cos a}{R}$$

$$\text{III.}\quad \cos(a+b)=\frac{\cos a\cos b-\sin a\sin b}{R}$$

$$\text{IV.}\quad \cos(a-b)=\frac{\cos a\cos b+\sin a\sin b}{R}$$

Or, multiplying both sides by R,

$$R\sin(a+b)=\sin a\cos b+\sin b\cos a,$$
$$R\sin(a-b)=\sin a\cos b-\sin b\cos a$$
$$R(\cos(a+b)=\cos a\cos b-\sin a\sin b$$
$$R\cos(a-b)=\cos a\cos b+\sin a\sin b$$

That is, the product of radius and the *sine* of the *sum* of two arcs, is equal to the product of the sine of the first arc into the cosine of the second + the product of the sine of the second into the cosine of the first.

The product of radius and the *sine* of the *difference* of two arcs, is equal to the product of the sine of the first arc into the cosine of the second —the product of the sine of the second into the cosine of the first.

The product of radius and the *cosine* of the *sum* of two arcs, is equal to the product of the cosines of the arcs — the product of their sines.

The product of radius and the *cosine* of the *difference* of two arcs, is equal to the product of the cosines of the arcs + the product of their sines.

These four equations may be considered as fundamental propositions, in what is called the *Arithmetic* of *Sines and Cosines, or Trigonometrical Analysis.*

Expressions for the sine and cosine of a DOUBLE *arc.*

209. When the sine and cosine of any arc are given, it is easy to derive from the equations in the preceding article, expressions for the sine and cosine of *double* that arc. As the two arcs a and b may be of any dimensions, they may be supposed to be *equal*. Substituting, then, a for its equal b, the first and the third of the four preceding equations will become,

$$R\ sin\ (a+a)=sin\ a\ cos\ a+sin\ a\ cos\ a$$
$$R\ cos\ (a+a)=cos\ a\ cos\ a-sin\ a\ sin\ a$$

That is, writing $sin^2 a$ for the square of the sine of a, and $cos^2 a$ for the square of the cosine of a,

$$\text{I.}\quad R\ sin\ 2a=2sin\ a\ cos\ a$$
$$\text{II.}\quad R\ cos\ 2a=cos^2 a-sin^2 a.$$

Expressions for the sine and cosine of HALF *a given arc.*

210. The arc in the preceding equations, not being necessarily limited to any particular value, may be *half* a, as well as a. Substituting then $\frac{1}{2}a$ for a, we have,

$$R\ sin\ a=2sin\ \tfrac{1}{2}a\ cos\ \tfrac{1}{2}a$$
$$R\ cos\ a=cos^2\ \tfrac{1}{2}a-sin^2\ \tfrac{1}{2}a$$

Putting the sum of the squares of the sine and cosine equal to the square of radius, (Art. 94.) and inverting the members of the last equation,

$$cos^2\tfrac{1}{2}a+sin^2\tfrac{1}{2}a=R^2$$
$$cos^2\tfrac{1}{2}a-sin^2\tfrac{1}{2}a=R\ cos\ a$$

If we *subtract* one of these from the other, the terms containing $cos^2\frac{1}{2}a$ will disappear ; and if we *add* them, the terms containing $sin^2\frac{1}{2}a$ will disappear : therefore,

$$2sin^2\tfrac{1}{2}a=R^2-R\ cos\ a$$
$$2cos^2\tfrac{1}{2}a=R^2+R\ cos\ a$$

Dividing by 2, and extracting the root of both sides,

$$\text{I. } sin\tfrac{1}{2}a=\sqrt{\tfrac{1}{2}R^2-\tfrac{1}{2}R\times cos\ a}$$
$$\text{II. } cos\tfrac{1}{2}a=\sqrt{\tfrac{1}{2}R^2+\tfrac{1}{2}R\times cos\ a}$$

Expressions for the sines and cosines of MULTIPLE *arcs.*

211. In the same manner, as expressions for the sine and cosine of a *double* arc, are derived from the equations in art. 208; expressions for the sines and cosines of other multiple arcs may be obtained, by substituting successively $2a$, $3a$, &c. for b, or for b and a both. Thus,

$$\text{I. } \begin{cases} R\,sin\ 3a=R\ sin(a+2a)=sin\ a\ cos\ 2a+sin\ 2a\ cos\ a \\ R\,sin\ 4a=R\ sin(a+3a)=sin\ a\ cos\ 3a+sin\ 3a\ cos\ a \\ R\,sin\ 5a=R\ sin(a+4a)=sin\ a\ cos\ 4a+sin\ 4a\ cos\ a \end{cases}$$
&c.

$$\text{II. } \begin{cases} R\ cos\ 3a=R\ cos(a+2a)=cos\ a\ cos\ 2a-sin\ a\ sin\ 2a \\ R\ cos\ 4a=R\ cos(a+3a)=cos\ a\ cos\ 3a-sin\ a\ sin\ 3a \\ R\ cos\ 5a=R\ cos(a+4a)=cos\ a\ cos\ 4a-sin\ a\ sin\ 4a \end{cases}$$
&c.

Expressions for the PRODUCTS *of sines and cosines.*

212. Expressions for the product of sines and cosines may be obtained, by adding and subtracting the four equations in art. 208, viz.

$$R\ sin(a+b)=sin\ a\ cos\ b+sin\ b\ cos\ a$$
$$R\ sin(a-b)=sin\ a\ cos\ b-sin\ b\ cos\ a$$
$$R\ cos(a+b)=cos\ a\ cos\ b-sin\ a\ sin\ b$$
$$R\ cos(a-b)=cos\ a\ cos\ b+sin\ a\ sin\ b$$

Adding the first and second,

$$R\ sin(a+b)+R\ sin(a-b)=2\ sin\ a\ cos\ b$$

Subtracting the second from the first,

$$R\ sin(a+b)-R\ sin(a-b)=2sin\ b\ cos\ a$$

Adding the third and fourth,

$$R\ cos(a-b)+R\ cos(a+b)=2cos\ a\ cos\ b$$

Subtracting the third from the fourth,

$$R\ cos(a-b)-R\ cos(a+b)=2sin\ a\ sin\ b$$

Inverting the members of each of these equations, and dividing by 2, we have,

I. $\sin a \cos b = \frac{1}{2}R \sin(a+b) + \frac{1}{2}R \sin(a-b)$
II. $\sin b \cos a = \frac{1}{2}R \sin(a+b) - \frac{1}{2}R \sin(a-b)$
III. $\cos a \cos b = \frac{1}{2}R \cos(a-b) + \frac{1}{2}R \cos(a+b)$
IV. $\sin a \sin b = \frac{1}{2}R \cos(a-b) - \frac{1}{2}R \cos(a+b)$

213. If b be taken equal to a, then $a+b=2a$, and $a-b=0$, the sine of which is 0; (Art. 201.) and the term in which this is a *factor*, is reduced to 0. (Alg. 112.) But the *cosine* of 0 is equal to radius, so that $R \times \cos 0 = R^2$. Reducing, then, the preceding equations,

The first becomes $\sin a \cos a = \frac{1}{2}R \sin 2a$
The third, $\cos^2 a = \frac{1}{2}R^2 + \frac{1}{2}R \cos 2a$
The fourth, $\sin^2 a = \frac{1}{2}R^2 - \frac{1}{2}R \cos 2a$

214. If s be the *sum*, and d the *difference* of two arcs, $\frac{1}{2}(s+d)$ will be equal to the greater, and $\frac{1}{2}(s-d)$ to the less. (Art. 153.) Substituting then, in the four equations in art. 212,

s for $a+b$ $\frac{1}{2}(s+d)$ for a
d for $a-b$, $\frac{1}{2}(s-d)$ for b, we have,

I. $\sin \frac{1}{2}(s+d(\cos \frac{1}{2}(s-d) = \frac{1}{2}R (\sin s + \sin d)$
II. $\sin \frac{1}{2}(s-d)\cos \frac{1}{2}(s+d) = \frac{1}{2}R (\sin s - \sin d)$
III. $\cos \frac{1}{2}(s+d)\cos \frac{1}{2}(s-d) = \frac{1}{2}R (\cos d + \cos s)$
IV. $\sin \frac{1}{2}(s+d)\sin \frac{1}{2}(s-d) = \frac{1}{2}R (\cos d - \cos s)$

Or, making $R=1$,

I. $\sin(a+b) + \sin(a-b) = 2 \sin a \cos b$
II. $\sin(a+b) - \sin(a-b) = 2 \sin b \cos a$
III. $\cos(a-b) + \cos(a+b) = 2 \cos a \cos b$
IV. $\cos(a-b) - \cos(a+b) = 2 \sin a \sin b$

215. If radius be taken equal to 1, the two first equations in art. 208, are,

$$\sin(a+b) = \sin a \cos b + \sin b \cos a$$
$$\sin(a-b) = \sin a \cos b - \sin b \cos a$$

Multiplying these into each other,

$$\sin(a+b) \times \sin(a-b) = \sin^2 a \cos^2 b - \sin^2 b \cos^2 a$$

But by art. 94, if radius is 1,

$$cos^2 b=1-sin^2 b, \text{ and } cos^2 a=1-sin^2 a$$

Substituting, then, for $cos^2 b$ and $cos^2 a$, their values, multiplying the factors, and reducing the terms, we have,

$$sin(a+b)\times sin(a-b)=sin^2 a-sin^2 b$$

Or, because the difference of the squares of two quantities is equal to the product of their sum and difference, (Alg. 235.)

$$sin(a+b)\times sin(a-b)=(sin\, a+sin\, b)\times(sin\, a-sin\, b)$$

That is, the product of the sine of the sum of two arcs, into the sine of their difference; is equal to the product of the sum of their sines, into the difference of their sines.

Expressions for the TANGENTS *of arcs.*

216. Expressions for the *tangents* of arcs may be derived from those already obtained for the sines and cosines. By art. 93, proportion 1st,

$$R : tan :: cos : sin$$

That is, $\frac{R}{tan}=\frac{cos}{sin}$, and $\frac{tan}{R}=\frac{sin}{cos}$, and $tan=\frac{R\times sin}{cos}$.

Thus $tan(a+b=\frac{R\, sin(a+b)}{cos(a+b)}$.

If, for $sin(a+b)$ and $cos(a+b)$ we substitute their values, as given in art. 208, we shall have,

$$tan(a+b)=\frac{R(sin\, a\, cos\, b+sin\, b\, cos\, a)}{cos\, a\, cos\, b-sin\, a\, sin\, b}.$$

217. Here the value of the tangent of the sum of two arcs is expressed, in terms of the *sines* and *cosines* of the arcs. To exchange these for terms of the *tangents* let the numerator and denominator of the second member of the equation be both divided by $cos\, a\, cos\, b$. This will not alter the value of the fraction. (Alg. 140.)

The *numerator*, divided by $cos\, a\, cos\, b$, is

$$\frac{R(sin\, a\, cos\, b+sin\, b\, cos\, a)}{cos\, a\, cos\, b}=R\left(\frac{sin\, a}{cos\, a}+\frac{sin\, b}{cos\, b}\right)=tan\, a+tan\, b$$

And the *denominator*, divided by $\cos a \cos b$, is

$$\frac{\cos a \cos b - \sin a \sin b}{\cos a \cos b} = 1 - \frac{\sin a}{\cos a} \times \frac{\sin b}{\cos b} = 1 - \frac{\tan a}{R} \times \frac{\tan b}{R}$$

$$\text{Therefore } \tan(a+b) = \frac{\tan a + \tan b}{1 - \frac{\tan a \tan b}{R^2}}$$

The denominator of the fraction may be cleared of the divisor R^2, by multiplying both the numerator and denominator into R^2. And if we proceed in a similar manner to find the tangent of $a-b$, we shall have,

$$218. \text{ I. } \tan(a+b) = \frac{R^2(\tan a + \tan b)}{R^2 - \tan a \tan b)}$$

$$\text{II. } \tan(a+b) = \frac{R^2(\tan a - \tan b)}{R^2 + \tan a \tan b)}$$

If the arcs a and b are *equal*, then substituting $\frac{1}{2}a$, a, $2a$, $3a$, &c. as in art. 210, 211,

$$\tan a = \tan(\tfrac{1}{2}a + \tfrac{1}{2}a) = \frac{R^2(2 \tan\frac{1}{2}a)}{R^2 - \tan^2\frac{1}{2}a}$$

$$\tan 2a = \tan(a+a) = \frac{R^2(2\tan a)}{R^2 - \tan^2 a}$$

$$\tan 3a = \tan(a+2a) = \frac{R^2(\tan a + \tan 2a)}{R^2 - \tan a \tan 2a}, \text{ &c.}$$

219. If we divide the first of the equations in art. 214, by the second; we shall have, after rejecting $\frac{1}{2}R^2$ from the numerator and denominator, (Alg. 140.)

$$\frac{\sin\frac{1}{2}(s+d)\cos\frac{1}{2}(s-d)}{\sin\frac{1}{2}(s-d)\cos\frac{1}{2}(s+d)} = \frac{\sin s + \sin d}{\sin s - \sin d}$$

But the first member of this equation, (Alg. 155,) is equal to

$$\frac{\sin\frac{1}{2}(s+d)}{\cos\frac{1}{2}(s+d)} = \frac{\cos\frac{1}{2}(s-d)}{\sin\frac{1}{2}(s-d)} = \frac{\tan\frac{1}{2}(s+d)}{R} \times \frac{R}{\tan\frac{1}{2}(s-d)} \cdot \text{(Art.216.)}$$

Therefore,

$$\frac{\sin s + \sin d}{\sin s - \sin d} = \frac{\tan\frac{1}{2}(s+d)}{\tan\frac{1}{2}(s-d)}$$

220. According to the notation in art. 214, s stands for the *sum* of two arcs, and d for their *difference*. But it is evident that arcs may be taken, whose sum shall be equal to *any* arc a, and whose difference shall be equal to any arc b, provided that a be *greater* than b. Substituting then, in the preceding equation, a for s, and b for d,

$$\frac{\sin a+\sin b}{\sin a-\sin b}=\frac{\tan \frac{1}{2}(a+b)}{\tan \frac{1}{2}(a-b)}. \quad \text{Or,}$$

$$\sin a+\sin b : \sin a-\sin b :: \tan \tfrac{1}{2}(a+b) : \tan \tfrac{1}{2}(a-b.)$$

That is, *The sum of the sines of two arcs or angles, is to the difference of those sines; as the tangent of half the sum of the arcs or angle, to the tangent of half their difference.*

By art. 143, the *sides of triangles* are as the sines of their opposite angles. It follows, therefore, from the preceding proposition, (Alg. 389.) that the sum of any two sides of a triangle, is to their difference; as the tangent of half the sum of the opposite angles, to the tangent of half their difference.

This is the second theorem applied to the solution of oblique angled triangles, which was *geometrically* demonstrated in art. 144.

Expressions for the *cotangents* may be obtained by putting

$$\cot = \frac{R^2}{\tan} \text{(Art. 93.)}$$

$$\text{Thus } \cot(a+b)=\frac{R^2}{\tan(a+b)}=\frac{R^2-\tan a \tan b}{\tan a+\tan b} \text{ (Art. 218.)}$$

Substituting $\frac{R^2}{\cot a}$ for $\tan a$, and $\frac{R^2}{\cot b}$ for $\tan b$,

$$\cot(a+b)=\frac{R^2-\frac{R^2}{\cot a}\times\frac{R^2}{\cot b}}{\frac{R^2}{\cot a}+\frac{R^2}{\cot b}}$$

Multiplying both the numerator and denominator by $\cot a$ $\cot b$, dividing by R^2, and proceeding in the same manner. for $\cot(a-b)$ we have,

$$\text{I. } \cot(a+b)=\frac{\cot a \cot b-R^2}{\cot b+\cot a}$$

$$\text{II. } \cot(a-b) = \frac{\cot a \cot b + R^2}{\cot b - \cot a}$$

220. *b*. By comparing the expressions for the sines, and cosines, with those for the tangents and cotangents, a great variety of formulæ may be obtained. Thus the tangent of the sum or the difference of two arcs, may be expressed in terms of the cotangent.

Putting radius $=1$, we have (Art. 93, 220.)

$$\text{I. } \tan(a+b) = \frac{1}{\cot(a+b)} = \frac{\cot b + \cot a}{\cot a \cot b - 1}$$

$$\text{II. } \tan(a-b) = \frac{1}{\cot(a-b)} = \frac{\cot b - \cot a}{\cot a \cot b + 1}$$

By art. 208,

$$\frac{\sin(a+b)}{\sin(a-b)} = \frac{\sin a \cos b + \sin b \cos a}{\sin a \cos b - \sin b \cos a}$$

Dividing the last member of the equation, in the first place by $\cos a \cos b$, as in art. 217, and then by $\sin a \sin b$, we have

$$\frac{\sin(a+b)}{\sin(a-b)} = \frac{\tan a + \tan b}{\tan a - \tan b} = \frac{\cot b + \cot a}{\cot b - \cot a}$$

In a similar manner, dividing the expressions for the cosines, in the first place by $\sin b \cos a$, and then by $\sin a \cos b$. we obtain

$$\frac{\cos(a+b)}{\cos(a-b)} = \frac{\cot b - \tan a}{\cot b + \tan a} = \frac{\cot a - \tan b}{\cot a + \tan b}$$

Dividing the numerator and denominator of the expression for the tangent of a, (Art. 218.) by $\tan \frac{1}{2}a$, we have

$$\tan a = \frac{2}{\cot \frac{1}{2}a - \tan \frac{1}{2}a}$$

These formulæ may be multiplied almost indefinitely, by combining the expressions for the sines, tangents, &c. The following are put down without demonstrations, for the exercise of the student.

$$\tan \tfrac{1}{2}a=\cot\tfrac{1}{2}a-2\cot a. \qquad \tan\tfrac{1}{2}a=\frac{1-\cos a}{\sin a}$$

$$\tan \tfrac{1}{2}a=\frac{\sin a}{1+\cos a} \qquad \tan^2\tfrac{1}{2}a=\frac{1-\cos a}{1+\cos a}$$

$$\sin a=\frac{2\tan\tfrac{1}{2}a}{1+\tan^2\tfrac{1}{2}a} \qquad \cos a=\frac{1-\tan^2\tfrac{1}{2}a}{1+\tan^2\tfrac{1}{2}a}$$

$$\cos a=\frac{\cot\tfrac{1}{2}a-\tan\tfrac{1}{2}a}{\cot\tfrac{1}{2}a+\tan\tfrac{1}{2}a} \qquad \sin a=\frac{2}{\cot\tfrac{1}{2}a+\tan\tfrac{1}{2}a}$$

$$\sin a=\frac{1}{\cot\tfrac{1}{2}a-\cot a} \qquad \sin a=\frac{1}{\cot a+\tan\tfrac{1}{2}a}$$

Expression for the *area* of a triangle, in terms of the sides.

221. Let the sides of the triangle ABC (Fig. 23.) be expressed by a, b, and c, the perpendicular CD by p, the segment AD by d, and the area by S.

Then $a^2=b^2+c^2-2cd$, (Euc. 13. 2.)

Transposing and dividing by $2c$

$$d=\frac{b^2+c^2-a^2}{2c}. \quad \text{Therefore } d^2=\frac{(b^2+c^2-a^2)^2}{4c^2}. \quad \text{(Alg. [223.)}$$

By Euc. 47. 1, $p^2=b^2-d^2=b^2-\dfrac{(b^2+c^2-a^2)^2}{4c^2}$

Reducing the fraction, (Alg. 150.) and extracting the root of both sides,

$$p=\frac{\qquad\qquad}{2c}\text{*}$$

This gives the length of the *perpendicular*, in terms of the sides of the triangle. But the *area* is equal to the product of the base into half the perpendicular height. (Alg. 518.) that is,

$$S=\tfrac{1}{2}cp=\tfrac{1}{4}\sqrt{4b^2c^2-(b^2+c^2-a^2)^2}$$

Here we have an expression for the area, in terms of the sides. But this may be reduced to a form much better adapted to arithmetical computation. It will be seen, that the quantities $4b^2c^2$, and $(b^2+c^2-a^2)^2$ are both *squares*; and that the whole expression under the radical sign is the *difference* of these squares. But the difference of two squares is equal to the product of the sum and difference of their roots. (Alg. 235.) Therefore $4b^2c^2-(b^2+c^2-a^2)^2$ may be resolved into the two factors,

$$\begin{cases} 2bc+(b^2+c^2-a^2) \text{ which is equal to } (b+c)^2-a^2 \\ 2bc-(b^2+c^2-a^2) \text{ which is equal to } a^2-(b-c)^2 \end{cases}$$

Each of these also, as will be seen in the expressions on the right, is the difference of two squares; and may, on the same principle, he resolved into factors, so that,

$$\begin{cases} (b+c)^2-a^2=(b+c+a)\times(b+c-a) \\ a^2-(b-c)^2=(a+b-c)\times(a-b+c) \end{cases}$$

Substituting, then, these four factors, in the place of the quantity which has been resolved into them, we have,

$$S=\tfrac{1}{4}\sqrt{(b+c+a)\times(b+c-a)\times(a+b-c)\times(a-b+c)}$$

---

* The expression for the perpendicular is the same, when one of the angles is *obtuse*, as in Fig. 24. Let $AD=d$.

Then $a^2=b^2+c^2+2cd$. (Euc. 12. 2.) And $d=\dfrac{-b^2-c^2+a^2}{2c}$

Therefore $d^2=\dfrac{(-b^2-c^2+a^2)^2}{4c^2}=\dfrac{(b^2+c^2-a^2)^2}{4c^2}$ (Al. 219.)

And $p=\dfrac{\sqrt{4b^2c^2-(b^2+c^2-a^2)^2}}{2c}$ as above

Here it will be observed, that all the three sides, a, b, and c, are in each of these factors.

Let $h=\frac{1}{2}(a+b+c)$ *half the sum* of the sides. Then

$$S=\sqrt{h\times(h-a)\times(h-b)\times(h-c)}$$

222. For finding the area of a triangle, then, when the three sides are given, we have this general rule ;

From half the sum of the sides, subtract each side severally ; multiply together the half sum and the three remainders ; and extract the square root of the product.

SECTION VIII.

COMPUTATION OF THE CANON.

ART. 223. THE trigonometrical canon is a set of tables containing the sines, cosines, tangents, &c. to every degree and minute of the quadrant. In the computation of these tables, it is common to find, in the first place, the sine and cosine of *one minute;* and then, by successive additions and multiplications, the sines, cosines, &c. of the larger arcs. For this purpose, it will be proper to begin with an arc, whose sine or cosine is a known portion of the radius. The cosine of 60° is equal to *half radius.* (Art. 96. Cor.) A formula has been given, (Art. 210.) by which, when the cosine of an arc is known, the cosine of *half* that arc may be obtained.

By successive bisections of 60°, we have the arcs

30°	0° 28′ 7′ 30‴
15°	0 14 3 45
7° 30′	0 7 1 52 30
3° 45′	0 3 30 56 15
1° 52′ 30″	0 1 45 28 7 30
0° 56′ 15″	0 0′ 52″ 44‴ 3⁗ 45′′′′′

By formula II, art. 210,

$$\cos \tfrac{1}{2}a=\sqrt{\tfrac{1}{2}R^2+\tfrac{1}{2}R\times\cos a}$$

If the radius be 1, and if $a=60°$, $b=30°$, $c=15°$, &c.; then

$$\cos b=\cos \tfrac{1}{2}a=\sqrt{\tfrac{1}{2}+\tfrac{1}{2}\times\tfrac{1}{2}}=0.8660254$$

$$\cos c=\cos \tfrac{1}{2}b=\sqrt{\tfrac{1}{2}+\tfrac{1}{2}\cos b}=0.9659258$$

$$\cos d=\cos \tfrac{1}{2}c=\sqrt{\tfrac{1}{2}+\tfrac{1}{2}\cos c}=0.9914449$$

$$\cos e=\cos \tfrac{1}{2}d=\sqrt{\tfrac{1}{2}+\tfrac{1}{2}\cos d}=0.9978589$$

Proceeding in this manner, by repeated extractions of the square root, we shall find the cosine of

$$0^\circ\ 0'\ 52''\ 44'''\ 3''''\ 45''''' \text{ to be } 0.99999996732$$

$$\text{And the sine (Art. 94.)}=\sqrt{1-cos^2}=0.000255663 46$$

This, however, does not give the sine of *one minute* exactly. The arc is a little *less* than a minute. But the ratio of very small arcs to each other, is so nearly equal to the ratio of their sines, that one may be taken for the other, without sensible error. Now the circumference of a circle is divided into 21600 parts, for the arc of 1′; and into 24576, for the arc of $0^\circ\ 0'\ 52''\ 44'''\ 3''''\ 45'''''$

Therefore,

$$21600 : 24576 :: 0.00025566346 : 0.0002908882.$$

which is the sine of 1 minute very nearly.*

$$\text{And the cosine }=\sqrt{1-sin^2}=0.9999999577$$

224. Having computed the sine and cosine of one minute, we may proceed, in a contrary order, to find the sines and cosines of *larger* arcs.

Making radius $=1$, and adding the two first equations in art. 208, we have

$$sin(a+b)+sin(a-b)=2sin\ a\ cos\ b$$

Adding the third and fourth,

$$cos(a+b)+cos(a-b)=2cos\ a\ cos\ b$$

Transposing $sin(a-b)$ and $cos(a-b)$

$$\text{I.}\quad sin(a+b)=2sin\ a\ cos\ b-sin(a-b)$$
$$\text{II.}\quad cos(a+b)=2cos\ a\ cos\ b-cos(a-b)$$

If we put $b=1'$, and $a=1'$, $2'$, $3'$, &c. successively, we shall have expressions for the sines and cosines of a series of arcs increasing regularly by one minute. Thus,

* See note H.

$$sin(1'+1')=2sin\,1'\times cos\,1'-sin\,0=0.0005817764,$$
$$sin(2'+1')=2sin\,2'\times cos\,1'-sin\,1'=0.0008726645,$$
$$sin(3'+1')=2sin\,3'\times cos\,1'-sin\,2'=0.0011635526,$$
&c. &c.

$$cos(1'+1')=2cos\,1'\times cos\,1'-cos\,0=09999998308$$
$$cos(2'+1')=2cos\,2'\times cos\,1'-cos\,1'=0.9999996192$$
$$cos(3'+1')=2cos\,3'\times cos\,1'-cos\,2'=0.9999993230$$
&c. &c.

The constant multiplier here, *cos* 1' is 0.9999999577, which is equal to 1—0.0000000423.

225. Calculating, in this manner, the sines and cosines from 1 minute up to 30 degrees, we shall have also the sines and cosines from 60° to 90°. For the sines of arcs between 0° and 30°, are the *cosines* of arcs between 60° and 90°. And the cosines of arcs between 0° and 30°, are the *sines* of arcs between 60 and 90°. (Art. 104.)

226. For the interval between 30° and 60°, the sines and cosines may be obtained by subtraction merely. As twice the sine of 30° is equal to radius; (Art. 96.) by making $a=$ 30°, the equation marked I, in article 224 will become

$$sin(30°+b)=cos\,b-sin(30°-b)$$

And putting $b=1', 2', 3'$, &c. successively.

$$sin(30°\,1')=cos\,1'-sin(29°\,59')$$
$$(30°\,2')=cos\,2'-sin(29°\,58')$$
$$(30°\,3')=cos\,3'-sin(29°\,57')$$
&c. &c.

If the *sines* be calculated from 30° to 60°, the *cosines* will also be obtained. For the sines of arcs between 30° and 45°, are the cosines of arcs between 45° and 60°. And the sines of arcs between 45° and 60°, are the cosines of arcs between 30° and 45°.* (Art. 96.

227. By the methods which have here been explained, the *natural* sines and cosines are found.

The *logarithms* of these, 10 being in each instance added to the index, will be the *artificial* sines and cosines, by which trigonometrical calculations are commonly made. (Art. 102, 3.)

228. The *tangents*, *cotangents*, *secants*, and *cosecants*, are easily derived from the sines and cosines. By art. 93,

* See note I.

$R : cos :: tan : sin$ $\quad$ $cos : R :: R : sec$

$R : sin :: cot : cos$ $\quad$ $sin : R :: R : cosec$

Therefore,

The tangent$=\dfrac{R \times sin}{cos}$ $\quad$ The secant$=\dfrac{R^2}{cos}$

The cotangent$=\dfrac{R \times cos}{sin}$ $\quad$ The cosecant$=\dfrac{R^2}{sin}$

Or, if the computations are made by *logarithms*,

The tangent$=10+sin-cos$, $\quad$ The secant$=20-cos$,

The cotangent$=10+cos-sin$, $\quad$ The cosecant$=20-sin$.

SECTION IX.

PARTICULAR SOLUTIONS OF TRIANGLES.*

ART. 231. ANY triangle whatever may be solved, by the theorems in sections III. IV. But there are other methods, by which, in certain circumstances, the calculations are rendered more expeditious, or more accurate results are obtained.

The differences in the *sines* of angles near 90°, and in the *cosines* of angles near 0°, are so small as to leave an uncertainty of several seconds in the result. The solutions should be varied, so as to avoid finding a very small angle by its cosine, or one near 90° by its sine.

The differences in the logarithmic *tangents* and *cotangents* are least at 45°, and increase towards each extremity of the quadrant. In no part of it, however, are they very small. In the tables which are carried to 7 places of decimals, the least difference for one second is 42. Any angle may be found within one second, by its tangent, if tables are used which are calculated to seconds.

But the differences in the logarithmic sines and tangents, within a few minutes of the beginning of the quadrant, and in cosines and tangents within a few minutes of 90°, though they are very large, are too *unequal* to allow of an exact determination of their corresponding angles, by taking *proportional parts* of the differences. Very small angles may be accurately found, from their sines and tangents, by the rules given in a note at the end.†

232. The following formulæ may be applied to *right angled* triangles, to obtain accurate results, by finding the sine or tangent of *half* an arc, instead of the whole.

In the triangle ABC (Fig. 20, Pl. II.) making AC radius,

$$AC : AB :: 1 : \text{Cos } A.$$

By conversion, (Alg. 389, 5.)

$$AC : AC - AB :: 1 : 1 - \text{Cos } A.$$

* Simson's, Woodhouse's, and Cagnoli's Trigonometry. † See Note K.

Therefore,

$$\frac{AC-AB}{AC}=1-\text{Cos } A=2\sin^2\tfrac{1}{2}A. \text{ (Art. 210.)}$$

Or,

$$\text{Sin } \tfrac{1}{2} A=\sqrt{\frac{AC-AB}{2AC}}$$

Again, from the first proportion, adding and subtracting terms, (Alg. 389, 7.)

$$AC+AB : AC-AB :: 1+\text{Cos } A : 1-\text{Cos } A.$$

Therefore,

$$\frac{AC-AB}{AC+AB}=\frac{1-\text{Cos } A}{1+\text{Cos } A}=\tan^2 \tfrac{1}{2}A. \text{ (Page. 120.)}$$

Or,

$$\text{Tan } \tfrac{1}{2}A=\sqrt{\frac{AC-AB}{AC+AB}}$$

233. Sometimes, instead of having two parts of a right angled triangle given, in addition to the right angle; we have only one of the parts, and the *sum* or *difference* of two others. In such cases, solutions may be obtained by the following proportions.

By the preceding formulæ, and Art. 140, 141,

1. $\text{Tan}^2 \tfrac{1}{2}A=\frac{AC-AB}{AC+AB}$

2. $BC^2=(AC-AB)\,(AC+AB)$

Multiplying these together, and extracting the root, we have,

$$\text{Tan } \tfrac{1}{2} A\times BC=AC-AB$$

Therefore,

$$\text{I. Tan}\tfrac{1}{2} A : 1 :: AC-AB : BC$$

That is, the tangent of half of one of the acute angles, is to 1, as the difference between the hypothenuse and the side at the angle, to the other side.

If, instead of multiplying, we *divide* the first equation above by the second, we have

$$\frac{\text{Tan } \tfrac{1}{2}A}{BC}=\frac{1}{AC+AB}$$

Therefore,

$$\text{II. } 1 : \tan \tfrac{1}{2}A :: AC+AB : BC$$

Again, in the triangle ABC, Fig. 20,

$$AB : BC : 1 :: \tan A$$

Therefore,

$$AB+BC : AB-BC :: 1+\tan A : 1-\tan A$$

Or,

$$AB+BC : AB-BC :: 1 : \frac{1-\tan A}{1+\tan A}$$

By art. 218, one of the arcs being A, and the other 45°, the tangent of which is equal to radius, we have,

$$\text{Tan } (45°-A)=\frac{1-\tan A}{1+\tan A}$$

Therefore,

$$\text{III. } 1 : \tan (45°-A) :: AB+BC : AB-BC.$$

That is, unity is to the tangent of the difference between 45° and one of the acute angles; as the sum of the perpendicular sides is to their difference.

Ex. 1. In a right angled triangle, if the difference of the hypothenuse and base be 64 feet, and the angle at the base $33°\frac{3}{4}$, what is the length of the perpendicular?

Ans. 211.

2. If the sum of the hypothenuse and base be 185.3, and the angle at the base 37°; what is the perpendicular?

Ans. 620.

3. Given the sum of the base and perpendicular 128.4, and the angle at the base $41°\frac{1}{4}$, to find the sides.

$1 : \tan(45°-41°\frac{1}{4}) :: 128.4 : 8.4$, the difference of the base and perpendicular. Half the difference added to, and subtracted from, the half sum, gives the base 68.4, and the perpendicular 60.

4. Given the sum of the hypothenuse and perpendicular 83, and the angle at the perpendicular 40°, to find the base.

5. Given the difference of the hypothenuse and perpendicular 16.5, and the angle at the perpendicular $37°\frac{1}{4}$, to find the base.

6. Given the difference of the base and perpendicular 35, and the angle at the perpendicular $27°\frac{1}{4}$, to find the sides.

234. The following solutions may be applied to the *third* and *fourth* cases of *oblique* angled triangles; in one of which, two sides and the included angle are given, and in the other. the three sides. See pages 87 and 88.

CASE III.

In astronomical calculations, it is frequently the case, that two sides of a triangle are given by their *logarithms*. By the following proposition, the necessity of finding the corresponding natural numbers is avoided.

THEOREM A. *In any plane triangle, of the two sides which include a given angle, the less is to the greater; as radius to the tangent of an angle greater than 45°:*

And radius is to the tangent of the excess of this angle above 45°; as the tangent of half the sum of the opposite angles, to the tangent of half their difference

In the triangle ABC, (Fig. 39.) let the sides AC and AB, and the angle A be given. Through A draw DH perpendicular to AC. Make AD and AF each equal to AC, and AH equal to AB. And let HG be perpendicular to a line drawn from C through F.

$$\text{Then AC : AB} :: \text{R : Tan ACH}$$

$$\text{And R : Tan(ACH} - 45°) :: \text{Tan}\tfrac{1}{2}(\text{ACB}+\text{B}) : \text{Tan}\tfrac{1}{2}(\text{ACB} - \text{B})$$

Demonstration.

In the right angled triangle ACD, as the acute angles are subtended by the equal sides AC and AD, each is 45°. For the same reason, the acute angles in the triangle CAF are each 45°. Therefore, the angle DCF is a right angle, the angles GFH and GHF are each 45°, and the line GH is equal to GF and parallel to DC.

In the triangle ACH, if AC be radius, AH which is equal to AB will be the tangent of ACH. Therefore,

$$\text{AC : AB} :: \text{R : Tan ACH.}$$

In the triangle CGH, if CG be radius, GH which is equal to FG will be the tangent of HCG. Therefore,

$$\text{R : Tan (ACH} - 45°) :: \text{CG : FG.}$$

And, as GH and DC are parallel, (Euc. 2. 6.)

CG : FG : : DH : FH.

But DH is, by construction, equal to the *sum*, and FH to the *difference* of AC and AB And by theorem II, [Art. 144.] the sum of the sides is to their difference ; as the tangent of half the sum of the opposite angles, to the tangent of half their difference. Therefore,

$$R : \text{Tan}\,(ACH-45°) : : \text{Tan}\,\tfrac{1}{2}(ACB+B) : \text{Tan}\,\tfrac{1}{2}(ACB \backsim B)$$

Ex. In the triangle ABC, (Fig. 30.) given the angle A= 26° 14′, the side AC=39, and the side AB=53.

AC	39	1.5910646	R	10.
AB	53	1.7242759	Tan 8° 39′ 9″	9.1823381
R		10.	Tan ½(B+C)76° 53′	10.6326181
Tan 53° 39′ 9″		10.1332113	Tan ½(B∽C)33°8′50″	9.8149562

The same result is obtained here, as by theorem II, p. 75.

To find the required *side* in this third case, by the theorems in section IV, it is necessary to find, in the first place, an *angle* opposite one of the given sides. But the required side may be obtained, in a different way, by the following proposition.

THEOREM B. *In a plane triangle, twice the product of any two sides, is to the difference between the sum of the squares of those sides, and the square of the third side, as radius to the cosine of the angle included between the two sides.*

In the triangle ABC, (Fig. 23.) whose sides are a, b, and c.

$$2bc : b^2+c^2-a^2 : : R : \text{Cos}\ A$$

For in the right angled triangle ACD, $b : d : : R : \text{Cos}\ A$
Multiplying by $2c$, $2bc : 2dc : : R : \text{Cos}\ A$
But, by Euclid 13. 2, $2dc=b^2+c^2-a^2$
Therefore, $2bc : b^2+c^2-a^2 : : R : \text{Cos}\ A.$

The demonstration is the same, when the angle A is *obtuse*, as in the triangle ABC, (Fig. 24.) except that a^2 is *greater* than b^2+c^2; (Euc. 12. 2.) so that the cosine of A is *negative*. See art. 194.

From this theorem are derived expressions, both for the *sides* of a triangle, and for the cosines of the *angles*. Converting the last proportion into an equation, and proceeding in the same manner with the other sides and angles, we have the following expressions;

For the angles.	For the sides.
$\text{Cos A}=\text{R}\times\frac{b^2+c^2-a^2}{2bc}$	$a=\sqrt{\left(b^2+c^2-\frac{2bc\,\text{Cos A}}{\text{R}}\right)}$
$\text{Cos B}=\text{R}\times\frac{a^2+c^2-b^2}{2ac}$	$b=\sqrt{\left(a^2+c^2-\frac{2ac\,\text{Cos B}}{\text{R}}\right)}$
$\text{Cos C}=\text{R}\times\frac{a^2+b^2-c^2}{2ab}$	$c=\sqrt{\left(a^2+b^2-\frac{2ab\,\text{Cos C}}{\text{R}}\right)}$

These formulæ are useful, in many trigonometrical investigations; but are not well adapted to logarithmic computation.

Case IV.

When the *three sides* of a triangle are given, the angles may be found, by either of the following theorems; in which a, b, and c are the sides, A, B, and C, the opposite angles, and h=half the sum of the sides.

Theorem C.

$$\begin{cases} \text{Sin A}=\frac{2\text{R}}{bc}\sqrt{h(h-a)(h-b)(h-c)} \\ \text{Sin B}=\frac{2\text{R}}{ac}\sqrt{h(h-a)(h-b)(h-c)} \\ \text{Sin C}=\frac{2\text{R}}{ab}\sqrt{h(h-a)(h-b)(h-c)} \end{cases}$$

The quantities under the radical sign are the same in all the equations.

In the triangle ACD, (Fig. 23.)

$\text{R} : b :: \text{Sin A} : p.$ Therefore, $\text{Sin A}\times b=\text{R}\times p.$

But $p=\frac{\sqrt{4b^2c^2-(b^2+c^2-a^2)^2}}{2c}$. (Art. 221. p. 105.)

This, by the reductions in page 106, becomes

$$p=\frac{\sqrt{2h\times2(h-a)\times2(h-b)\times2(h-c)}}{2c}$$

Substituting this value of p, and reducing,

$$\text{Sin A}=\frac{2R}{bc}\sqrt{h(h-a)(h-b)(h-c)}$$

The arithmetical calculations may be made, by adding the logarithms of the factors under the radical sign, dividing the sum by 2, and to the quotient, adding the logarithms of radius and 2, and the arithmetical complements of the logarithms of b and c. (Arts. 39, 47, 59.)

Ex. Given $a=134$, $b=108$, and $c=80$, to find A, B, and C.

For the angle A.

h	161	log.	2.2068259
$h-a$	27	log.	1.4313638
$h-b$	53	log.	1.7242759
$h-c$	81	log.	1.9084850
			2)7.2709506
			*3.6354753
RX2		log.	10.3010300
			13.9365053
b	108	a. c.	7.9665762
c	80	a. c.	8.0969100
Sin A.			9.9999915

A=89° 38′ 31″

For the angle B.

			13.9365053
a	134	a. c.	7.8728952
c	80	a. c.	8.0969100
Sin B.			9.9063105

B=53° 42′ 9″

For the angle C.

			13.9365053
a	134	a. c.	7.8728952
b	108	a. c.	7.9665762
Sin C.			9.7759767

C=36° 39′ 20″

THEOREM D.

$$\begin{cases}\text{Sin }\tfrac{1}{2}\text{A}=\text{R}\sqrt{\dfrac{(h-b)(h-c)}{bc}}\\ \text{Sin }\tfrac{1}{2}\text{B}=\text{R}\sqrt{\dfrac{(h-a)(h-c)}{ac}}\\ \text{Sin }\tfrac{1}{2}\text{C}=\text{R}\sqrt{\dfrac{(h-a)(h-b)}{ab}}\end{cases}$$

By art. 210, $2\text{ Sin}^2\tfrac{1}{2}\text{A}=\text{R}^2-\text{R}\times \cos\text{ A}$.

Substituting for *cos* A, its value, as given in page 132,

$$2\text{Sin}^2\tfrac{1}{2}\text{A}=\text{R}^2-\text{R}^2\times\frac{b^2+c^2-a^2}{2bc}$$

* This is the logarithm of the *area* of the triangle. (Art. 222.)

But $R^2 = R^2 \times \frac{2bc}{2bc}$. And $-R^2 \times \frac{b^2+c^2-a^2}{2bc} = R^2 \times \frac{a^2-b^2-c^2}{2bc}$

Therefore $2\text{Sin}^2 \tfrac{1}{2}A = R^2 \times \frac{2bc+a^2-b^2-c^2}{2bc}$

But $2bc+a^2-b^2-c^2 = a^2-(b-c)^2 = (a+b-c)(a-b+c)$ (Alg. 235.).

Putting then $h = \tfrac{1}{2}(a+b+c)$, reducing, and extracting,

$$\text{Sin}\ \tfrac{1}{2}A = R\sqrt{\frac{(h-b)(h-c)}{bc}}$$

Ex. Given a, b, and c, as before, to find A and B.

For the angle A.

$h-b$	53	1.7242759
$h-c$	81	1.9084850
b	108 a. c.	7.9665762
c	80 a. c.	8.0969100
		2)19.6962471
Sin ½A		9.8481235

A = 89° 38′ 31″

For the angle B.

$h-a$	27	1.4313638
$h-c$	81	1.9084850
a	134 a. c.	7.8728952
c	80 a. c.	8.0969100
		2)19.3096540
Sin ½B		9.6548270

B = 53° 42′ 9″

THEOREM E.
$$\begin{cases} \text{Cos}\ \tfrac{1}{2}A = R\sqrt{\frac{h(h-a)}{bc}} \\ \text{Cos}\ \tfrac{1}{2}B = R\sqrt{\frac{h(h-b)}{ac}} \\ \text{Cos}\ \tfrac{1}{2}C = R\sqrt{\frac{h(h-c)}{ab}} \end{cases}$$

By art. 210, $2\text{Cos}^2 \tfrac{1}{2}A = R^2 + R \times \cos A$,

Substituting and reducing, as in the demonstration of the last theorem,

$$2\text{Cos}^2 \tfrac{1}{2}A = R^2 \times \frac{2bc+b^2+c^2-a^2}{2bc} = R^2 \times \frac{(b+c+a)(b+c-a)}{2bc}$$

Putting $h = \tfrac{1}{2}(a+b+c)$ reducing and extracting,

$$\text{Cos}\ \tfrac{1}{2}A = R\sqrt{\frac{h(h-c)}{bc}}$$

Ex. Given the sides 134, 108, 80; to find B and C.

For the angle B.			
h	161		2.2068259
$h-b$	53		1.7242759
a	134	a. c.	7.8728952
c	80	a. c.	8.0969100
			2)19.9009070
Cos ½B			9.9504535
B=53° 42′ 9″			

For the angle C.			
h	161		2.2068259
$h-c$	81		1.9084850
a	134	a. c.	7.8728952
b	108	a. c.	7.9665762
			2)19.9547823
Cos ½C			9.9773911
C=36° 39′ 20″			

THEOREM F.

$$\left\{\begin{array}{l}\text{Tan }\tfrac{1}{2}A=R\sqrt{\dfrac{(h-b)(h-c)}{h(h-a)}}\\ \text{Tan }\tfrac{1}{2}B=R\sqrt{\dfrac{(h-a)(h-c)}{h(h-b)}}\\ \text{Tan }\tfrac{1}{2}C=R\sqrt{\dfrac{(h-a)(h-b)}{h(h-c)}}\end{array}\right.$$

The tangent is equal to the product of radius and the sine, divided by the cosines. (Art. 216.) By the last two theorems, then,

$$\text{Tan }\tfrac{1}{2}A=\frac{R\sin\tfrac{1}{2}A}{\cos\tfrac{1}{2}a}=R^2\sqrt{\frac{(h-b)(h-c)}{bc}}\div R\sqrt{\frac{h(h-a)}{bc}}$$

$$\text{That is, Tan }\tfrac{1}{2}A=R\sqrt{\frac{(h-b)(h-c)}{h(h-a)}}$$

Ex. Given the sides as before, to find A and C.

For the angle A.			
$h-b$	53		1.7242759
$h-c$	81		1.9084850
$h-a$	27	a. c.	8.5686362
h	161	a. c.	7.7931741
			2)19.9945712
Tan ½A			9.9972856
A=89° 38′ 31′			

For the angle B.			
$h-a$	27		1.4313638
$h-b$	53		1.7242759
$h-c$	81	a. c.	8.0915150
h	161	a. c.	7.7931741
			2)19.0403288
Tan ½C			9.5201644
C=36° 39′ 20″			

The three last theorems give the angle required, *without ambiguity*. For the *half* of any angle must be less than 90.°

Of these different methods of solution, each has its advantages in particular cases. It is expedient to find an angle, sometimes by its sine, sometimes by its cosine, and sometimes by its tangent.

By the first of the four preceding theorems marked C, D, E, and F, the calculation is made for the *sine* of the *whole* angle; by the others, for the *sine*, *cosine*, or *tangent*, of *half* the angle. For finding an angle near 90°, each of the three last theorems is preferable to the first. In the example above, A would have been uncertain to several seconds, by theorem C, if the other two angles had not been determined also.

But for a very *small* angle, the first method has an advantage over the others. The third, by which the calculation is made for the *cosine* of half the required angle, is in this case the most defective of the four. The second will not answer well for an angle which is almost 180°. For the *half* of this is almost 90°; and near 90°, the differences of the sines are very small.

NOTES.

Note A. Page 1.

THE *name* Logarithm is from λόγος *ratio*, and ἀριθμὸς *number*. Considering the ratio of a to 1 as a *simple* ratio, that of a^2 to 1 is a *duplicate* ratio, of a^3 to 1 a *triplicate* ratio, &c. (Alg. 354.) Here the *exponents* or *logarithms* 2, 3, 4, &c. show how many times the simple ratio is *repeated as a factor*, to form the compound ratio. Thus the ratio of 100 to 1, is the *square* of the ratio of 10 to 1; the ratio of 1000 to 1, is the *cube* of the ratio of 10 to 1, &c. On this account, logarithms are called the *measures* of ratios; that is of the ratios which different numbers bear to unity. See the Introduction to Hutton's Tables, and Mercator's Logarithmo-Technia, in Masseres' Scriptores Logarithmici.

Note B. p. 4.

If 1 be added to $-.09691$, it becomes $1-.09691$, which is equal to $+.90309$. The decimal is here rendered positive, by *subtracting* the figures from 1. But it is made 1 too great. This is compensated, by adding -1 to the *integral* part of the logarithm. So that $-2-.09691=-3+.90309$.

In the same manner, the decimal part of any logarithm which is wholly negative, may be rendered positive, by subtracting it from 1, and adding -1 to the index. The subtraction is most easily performed, by taking the right hand significant figure from 10, and each of the other figures from 9. (Art. 55.)

On the other hand, if the index of a logarithm be negative, while the decimal part is positive; the whole may be rendered negative, by subtracting the decimal part from 1, and taking -1 from the index.

NOTE C. p. 7.

It is common to *define* logarithms to be a series of numbers in arithmetical progression, corresponding with another series in geometrical progression. This is calculated to perplex the learner, when, upon opening the tables, he finds that the natural numbers, as they stand there, instead of being in *geometrical*, are in *arithmetical* progression; and that the logarithms are *not* in arithmetical progression.

It is true, that a geometrical series may be obtained, by taking out, here and there, a few of the natural numbers; and that the logarithms of these will form an arithmetical series. But the definition is not applicable to the whole of the numbers and logarithms, as they stand in the tables.

The supposition that positive and negative numbers have the same series of logarithms, (p. 7.) is attended with some theoretical difficulties. But these do not affect the practical rules for calculating by logarithms.

NOTE D. p. 43.

To revert a series, of the form

$$x=an+bn^2+cn^3+dn^4+en^5+\&c.$$

that is, to find the value of n, in terms of x, *assume* a series, with indeterminate co-efficients, (Alg. 490. *b*.)

$$\text{Let } n=Ax+Bx^2+Cx^3+Dx^4+Ex^5+\&c.$$

Finding the powers of this value of n, by multiplying the series into itself, and arranging the several terms according to the powers of x; we have

$$n^2=A^2x^2+2ABx^3+\left.\begin{matrix}+2AC\\+B^2\end{matrix}\right\}x^4\left.\begin{matrix}+2BC\\+2AD\end{matrix}\right\}x^5+\&c.$$

$$n^3=A^3x^3+3A^2Bx^4\left.\begin{matrix}+3A^2C\\+3AB^2\end{matrix}\right\}x^5+\&c.$$

$$n^4=A^4x^4+4A^3Bx^5+\&c.$$

$$n^5=A^5x^5+\&c.$$

Substituting these values, for n and its powers, in the first series above, we have

$$x=\left\{\begin{array}{llll} aA\,x+\left.\begin{array}{l} aB \\ +bA^2 \end{array}\right\} x^2 & +\left.\begin{array}{l} aC \\ +2bAB \\ +cA^3 \end{array}\right\} x^3 & +\left.\begin{array}{l} aD \\ +2bAC \\ +bB^2 \\ +3cA^2B \\ +dA^4 \end{array}\right\} x^4 & +\left.\begin{array}{l} aE \\ +2bBC \\ +2bAD \\ +3cA^2C \\ +3cAB^2 \\ +4dA^3B \\ +eA^5 \end{array}\right\} x^5 \end{array}\right.$$

Transposing x, and making the co-efficients of the several powers of x each equal to 0, we have

$$aA-1=0,$$
$$aB+bA^2=0,$$
$$aC+2bAB+cA^3=0,$$
$$aD+2bAC+bB^2+3cA^2B+dA^4=0,$$
$$aE+2bBC+2bAD+3cA^2C+3cAB^2+4dA^3B+eA^5=0.$$

And reducing the equations,

$$A=\tfrac{1}{2}$$
$$B=-\frac{b}{a^3}$$
$$C=\frac{2b^2-ac}{a^5}$$
$$D=-\frac{5b^3-5abc+a^2d}{a^7}$$
$$E=b\frac{14b^4-21ab^2c+3a^2c^2+6a^2bd-a^3e}{a^9}$$

These are the values of the co-efficients A, B, C, &c. in the assumed series

$$n=Ax+Bx^2+Cx\ +Dx^4+Ex^5+\&c.$$

Applying these results to the logarithmic series. (Art. 66. p. 43.)

$$n=n-\tfrac{1}{2}n^2+\tfrac{1}{3}n^3-\tfrac{1}{4}n^4+\tfrac{1}{5}n^5+\&c.$$

in which

$a=1, b=-\frac{1}{2}\ c=\frac{1}{3}, d=-\frac{1}{4}, e=\frac{1}{5},$

we have, in the inverted series

$n=Ax+Bx^2+Cx^3+Dx^4+Ex^6+$ &c.

$A=\frac{1}{a}=1 \qquad D=\frac{1}{2.3.4}$

$B=-b=\frac{1}{2}$

$C=2b^2-ac=\frac{1}{2.3} \qquad E=\frac{1}{2.3.4.5}$

Therefore

$$n=x+\frac{x^2}{2}+\frac{x^3}{2.3}+\frac{x^4}{2.3.4}+\frac{x^4}{2.3.4.5}+\&c.$$

Note E. p. 50.

According to the scheme lately introduced into France, of dividing the denominations of weights, measures, &c. into tenths, hundredths, &c. the fourth part of a circle is divided into 100 degrees, a degree into 100 minutes, a minute into 100 seconds, &c. The whole circle contains 400 of these degrees; a plane triangle 200. If a right angle be taken for the measuring *unit;* degrees, minutes. and seconds, may be written as decimal fractions. Thus 36° 5′ 49″ is 0.360549.

According to the French division $\left\{\begin{array}{c}10^\circ=9^\circ\\100'=54'\\1000''=324''\end{array}\right\}$ English.

Note F. p. 82.

If the perpendicular be drawn from the angle opposite the longest side, it will always fall *within* the triangle; because the other two angles must, of course, be acute. But if one of the angles at the base be *obtuse*, the perpendicular will fall *without* the triangle, as CP, (Fig. 38.)

In this case, the side on which the perpendicular falls, is to the sum of the other two; as the difference of the latter, to the *sum* of the segments made by the perpendicular.

The demonstration is the same, as in the other case, except that AH=BP+PA, instead of BP−PA.

Thus in the circle BDHL (Fig. 38.) of which C is the centre,

$$AB \times AH = AL \times AD \text{; therefore } AB : AD :: AL : AH.$$

$$\text{But } AD = CD + CA = CB + CA$$
$$\text{And } AL = CL - CA = CB - CA$$
$$\text{And } AH = HP + PA = BP + PA$$

Therefore

$$AB : CB + CA :: CB - CA : BP + PA.$$

When the three sides are given, it may be known whether one of the angles is obtuse. For any angle of a triangle is obtuse or acute, according as the square of the sine subtending the angle is *greater*, or *less*, than the sum of the squares of the sides containing the angle. (Euc. 12, 13. 2.)

Note G. p. 104.

Gunter's *Sliding Rule* is constructed upon the same principle as his scale, with the addition of a slider, which is so contrived as to answer the purpose of a pair of compasses, in working proportions, multiplying, dividing, &c. The lines on the *fixed part* are the same as on the scale. The *slider* contains two lines of numbers, a line of logarithmic sines, and a line of logarithmic tangents.

To *multiply* by this, bring 1 on the slider, against one of the factors on the fixed part; and against the other factor on the slider, will be the product on the fixed part. To divide, bring the divisor on the slider, against the dividend on the fixed part; and against 1 on the slider, will be the quotient on the fixed part. To work a *proportion*, bring the first term on the slider, against one of the middle terms on the fixed part; and against the other middle term on the slider, will be the fourth term on the fixed part. Or the first term may be taken on the fixed part; and then the fourth term will be found on the slider.

Another instrument frequently used in trigonometrical constructions, is

The Sector.

This consists of two equal scales, moveable about a point as a centre. The lines which are drawn on it are of two kinds; some being parallel to the sides of the instrument, and others diverging from the central point, like the radii of a circle. The latter are called the *double* lines, as each is repeated upon the two scales. The *single* lines are of the same nature, and have the same use, as those which are put upon the common scale; as the lines of equal parts, of chords, of latitude, &c. on one face; and the logarithmic lines of numbers, of sines, and of tangents, on the other.

The *double* lines are

A line of *Lines*, or equal parts, marked	Lin. or L.
A line of *Chords*,	Cho. or C.
A line of natural *Sines*,	Sin. or S.
A line of natural *Tangents* to 45°.	Tan. or T.
A line of tangents *above* 45°,	Tan. or T.
A line of natural *Secants*,	Sec. or S.
A line of *Polygons*,	Pol. or P.

The double lines of *Chords*, of *sines*, and of *tangents* to 45°, are all of the same radius; beginning at the central point, and terminating near the other extremity of each scale; the chords at 60°, the sines at 90°, and the tangents at 45°. (See art. 95.) The line of *lines* is also of the same length, containing ten equal parts which are numbered, and which are again subdivided. The radius of the lines of secants, and of tangents above 45°, is about one fourth of the length of the other lines. From the end of the radius, which for the secants is at 0, and for the tangents at 45°, these lines extend to between 70° and 80°. The line of polygons is numbered 4, 5, 6, &c. from the extremity of each scale, towards the centre.

The simple principle on which the utility of these several pairs of lines depends is this, that *the sides of similar triangles are proportional.* (Euc. 4. 6.) So that sines, tangents, &c. are furnished *to any radius*, within the extent of the opening of the two scales. Let AC and AC′ (Fig. 40.) be any pair of lines on the sector, and AB and AB′ equal portions of these lines. As AC and AC′ are equal, the triangle ACC′ is isosceles, and similar to ABB′. Therefore,

$$AB : AC :: BB' : CC'.$$

Distances measured from the centre on either scale, as AB and AC, are called *lateral distances*. And the distances between corresponding points of the two scales, as BB′ and CC′ are called *transverse distances*.

Let AC and CC′ be radii of two circles. Then if AB be the chord, sine, tangent, or secant, of any number of degrees in one; BB′ will be the chord, sine, tangent, or secant, of the same number of degrees in the other. (Art. 119.) Thus, to find the *chord* of 30°, to a radius of four inches, open the sector so as to make the transverse distance from 60 to 60, on the lines of chords, four inches; and the distance from 30 to 30, on the same lines, will be the chord required. To find the *sine* of 28°, make the distance from 90 to 90, on the lines of sines, equal to radius; and the distance from 28 to 28 will be the sine. To find the *tangent* of 37°, make the distance from 45 to 45, on the lines of tangents, equal to radius; and the distance from 37 to 37 will be the tangent. In finding *secants*, the distance from 0 to 0 must be made radius. (Art. 201.)

To lay down an *angle* of 34°, describe a circle, of any convenient radius, open the sector, so that the distance from 60 to 60 on the lines of chords shall be equal to this radius, and to the circle apply a chord equal to the distance from 34 to 34. (Art. 161.) For an angle above 60°, the chord of *half* the number of degrees may be taken, and applied *twice* on the arc, as in art. 161.

The line of *polygons* contains the chords of arcs of a circle which is divided into equal portions. Thus the distances from the centre of the sector to 4, 5, 6, and 7, are the chords of $\frac{1}{4}$, $\frac{1}{5}$, $\frac{1}{6}$, and $\frac{1}{7}$ of a circle, The distance 6 is the radius. (Art. 95.) This line is used to make a regular polygon, or to inscribe one in a given circle. Thus, to make a *pentagon* with the transverse distance from 6 to 6 for radius, describe a circle, and the distance from 5 to 5 will be the length of one of the sides of a pentagon inscribed in that circle.

The line of *lines* is used to divide a line into equal or proportional parts, to find fourth proportionals, &c. Thus, to divide a line into 7 equal parts, make the length of the given line the transverse distance from 7 to 7, and the distance from 1 to 1 will be one of the parts. To find $\frac{3}{5}$ of a line, make the transverse distance from 5 to 5 equal to the given line; and the distance from 3 to 3 will be $\frac{3}{5}$ of it.

In working the *proportions in trigonometry* on the sector, the lengths of the sides of triangles are taken from the line of lines, and the degrees and minutes from the lines of sines, tangents, or secants. Thus in art. 135, ex. 1,

$$35 : R :: 26 : \text{Sin } 48^\circ.$$

To find the fourth term of this proportion by the sector, make the lateral distance 35 on the line of lines, a transverse distance from 90 to 90 on the lines of sines; then the lateral distance 26 on the line of lines, will be the transverse distance from 48 to 48 on the lines of sines.

For a more particular account of the construction and uses of the Sector, see Stone's edition of Bion on Mathematical Instruments, Hutton's Dictionary, and Robertson's Treatise on Mathematical Instruments.

Note H. p. 124.

The errour in supposing that arcs less than 1 minute are proportional to their sines, can not affect the first ten places of decimals. Let AB and AB′ (Fig. 41.) each equal 1 minute. The tangents of these arcs BT and B′T are equal, as are also the sines BS and B′S. The arc BAB′ is greater than BS+B′S, but less than BT+B′T. Therefore BA is greater than BS, but less than BT: that is, *the difference between the sine and the arc is less than the difference between the sine and the tangent.*

Now the sine of 1 minute is	0.000290888216
And the tangent of 1 minute is	0.000290888204
The difference is	0.000000000012

The difference between the sine and the arc of 1 minute is less than this; and the errour in supposing that the sines of 1′, and of 0′ 52″ 44‴ 3⁗ 45′′′′′ are proportional to their arcs, as in art. 223, is still less.

Note I. p. 125.

There are various ways in which sines and cosines may be more *expeditiously* calculated, than by the method which is given here. But as we are already supplied with accurate trigonometrical tables, the computation of the canon is, to the great body of our students, a subject of speculation, rather than of practical utility. Those who wish to enter into a minute examination of it, will of course consult the treatises in which it is particularly considered.

There are also numerous formulæ of *verification*, which are used to detect the errours with which any part of the calculation is liable to be affected. For these, see Legendre's and Woodhouse's Trigonometry, Lacroix's Differential Calculus, and particularly Euler's Analysis of Infinites.

NOTE K. p. 127.

The following rules for finding the sine or tangent of a very small arc, and, on the other hand, for finding the arc from its sine or tangent, are taken from Dr. Maskelyne's Introduction to Taylor's Logarithms.

To find the logarithmic SINE *of a very small arc.*

From the sum of the constant quantity 4.6855749, and the logarithm of the given arc reduced to seconds and decimals, subtract one third of the arithmetical complement of the logarithmic cosine.

To find the logarithmic TANGENT *of a very small arc.*

To the sum of the constant quantity 4.6855749, and the logarithm of the given arc reduced to seconds and decimals, add two thirds of the arithmetical complement of the logarithmic cosine.

To find a small arc from its logarithmic SINE.

To the sum of the constant quantity 5.3144251, and the given logarithmic sine, add one third of the arithmetical complement of the logarithmic cosine. The remainder diminished by 10, will be the logarithm of the number of seconds in the arc.

To find a small arc from its logarithmic TANGENT.

From the sum of the constant quantity 5.3144251, and the given logarithmic tangent, subtract two thirds of the arithmetical complement of the logarithmic cosine. The remainder diminished by 10, will be the logarithm of the number of seconds in the arc.

For the demonstration of these rules, see Woodhouse's Trigonometry, p. 189.

A TABLE OF

NATURAL SINES AND TANGENTS;

TO EVERY TEN MINUTES OF A DEGREE.

IF the given angle is less than 45°, look for the title of the column, at the *top* of the page; and for the degrees and minutes, on the *left*. But if the angle is between 45° and 90°, look for the title of the column, at the *bottom*; and for the degrees and minutes, on the *right*.

D. M.	Sine	Tangent	Cotangent	Cosine	D. M.
0° 0′	0.0000000	0.0000000	Infinite	1.0000000	90° 0′
10	0029089	0029089	343.77371	0.9999958	50
20	0058177	0058178	171.88540	9999831	40
30	0087265	0087269	114.58865	9999619	30
40	0116353	0116361	85.939791	9999323	20
0° 50′	0145439	0145454	68.750087	9998942	89° 10
1° 0′	0.0174524	0.0174551	57.289962	0.9998477	89° 0
10	0203608	0203650	49.103881	9997927	50
20	0232690	0232753	42.964077	9997292	40
30	0261769	0261859	38.188459	9996573	30
40	0290847	0290970	34.367771	9995770	20
1° 50′	0319922	0320086	31.241577	9994881	88° 10′
2° 0′	0.0348995	0 0349208	28.636253	0.9993908	88° 0′
10	0378065	0378335	26.431600	9992851	50
20	0407131	0407469	24.541758	9991709	40
30	0436194	0436609	22.903766	9990482	30
40	0465253	0465757	21.470401	9989171	20
2° 50′	0494308	0494913	20.205553	9987775	87° 10′
3° 0′	0.0523360	0.0524078	19.081137	0.9986295	87° 0′
10	0552406	0553251	18.074977	9984731	50
20	0581448	0582434	17.169337	9983082	40
30	0610485	0611626	16.349855	9981348	30
40	0639517	0640829	15.604784	9979530	20
3° 50′	0668544	0670043	14.924417	9977627	86° 10′
4° 0′	00697565	0.0699268	14.300666	0.9975641	86° 0′
10	0726580	0728505	13.726738	9973569	50
20	0755589	·0757755	13.196883	9971413	40
30	0784591	0787017	12.706205	9969173	30
40	0813587	0816293	12.250505	9966849	20
4° 50′	0842576	0848583	11.826167	9964440	85° 10′
5° 0′	0.0871557	0.0874887	11.430052	0.9961947	85° 0
10	0900532	0904206	11.059431	9959370	50
20	0929499	0933540	10.711913	9956708	40
30	0958458	0962890	16.385397	9953962	30
40	0987408	0992257	10.078031	9951132	20
5° 50′	1016351	1021641	9.7881732	9948217	84° 10
D. M.	Cosine	Cotangent	Tangent	Sine	D. M.

D. M.	Sine	Tangent.	Cotangent	Cosine	D. M.
6° 0′	0.1045285	0.1051042	9 5143645	0.9945219	84° 0′
10	1074210	1080462	9.2553035	9942136	50
20	1103126	1109899	9.0098261	9938969	40
30	1132032	1139356	8.7768874	9935719	30
40	1160929	1168832	8.5555468	9932384	20
6° 50′	1189816	1198329	8.3449558	9928065	83° 10′
7° 0′	0.1218693	0.1227846	8.1443464	0.9925462	83° 0′
10	1247560	1257384	7.9530224	9921874	50
20	1276416	1286943	7.7703506	9918204	40
30	1305262	1316525	7.5957541	9914449	30
40	1334096	1346129	7.4287064	9910610	20
7° 50′	1362919	1375757	7.2687255	9906687	82° 10′
8° 0′	0.1391731	0.1404085	7.1153697	0.9902681	82° 0′
10	1420531	1435084	6.9682335	9898590	50
20	1449319	1464784	6.8269437	9894416	40
30	1478094	1494510	6.6911562	9890159	30
40	1506857	1524262	6.5605538	9885817	20
8° 50′	1535607	1554040	6.4348428	9881392	81° 10′
9° 0′	0.1564345	0.1583844	6.3137515	0.9876883	81° 0′
10	1593069	1613677	6·1970279	9872291	50
20	1621779	1643537	6.0844381	9867615	40
30	1650476	1673426	5.9757644	9862856	30
40	1679159	1703344	5.8708042	9858013	20
9° 50′	1707828	1733292	5.7693688	9853087	80° 10′
10° 0′	0.1736482	0.1763270	5.6712818	0.9848078	80° 0′
10	1765121	1793279	5.5763786	9842985	50
20	1793746	1823319	5.4845052	9837808	40
30	1822355	1853390	5.3955172	9832549	30
40	1850949	1883495	5.3092793	9827206	20
10° 50′	1879528	1913632	5.2256647	9821781	79° 10′
11° 0′	0.1908090	0.1943803	5.1445540	0.9816272	79° 0′
10	1936636	1974008	5.0658352	9810680	50
20	1965166	2004248	4.9894027	9805005	40
30	1993679	2034523	4.9151570	9799247	30
40	2022176	2064834	4.8430045	9793406	20
11° 50′	2050655	2095181	4.7728568	9787483	78° 10′
D. M.	Cosine	Cotangent	Tangent	Sine	D. M.

D. M.	Sine	Tangent	Cotangent	Cosine	D. M.
12° 0′	0.2079117	0.2125566	4.7046301	0.9781476	78° 0′
10	2107561	2155988	4.6382457	9775387	50
20	2135988	2186448	4.5736287	9769215	40
30	2164396	2216947	4.5107085	9762960	30
40	2192786	2247485	4.4494181	9756623	20
12° 50′	2221158	2278063	4.3896940	9750203	77° 10
13° 0′	0.2249511	0.2308682	4.3314759	0.9743701	77° 0′
10	2277844	2339342	4.2747066	9737116	50
20	2306159	2370044	4.2193318	9730449	40
30	2334454	2400788	4.1652998	9723699	30
40	2362729	2431575	4.1125614	9716867	20
13° 50′	2390984	2462405	4.0610700	9709953	76° 10′
14° 0′	0.2419219	0.2493280	4.0107809	0.9702957	76° 0′
10	2447433	2524200	3.9616518	9695879	50
20	2475627	2555165	3.9136420	9688719	40
30	2503800	2586176	3.8667131	9681476	30
40	2531952	2617234	3.8208281	9674152	20
14° 50′	2560082	2648339	3.7759519	9666746	75° 10′
15° 0′	0.2588190	0.2679492	3.7320508	0.9659258	75° 0′
10	2616277	2710694	3.6890927	9651689	50
20	2644342	2741945	3.6470467	9644037	40
30	2672384	2773245	3.6058835	9636305	30
40	2700403	2804597	3.5655749	9628490	20
15° 50′	2728400	2835999	3.5260938	9620594	74° 10′
16° 0′	0.2756374	0.2867454	3.4874144	0.9612617	74° 0′
10	2784324	2898961	3.4495120	9604558	50
20	2812251	2930521	3.4123626	9596418	40
30	2840153	2962135	3.3759434	9588197	30
40	2868032	2993803	3.3402326	9579895	20
16° 50′	2895887	3025527	3.3052091	9571512	73° 10′
17° 0′	0.2923717	0.3057307	3.2708526	0.9563048	73° 0′
10	2951522	3089143	3.2371438	9554502	50
20	2979303	3121036	3.2040638	9545876	40
30	3007058	3152988	3.1715948	9537170	30
40	3034788	3184998	3.1397194	9528382	20
17° 50′	3062492	3217067	3.1084210	9519514	72° 10′
D. M.	Cosine	Cotangent	Tangent	Sine	D. M.

D. M.	Sine	Tangent	Cotangent	Cosine	D. M.
18° 0′	0.3090170	0.3249197	3.0776835	0.9510565	72° 0′
10	3117822	3281387	3.0474915	9501536	50
20	3145448	3313639	3.0178301	9492426	40
30	3173047	3345953	2.9886850	9483237	30
40	3200619	3378330	2.9600422	9473966	20
18° 50′	3228164	3410771	2.9318885	9464616	71° 10′
19° 0′	0.3255682	0.3443276	2.9042109	0.9455186	71° 0′
10	3283172	3475846	2.8769970	9445675	50
20	3310634	3508483	2.8502349	9436085	40
30	3338069	3541186	2.8239129	9426415	30
40	3365475	3573956	2.7980198	9416665	20
19° 50′	3392852	3606795	2.7725448	9406835	70° 10′
20° 0′	0.3420201	0.3639702	2.7474774	0.9396926	70° 0′
10	3447521	3672680	2.7228076	9386938	50
20	3474812	3705728	2.6985254	9376869	40
30	3502074	3738847	2.6746215	9366722	30
40	3529306	3772038	2.6510867	9356495	20
20° 50′	3556508	3805302	2 6279121	9346189	69° 10′
21° 0′	0.3583679	0.3838640	2.6050891	0.9335804	69° 0′
10	3610821	3872053	2.5826094	9325340	50
20	3637932	3905541	2.5604649	9314797	40
30	3665012	3939105	2.5386479	9304176	30
40	3692061	3972746	2.5171507	9293475	20
21° 50′	3719079	4006465	2.4959661	9282696	68° 10′
22° 0′	0.3746066	0.4040262	2.4750869	0.9271839	68° 0′
10	3773021	4074139	2.4545061	9260902	50
20	3799944	4108097	2.4342172	9249888	40
30	3826834	4142136	2.4142136	9238795	30
40	3853693	4176257	2.3944889	9227624	20
22° 50′	3880518	4210460	2.3750372	9216375	67° 10′
23° 0′	0.3907311	0.4244748	2 3558524	0.9205049	67° 0′
10	3934071	4279121	2.3369287	9193644	50
20	3960798	4313579	2.3182606	9182161	40
30	3987491	4348124	2.2998425	9170601	30
40	4014150	4382756	2.2816693	9158963	20
23° 50′	4040775	4417477	2.2637357	9147247	66° 10′
D. M.	Cosine	Cotangent	Tangent	Sine	D. M.

D. M.	Sine	Tangent	Cotangent	Cosine	D. M.
24° 0′	0.4067366	0.4452287	2.2460368	0.9135455	66° 0′
10	4093923	4487187	2.2285676	9123584	50
20	4120445	4522179	2.2113234	9111637	40
30	4146932	4557263	2.1942997	9099613	30
40	4173385	4592439	2.1774920	9087511	20
24° 50′	4199801	4627710	2.1608958	9075333	65° 10′
25° 0′	0.4226183	0.4663077	2.1445069	0.9063078	65° 0′
10	4252528	4698539	2.1283213	9050746	50
20	4278838	4734098	2.1123348	9038338	40
30	4305111	4769755	2.0965436	9025853	30
40	4331348	4805512	2.0809438	9013292	20
25° 50′	4357548	4841368	2.0655318	9000654	64° 10′
26° 0′	0.4383711	0.4877326	2.0503038	0.8987940	64° 0′
10	4409838	4913386	2.0352565	8975151	50
20	4435927	4949549	2.0203862	8962285	40
30	4461978	4985816	2.0056897	8949344	30
40	4487992	5022189	1.9911637	8936326	20
26° 50′	4513967	5058668	1.9768050	8923234	63° 10′
27° 0′	0.4539905	0.5095254	1.9626105	0.8910065	63° 0′
10	4565805	5131950	1.9485772	8896822	50
20	4591665	5168755	1.9347020	8883503	40
30	4617486	5205671	1.9209821	8870108	30
40	4643269	5242698	1.9074147	8856639	20
27° 50′	4669012	5279889	1.8939971	8843095	62° 10′
28° 0′	0.4694716	0.5317094	1.8807265	0.8829476	62° 0′
10	4720380	5354465	1.8676003	8815782	50
20	4746004	5391953	1.8546159	8802014	40
30	4771588	5429557	1.8417709	8788171	30
40	4797131	5467281	1.8290628	8774254	20
28° 50′	4822634	5505125	1.8164892	8760263	61° 10′
29° 0′	0.4848096	0.5543091	1.8040478	0.8746197	61° 0′
10	4873517	5581179	1.7917362	8732058	50
20	4898897	5619391	1.7795524	8717844	40
30	4924236	5657728	1.7674940	8703557	30
40	4949532	5696191	1.7555590	8689196	20
29° 50′	4974787	5734783	1.7437453	8674762	60° 10′
D. M.	Cosine	Cotangent	Tangent	Sine	D. M.

D. M.	Sine	Tangent	Cotangent	Cosine	D. M.
30° 0′	0.5000000	0.5773503	1.7320508	0.8660254	60° 0′
10	5025170	5812353	1.7204736	8645673	50
20	5050298	5851335	1.7090116	8631019	40
30	5075384	5890450	1.6976631	8616292	30
40	5100426	5929699	1.6864261	8601491	20
30° 50′	5125425	5969084	1.6752988	8586619	59° 10′
31° 0′	0.5150381	0.6008606	1.6642795	0.8571673	59° 0′
10	5175293	6048266	1.6533663	8556655	50
20	5200161	6088067	1.6425576	8541564	40
30	5224986	6128008	1.6318517	8526402	30
40	5249766	6168092	1.6212469	8511167	20
31° 50′	5274502	6208320	1.6107417	8495860	58° 10′
32° 0′	0.5299193	0.6248694	1.6003345	0.8480481	58° 0′
10	5323839	6289214	1.5900238	8465030	50
20	5348440	6329883	1.5798079	8449508	40
30	5372996	6370703	1.5696856	8433914	30
40	5397507	6411673	1.5596552	8418249	20
32° 50′	5421971	6452797	1.5497155	8402513	57° 10′
33° 0′	0.5446390	0.6494076	1.5398650	0.8386706	57° 0′
10	5470763	6535511	1.5301023	8370827	50
20	5495090	6577103	1.5204261	8354878	40
30	5519370	6618856	1.5108352	8338858	30
40	5543603	6660769	1 5013282	8322768	20
33° 50′	5567790	6702845	1.4919039	8306607	56° 10′
34° 0′	0.5591929	0.6745085	1.4825610	0.8290376	56° 0′
10	5616021	6787492	1.4732983	8274074	50
20	5640066	6830066	1.4641147	8257703	40
30	5664062	6872810	1.4550090	8241262	30
40	5688011	6915725	1.4459801	8224751	20
34° 50′	5711912	6958813	1.4370268	8208170	55° 10′
35° 0′	0.5735764	0.7002075	1.4281480	0.8191520	55° 0′
10	5759568	7045515	1.4193427	8174801	50
20	5783323	7089133	1.4106098	8158013	40
30	5807030	7132931	1.4019483	8141155	30
40	5830687	7176911	1.3933571	8124229	20
35° 50′	5854294	7221075	1.3848353	8107234	54° 10′
D. M.	Cosine	Cotangent	Tangent	Sine	D. M.

D. M.	Sine	Tangent.	Cotangent	Cosine	D. M.
36° 0′	0.5877853	0.7265425	1.3763819	0.8090170	54° 0′
10	5901361	7309963	1.3679959	8073038	50
20	5924819	7354691	1.3596764	8055837	40
30	5948228	7399611	1.3514224	8038569	30
40′	5971586	7444724	1.3432331	8021232	20
36° 50	5994893	7490033	1.3351075	8003827	53° 10′
37° 0′	0.6018150	0.7535541	1.3270448	0.7986355	53° 0′
10	6041356	7581248	1.3190441	7968815	50
20	6064511	7627157	1.3111046	7951208	40
30	6087614	7673270	1.3032254	7933533	30
40	6110666	7719589	1.2954057	7915792	20
37° 50′	6133666	7766118	1.2876447	7897983	52° 10′
38° 0′	0.6156615	0.7812856	1.2799416	0.7880108	52° 0′
10	6179511	7859808	1.2722957	7862165	50
20	6202355	7906975	1.2647062	7844157	40
30	6225146	7954359	1.2571723	7826082	30
40	6247885	8001963	1.2496933	7807940	20
38° 50′	6270571	8049790	1.2422685	7789733	51° 10′
39° 0′	0 6293204	0.8097840	1.2348972	0.7771460	51° 0′
10	6315784	8146118	1.2275786	7753121	50
20	6338310	8194625	1.2203121	7734716	40
30	6360782	8243364	1.2130970	7716246	30
40	6383201	8292337	1.2059327	7697710	20
39° 50	6405566	8341547	1.1988184	7679110	50° 10′
40° 0′	0.6427876	0.8390996	1.1917536	0.7660444	50° 0′
10	6450132	8440688	1.1847376	7641714	50
20	6472334	8490624	1.1777698	7622919	40
30	6494480	8540807	1.1708496	7604060	30
40	6516572	8591240	1.1639763	7585136	20
40° 50′	6538609	8641926	1.1571495	7566148	49° 10′
41° 0	0 6560590	0.8692867	1.1503684	0.7547096	49° 0′
10	6582516	8744067	1.1436326	7527980	50
20	6604386	8795528	1.1369414	7508800	40
30	6626200	8847253	1.1302944	7489557	30
40	6647959	8899244	1.1236909	7470251	20
41° 50′	6669661	8951506	1.1171305	7450881	48° 10′
D. M.	Cosine	Cotangent	Tangent	Sine	D. M.

D. M.	Sine	Tangent	Cotangent	Cosine	D. M.
42° 0′	0.6691306	0.9004040	1.1106125	0.7431448	48° 0′
10	6712895	9056851	1.1041365	7411953	50
20	6734427	9109940	1.0977020	7392394	40
30	6755902	9163312	1.0913085	7372773	30
40	6777320	9216969	1.0849554	7353090	20
42° 50′	6798681	9270914	1.0786423	7333345	47° 10
43° 0′	0.6819984	0.9325151	1.0723687	0.7313537	47° 0′
10	6841229	9379683	1.0661341	7293668	50
20	6862416	9434513	1.0599381	7273736	40
30	6883546	9489646	1.0537801	7253744	30
40	6904617	9545083	1.0476598	7233690	20
43° 50′	6925630	9600829	1.0415767	7213574	46° 10′
44° 0′	0.6946584	0.9656888	1.0355303	0.7193398	46° 0′
10	6967479	9713262	1.0295203	7173161	50
20	6988315	9769956	1.0235461	7152863	40
30	7009093	9826973	1 0176074	7132504	30
40	7029811	9884316	1.0117038	7112086	20
44° 50′	7050469	9941991	1.0058348	7091607	45° 10′
45° 0′	0.7071068	1.0000000	1.0000000	0.7071068	45° 0′
D. M.	Cosine	Cotangent	Tangent	Sine	D. M.

The *Secants and Cosecants*, which are not inserted in this table, may be easily supplied. If 1 be divided by the cosine of an arc, the quotient will be the secant of that arc. (Art. 228.) And if 1 be divided by the sine, the quotient will be the cosecant.

D.	M.	Sine.	Tangent.	Cotangent.	Cosine.	D.	M.
42°	0	0.6691306	0.9004040	1.1106125	0.7431448	[illegible]	0
	10	[illegible]	[illegible]	[illegible]	[illegible]		50
	20	[illegible]	[illegible]	[illegible]	[illegible]		40
	30	[illegible]	[illegible]	[illegible]	[illegible]		30
	40	[illegible]	[illegible]	[illegible]	[illegible]		20
	50	[illegible]	[illegible]	[illegible]	[illegible]		10
43°	0	[illegible]	[illegible]	[illegible]	[illegible]	[illegible]	0
	10	[illegible]	[illegible]	[illegible]	[illegible]		50
	20	[illegible]	[illegible]	[illegible]	[illegible]		40
	30	[illegible]	[illegible]	[illegible]	[illegible]		30
	40	[illegible]	[illegible]	[illegible]	[illegible]		20
	50	[illegible]	[illegible]	[illegible]	[illegible]		10
44°	0	[illegible]	[illegible]	[illegible]	[illegible]	[illegible]	0
	10	[illegible]	[illegible]	[illegible]	[illegible]		50
	20	[illegible]	[illegible]	[illegible]	[illegible]		40
	30	[illegible]	[illegible]	[illegible]	[illegible]		30
	40	[illegible]	[illegible]	[illegible]	[illegible]		20
	50	[illegible]	[illegible]	[illegible]	[illegible]		10
45°	0	[illegible]	[illegible]	[illegible]	[illegible]	45°	0
D.	M.	Cosine.	Cotangent.	Tangent.	Sine.	D.	M.

The Secants and Cosecants, which are not inserted in this table, may be easily supplied. If 1 be divided by the cosine of an arc, the quotient will be the secant of that arc. (Art. 228.) And if 1 is divided by the sine, the quotient will be the cosecant.

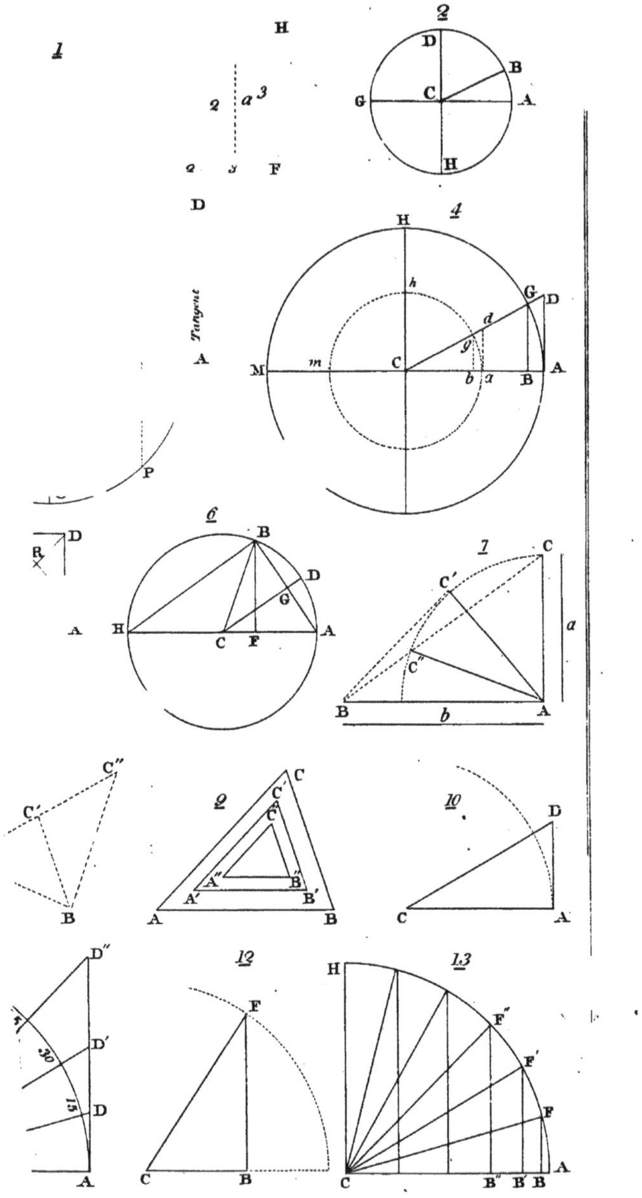

N. Jocelin Sc. N. H.

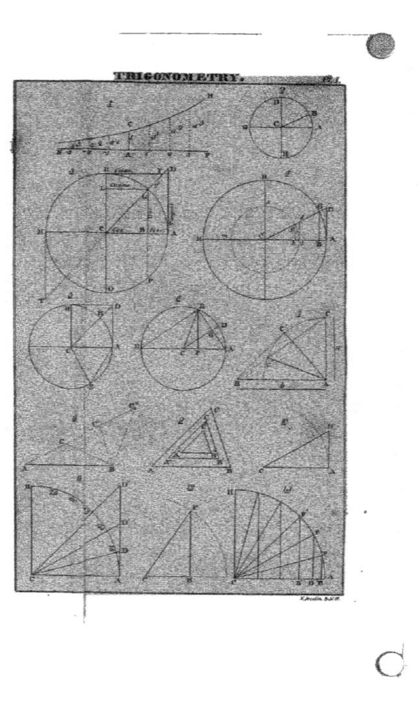
TRIGONOMETRY.

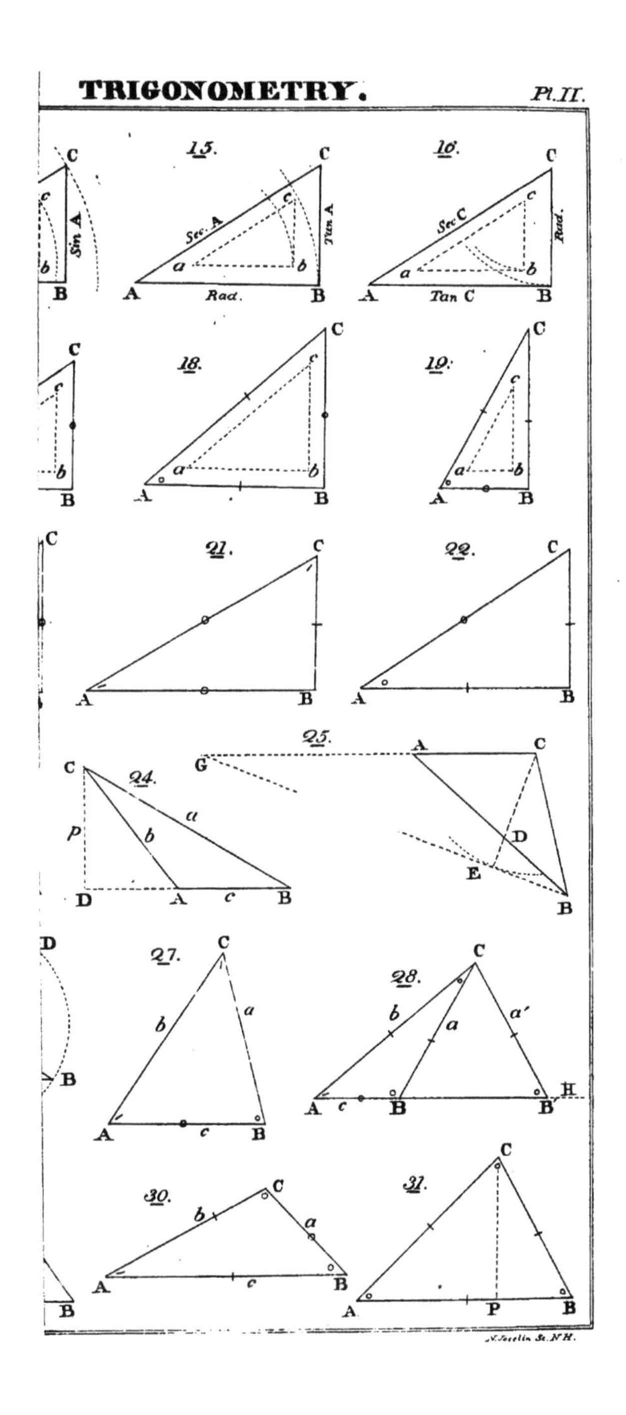
15.
16.
18.
19.
21.
22.
24.
25.
27.
28.
30.
31.
Sec. A
Tan A
Rad.
Sec C
Tan C
Sin A
A
B
C
a
b
c
D
E
G
H
P
p
a'
N. Jocelin Sc. N.H.

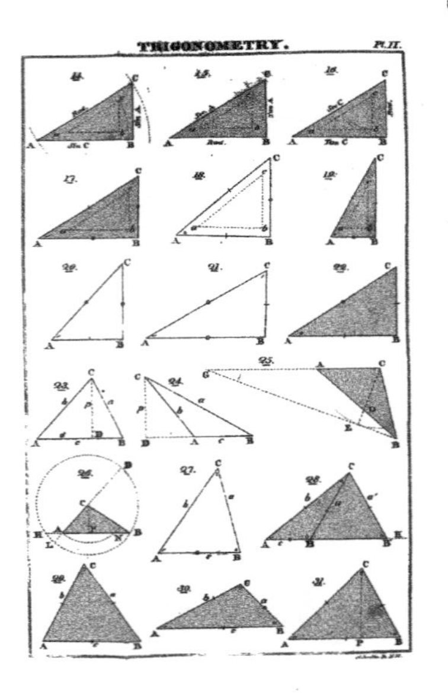
TRIGONOMETRY.
Pl. II.

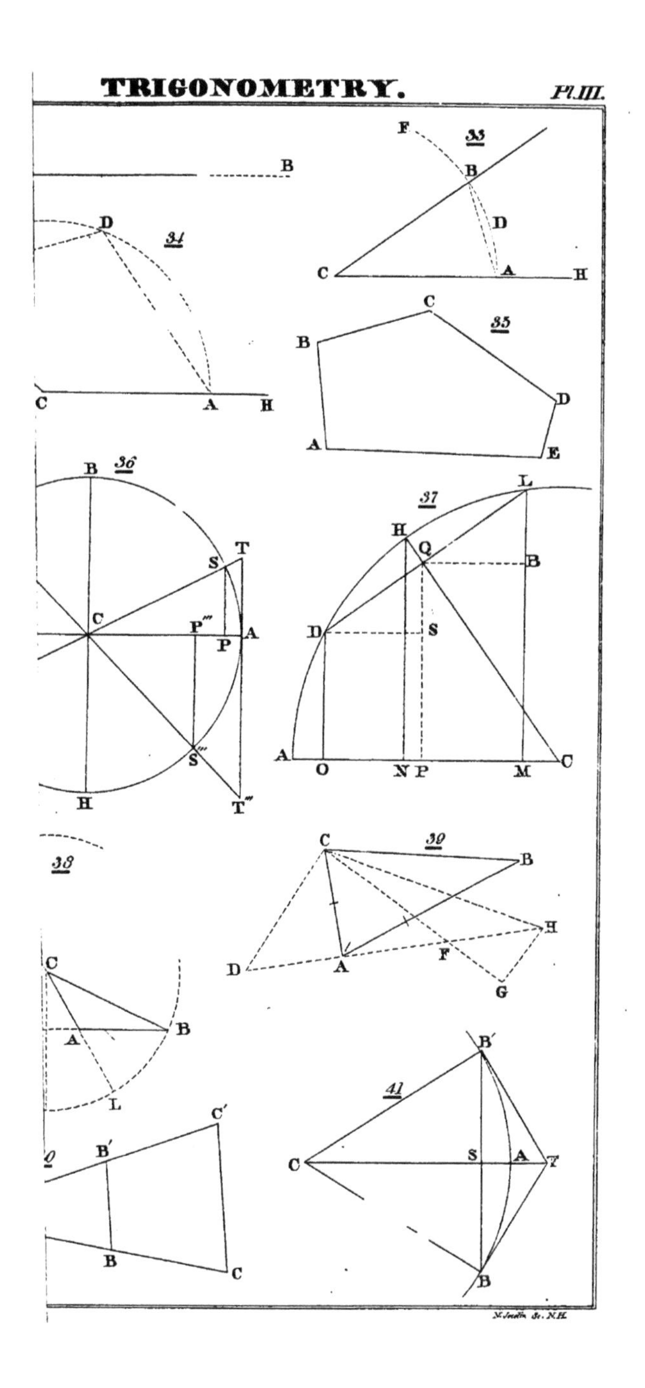
33
F
B
D
C
A
H
34
D
C
A
H
B
35
A
B
C
D
E
36
B
S
T
C
P'''
P
A
S'''
H
T'''
37
L
H
Q
B
D
S
A
O
N
P
M
C
38
C
A
B
L
39
C
B
H
D
A
F
G
40
B'
C'
B
C
41
B'
C
S
A
T
B
N. Jocelin Sc. N.H.

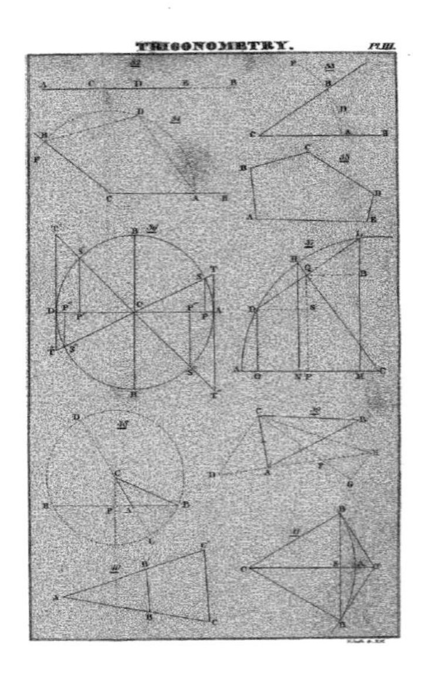
TRIGONOMETRY.
Pl. III.

A PRACTICAL APPLICATION

OF

THE PRINCIPLES OF GEOMETRY

TO THE

MENSURATION

OF

SUPERFICIES AND SOLIDS:

BEING

THE THIRD PART

OF

A COURSE OF MATHEMATICS,

ADAPTED TO THE METHOD OF INSTRUCTION IN

THE AMERICAN COLLEGES.

By JEREMIAH DAY, D. D. LL. D.

President of Yale College.

THE SECOND EDITION.

NEW-HAVEN:

PRINTED AND PUBLISHED BY S. CONVERSE.

1825.

The following short Treatise contains little more than an application of the principles of Geometry, to the numerical calculation of the superficial and solid contents of such figures as are treated of in the Elements of Euclid. As the plan proposed for the work of which this number is a part, does not admit of introducing rules and propositions which are not demonstrated; the particular consideration of the areas of the Conic Sections and other curves, with the contents of solids produced by their revolution, is reserved for succeeding parts of the course. The student would be little profited by applying arithmetical calculation, in a mechanical way, to figures of which he has not yet learned even the definitions. But as this number may fall into the hands of some who will not read those which are to follow, the principal rules for conic areas and solids, and for the gauging of casks, are given, without demonstrations, in the appendix. Those who wish to take a complete view of Mensuration, in all its parts, are referred to the valuable treatise of Dr. Hutton on the subject.

CONTENTS.

SECTION I.

AREAS OF FIGURES BOUNDED BY RIGHT LINES.

ART. 1. THE following definitions, which are nearly the same as in Euclid, are inserted here for the convenience of reference.

I. *Four-sided* figures have different names, according to the relative position and length of the sides. A *parallelogram* has its opposite sides equal and parallel; as ABCD. (Fig. 2.) A *rectangle*, or *right parallelogram*, has its opposite sides equal, and all its angles right angles; as AC. (Fig 1.) A *square* has all its sides equal, and all its angles right angles; as AB GH. (Fig. 3.) A *rhombus* has all its sides equal, and its angles oblique: as ABCD. (Fig. 3.) A *rhomboid* has its opposite sides equal, and its angles oblique; as ABCD. (Fig. 2.) A *trapezoid* has only two of its sides parallel; as ABCD. (Fig. 4.) Any other four sided figure is called a *trapezium*.

II. A figure which has more than four sides is called a *polygon*. A *regular* polygon has all its sides equal, and all its angles equal.

III. The *height* of a *triangle* is the length of a perpendicular, drawn from one of the angles to the opposite side; as CP. (Fig. 5.) The *height* of a *four-sided* figure is the perpendicular distance between two of its parallel sides; as CP. (Fig. 4.)

IV. The *area* or *superficial contents* of a figure is the *space* contained within the line or lines by which the figure is bounded.

2. In calculating areas, some particular portion of surface is fixed upon, as the *measuring unit*, with which the given figure is to be compared. This is commonly a *square;* as a square inch, a square foot, a square rod, &c. For this reason, determining the quantity of surface in a figure is called *squaring it*, or finding its *quadrature;* that is, finding a square or number of squares to which it is equal.

3. The *superficial* unit has generally the same name, as the *linear* unit which forms the side of the square.

The side of a square inch is a linear inch;
of a square foot, a linear foot;
of a square rod, a linear rod, &c.

There are some superficial measures, however, which have no corresponding denominations of length. The *acre*, for instance, is not a square which has a line of the same name for its side.

The following tables contain the linear measures in common use, with their corresponding square measures.

Linear Measures.				*Square Measures.*			
12	Inches	=1	Foot.	144	Inches	=1	Foot.
3	Feet	=1	Yard.	9	Feet	=1	Yard.
6	Feet	=1	Fathom.	36	Feet	=1	Fathom.
$16\frac{1}{2}$	Feet	=1	Rod.	$272\frac{1}{4}$	Feet	=1	Rod.
$5\frac{1}{2}$	Yards	=1	Rod.	$30\frac{1}{4}$	Yards	=1	Rod.
4	Rods	=1	Chain.	16	Rods	=1	Chain.
40	Rods	=1	Furlong.	1600	Rods	=1	Furlong.
320	Rods	=1	Mile.	102400	Rods	=1	Mile.

An *acre* contains 160 square rods, or 10 square chains.

By reducing the denominations of square measure, it will be seen that

1 sq. mile=640 acres=102400 rods=27878400 ft.=4014489600 inches.
1 acre=10 chains=160 rods=43560 feet=6272640 inches.

The fundamental problem in the mensuration of superficies is the very simple one of determining the area of a *right parallelogram.* The contents of other figures, particularly those which are rectilinear, may be obtained by finding parallelograms which are equal to them, according to the principles laid down in Euclid.

PROBLEM I.

To find the area of a PARALLELOGRAM, *square, rhombus, or rhomboid.*

4. MULTIPLY THE LENGTH BY THE PERPENDICULAR HEIGHT OR BREADTH.

It is evident that the number of *square* inches in the parallelogram AC (Fig. 1.) is equal to the number of *linear* inches in the length AB, repeated as many times as there are inches in the breadth BC. For a more particular illustration of this, see Alg. 511—514.

The oblique parallelogram or rhomboid ABCD (Fig. 2.) is equal to the right parallelogram GHCD. (Euc. 36. 1.) The area, therefore, is equal to the length AB multiplied into the perpendicular height HC. And the rhombus ABCD (Fig. 3.) is equal to the *parallelogram* ABGH. As the sides of a *square* are all equal, its area is found, by *multiplying one of the sides into itself.*

Ex. 1. How many square feet are there in a floor $23\frac{1}{2}$ feet long, and 18 feet broad? Ans. $23\frac{1}{2} \times 18 = 423$.

2. What are the contents of a piece of ground which is 66 feet square? Ans. 4356 sq. feet=16 sq. rods.

3. How many square feet are there in the four sides of a room which is 22 feet long, 17 feet broad, and 11 feet high? Ans. 858.

Art. 5. If the sides and angles of a parallelogram are given, the perpendicular height may be easily found by trigonometry. Thus CH (Fig. 2.) is the perpendicular of a right angled triangle, of which BC is the hypothenuse. Then (Trig. 134.)

$$R : BC :: \text{Sin } B : CH.$$

The area is obtained by multiplying CH thus found, into the length AB.

Or, to reduce the two operations to one,

As radius,
To the sine of any angle of a parallelogram;
So is the product of the sides including that angle,
To the area of the parallelogram.

For the area$= AB \times CH$ (Fig. 2.) But $CH = \dfrac{BC \times \text{Sin } B}{R}$.

Therefore,

The area$= \dfrac{AB \times BC + \text{Sin } B}{R}$. Or $R : \text{Sin}B :: AB \times BC :$ *the area.*

Ex. If the side AB be 58 rods, BC 42 rods, and the angle B 63°, what is the area of the parallelogram?

As radius		10.00000
To the sine of B	63°	9.94988
{ So is the product of AB	58	1.76343
{ Into BC (Trig. 39,)	42	1.62325
To the area	2170.5 sq. rods	3.33656

2. If the side of a rhombus is 67 feet, and one of the angles 73°, what is the area? Ans. 4292.7 feet.

6. When the dimensions are given in feet and inches, the multiplication may be conveniently performed by the arithmetical rule of *Duodecimals;* in which each inferior denomination is one twelfth of the next higher. Considering a foot as the measuring *unit*, a prime is the twelfth part of a foot; a second, the twelfth part of a prime, &c. It is to be observed, that, in measures of *length, inches* are *primes;* but in *superficial* measure they are *seconds*. In both, a prime is $\frac{1}{12}$ of a foot. But $\frac{1}{12}$ of a *square* foot is a parallelogram a foot long and an inch broad. The twelfth part of this is a square inch, which is $\frac{1}{144}$ of a square foot.

Ex. 1. What is the surface of a board 9 feet 5 inches, by 2 feet 7 inches.

F		
9	5′	
2	7	
18	10	
5	5	11
24	3	11″, or 24 feet 47 inches.

2. How many feet of glass are there in a window 4 feet 11 inches high, and 3 feet 5 inches broad?

Ans. 16F. 9′ 7″, or 16 feet 115 inches.

7. If the area and one side of a parallelogram be given, the other side may be found by *dividing the area by the given side.* And if the area of a *square* be given, the side may be found by *extracting the square root of the area.* This is merely reversing the rule in art. 4. See Alg. 520, 521.

Ex. 1. What is the breadth of a piece of cloth which is 36 yds. long, and which contains 63 square yds. Ans. $1\frac{3}{4}$ yds.

2. What is the side of a square piece of land containing 289 square rods?

3. How many yards of carpeting $1\frac{1}{4}$ yard wide, will cover a floor 30 feet long and $22\frac{1}{2}$ broad?

Ans. $30\times22\frac{1}{2}$ feet$=10\times7\frac{1}{2}=75$ yds. And $75\div1\frac{1}{4}=60$.

4. What is the side of a square which is equal to a parallelogram 936 feet long and 104 broad?

5. How many panes of 8 by 10 glass are there, in a window 5 feet high, and 2 feet 8 inches broad?

PROBLEM II.

To find the area of a TRIANGLE.

8. RULE I. MULTIPLY ONE SIDE BY HALF THE PERPENDICULAR FROM THE OPPOSITE ANGLE. Or, multiply half the side by the perpendicular. Or, multiply the whole side by the perpendicular, and take half the product.

The area of the triangle ABC (Fig. 5.) is equal to $\frac{1}{2}$ PC$\times$AB, because a parallelogram of the same base and height is equal to PC$\times$AB, (Art. 4.) and by Euc. 41. 1, the triangle is half the parallelogram.

Ex. 1. If AB (Fig. 5.) be 65 feet, and PC 31.2, what is the area of the triangle? Ans. 1014 square feet.

2. What is the surface of a triangular board, whose base is 3 feet 2 inches, and perpendicular height 2 feet 9 inches?

Ans. 4F. 4′ 3″, or 4 feet 51 inches.

9. If two sides of a triangle and the included angle, are given, the perpendicular on one of these sides may be easily found by rectangular trigonometry. And the area may be calculated in the same manner as the area of a parallelogram in art. 5. In the triangle ABC (Fig. 2.)

R : BC : : Sin B : CH

And because the triangle is half the parallelogram of the same base and height,

As radius,
To the sine of any angle of a triangle;
So is the product of the sides including that angle,
To twice the area of the triangle. (Art. 5.)

Ex. If AC (Fig. 5.) be 39 feet, AB 65 feet, and the angle at A 53° 7′ 48″, what is the area of the triangle?

Ans. 1014 square feet.

9. *b*. If *one side* and the *angles* are given; then
As the product of radius and the sine of the angle opposite the given side,
To the product of the sines of the two other angles;
So is the square of the given side,
To twice the area of the triangle.

If PC (Fig. 5.) be perpendicular to AB.

$$R : \text{Sin } B :: BC : CP$$

$$\text{Sin } ACB : \text{Sin } A :: AB : BC$$

Therefore (Alg. 390, 382.)

$R \times \text{Sin } ACB : \text{Sin } A \times \text{Sin } B :: AB \times BC : CP \times BC :: \overline{AB}^2 : AB \times CP =$ twice the area of the triangle.

Ex. If one side of a triangle be 57 feet, and the angles at the ends of this side 50° and 60°, what is the area?

Ans. 1147 sq. feet.

10. If the *sides* only of a triangle are given, an angle may be found, by oblique trigonometry, Case IV, and then the perpendicular and the area may be calculated. But the area may be more directly obtained, by the following method.

Rule II. When the three sides are given, *from half their sum subtract each side severally, multiply together the half sum and the three remainders, and extract the square root of the product.*

If the sides of the triangle are a, b, and c, and if h=half their sum, then

$$\textit{The area} = \sqrt{h \times (h-a) \times (h-b) \times (h-c)}$$

For the demonstration of this rule, see Trigonometry, Art. 221.

If the calculation be made by *logarithms*, add the logarithms of the several factors, and half their sum will be the logarithm of the area. (Trig. 39, 47.)

Ex. I. In the triangle ABC (Fig. 5.) given the sides a 52 feet, b 39, and c 65; to find the side of a square which has the same area as the triangle.

$\frac{1}{2}(a+b+c)=h=78$ $\qquad h-b=39$

$h-a=26$ $\qquad h-c=13$

Then the area $=\sqrt{78 \times 26 \times 39 \times 13} = 1014$ square feet.

By logarithms.

The half sum	=78	1.89209
First remainder	=26	1.41497
Second do.	=39	1.59106
Third do.	=13	1.11394
		2)6.01206
The area required	=1014	2)3.00603
Side of the square	=31.843 (Trig. 47)	1.50301

2. If the sides of a triangle are 134, 108, and 80 rods, what is the area? Ans. 4319.

3. What is the area of a triangle whose sides are 371, 264, and 225 feet?

11. In an *equilateral* triangle, one of whose sides is a, the expression for the area becomes

$$\sqrt{h\times(h-a)\times(h-a)\times(h-a)}$$

But as $h=\frac{3}{2}a$, and $h-a=\frac{3}{2}a-a=\frac{1}{2}a$, the area is

$$\sqrt{\tfrac{3}{2}a\times\tfrac{1}{2}a\times\tfrac{1}{2}a\times\tfrac{1}{2}a}=\sqrt{\tfrac{3}{16}a^4}=\tfrac{1}{4}a^2\sqrt{3} \text{ (Alg. 271.)}$$

That is, the area of an equilateral triangle is equal to $\frac{1}{4}$ the square of one of its sides, multiplied into the square root of 3, which is 1.732.

Ex. 1. What is the area of a triangle whose sides are each 34 feet? Ans. $500\frac{1}{2}$ feet.

2. If the sides of a triangular field are each 100 rods, how many acres does it contain?

PROBLEM III.

To find the area of a TRAPEZOID.

21. MULTIPLY HALF THE SUM OF THE PARALLEL SIDES INTO THEIR PERPENDICULAR DISTANCE.

The area of the trapezoid ABCD (Fig. 4.) is equal to half the sum of the sides AB and CD, multiplied into the perpendicular distance PC or AH. For the whole figure is made up of the two triangles ABC and ADC; the area of the first of which is equal to the product of half the base AB into the perpendicular PC, (Art. 8.) and the area of the other is equal to the product of half the base DC into the perpendicular AH or PC.

Ex. If AB (Fig. 4.) be 46 feet, BC 31, DC 38, and the angle B 70°, what is the area of the trapezoid?

R : BC : : Sin B : PC=29.13. And $42\times29.13=1223\frac{1}{4}$.

2. What are the contents of a field which has two parallel sides 65 and 38 rods, distant from each other 27 rods?

PROBLEM IV.

To find the area of a TRAPEZIUM, *or of an irregular* POLYGON.

13. DIVIDE THE WHOLE FIGURE INTO TRIANGLES, BY DRAWING DIAGONALS, AND FIND THE SUM OF THE AREAS OF THESE TRIANGLES. (Alg. 519.)

If the perpendiculars in two triangles fall upon the *same diagonal*, the area of the trapezium formed of the two triangles, is equal to half the product of the diagonal into the sum of the perpendiculars.

Thus the area of the trapezium ABCH (Fig. 6.) is

$$\tfrac{1}{2}BH\times AL+\tfrac{1}{2}BH\times CM=\tfrac{1}{2}BH\times(AL+CM)$$

Ex. In the irregular poloygon ABCDH (Fig. 6.)

if the diagonals $\begin{cases}BH=36,\\CH=32,\end{cases}$ and the perpendiculars $\begin{cases}AL=5.3\\CM=9.4\\DN=7.3\end{cases}$

The area $=18\times14.6+16\times7.3=379.6$

14. If the diagonals of a *trapezium* are given, the area may be found, nearly in the same manner as the area of a parallelogram in Art. 5, and the area of a triangle in Art. 9.

In the trapezium ABCD (Fig. 8.) the sines of the four angles at N, the point of intersection of the diagonals, are all equal. For the two acute angles are *supplements* of the other two, and therefore have the same sine. (Trig. 90.) Putting, then, Sin N for the sine of each of these angles, the areas of the four triangles of which the trapezium is composed, are given by the following proportions; (Art. 9.)

$$R : \text{Sin } N :: \begin{cases}BN\times AN : 2\ \textit{area}\ ABN\\BN\times CN : 2\ \textit{area}\ BCN\\DN\times CN : 2\ \textit{area}\ CDN\\DN\times AN : 2\ \textit{area}\ ADN\end{cases}$$

And by addition, (Alg. 388, Cor. 1.*)

R : Sin N : : BN × AN + BN × CN + DN × CN + DN × AN : 2 *area* ABCD.

The 3d term = (AN + CN) × (BN + DN) = AC × BD, by the figure.

Therefore, R : Sin N : : AC × BD : 2 *area* ABCD. That is,

As Radius,
To the sine of the angle, at the intersection of the diagonals of a trapezium ;
So is the product of the diagonals,
To twice the area of the trapezium.

It is evident that this rule is applicable to a parallelogram, as well as to a trapezium.

If the diagonals intersect at *right angles*, the sine of N is equal to radius ; (Trig. 95.) and therefore the product of the diagonals is equal to twice the area. (Alg. 395.†)

Ex. 1. If the two diagonals of a trapezium are 37 and 62, and if they intersect at an angle of 54°, what is the area of the trapezium ? Ans. 928.

2. If the diagonals are 85 and 93, and the angle of intersection, 74°, what is the area of the trapezium ?

14. *b.* When a trapezium can be *inscribed in a circle*, the area may be found by either of the following rules.

I. *Multiply together any two adjacent sides, and also the two other sides ; then multiply half the sum of these products by the sine of the angle included by either of the pairs of sides multiplied together.*

Or,

II. *From half the sum of all the sides, subtract each side severally, multiply together the half sum and the four remainders, and extract the square root of the product.*

If the sides are a, b, c, and d ; and if h = half their sum ;

$$\text{The area} = \sqrt{(h-a) \times (h-b) \times (h-c) \times (h-d)}$$

* Euclid 2, 5. Cor. † Euc. 14. 5.

If the trapezium ABCD (Fig. 33) can be inscribed in a circle, the sum of the opposite angles BAD and BCD is 180° (Euc. 22. 3) Therefore the *sine* of BAD is equal to that of BCD or P′CD.

If s=the sine of either of these angles, radius being 1, and if

$$AB=a, \quad BC=b, \quad CD=c, \quad AD=d;$$

The triangle $BAD=\frac{1}{2}ad\times s$, And $BCD=\frac{1}{2}bc\times s$; (Art. 9.)

Therefore,

I. *The area of* $ABCD=\frac{1}{2}(ad+bc)\times s$.

To obtain the value of s, in terms of the sides of the trapezium, draw DP and DP′ perpendicular to BA and BC.

Then $\quad \text{Rad.} : s :: AD : DP :: CD : DP'$.

Also $\quad AP^2=AD^2-DP^2$, and $CP'^2=CD^2-DP'^2$.

So that $\begin{cases} DP=AD\times s=ds \\ DP'=CD\times s=cs \end{cases}$ And $\begin{cases} AP=\sqrt{d^2-d^2s^2}=d\sqrt{1-s^2} \\ CP'=\sqrt{c^2-c^2s^2}=c\sqrt{1-s^2} \end{cases}$

But by the figure $\begin{cases} BP=AB-AP=a-d\sqrt{1-s^2} \\ BP'=BC+CP'=b+c\sqrt{1-s^2} \end{cases}$

$$\text{And } \overline{BP}^2+\overline{DP}^2=\overline{DB}^2=\overline{BP'}^2+\overline{DP'}^2$$

$$\text{That is } a^2-2ad\sqrt{1-s^2}+d^2=b^2+2bc\sqrt{1-s^2}+c^2$$

Reducing the equation, we have

$$s^2=1-\frac{(b^2+c^2-a^2-d^2)^2}{(2ad+2bc)^2}, \text{ and}$$

$$s=\frac{\sqrt{(2ad+2bc)^2-(b^2+c^2-a^2-d^2)^2}}{2ad+2bc}$$

Substituting for s in the first rule, the value here found, we have the area of the trapezium, equal to

$$\tfrac{1}{4}\sqrt{(2ad+2bc)^2-(b^2+c^2-a^2-d^2)^2}$$

The expression under the radical sign is the difference of two squares, and may be resolved, as in Trig. 221, into the factors

$$(\overline{b+c}^2-\overline{a-d}^2\times(\overline{a+d}^2-\overline{b-c}^2)$$

and these again into

$$(a+b+c-d)(b+c+d-a)(a+b+d-c)(a+d+c-b)$$

If then h=half the sum of the sides of the trapezium,

II. *The area* $=\sqrt{(h-a)\times(h-b)\times(h-c)\times(h-d)}$

If one of the sides, as d, is supposed to be diminished, till it is reduced to nothing; the figure becomes a *triangle*, and the expression for the area is the same as in art. 10. See Hutton's Mensuration.

PROBLEM V.

To find the area of a REGULAR POLYGON.

15. MULTIPLY ONE OF ITS SIDES INTO HALF ITS PERPENDICULAR DISTANCE FROM THE CENTRE, AND THIS PRODUCT INTO THE NUMBER OF SIDES.

A regular polygon contains as many equal triangles as the figure has sides. Thus the hexagon ABDFGH (Fig. 7.) contains six triangles, each equal to ABC. The area of one of them is equal to the product of the side AB, into half the perpendicular CP (Art. 8.) The area of the whole, therefore, is equal to this product multiplied into the *number* of sides.

Ex. 1. What is the area of a regular octagon, in which the length of a side is 60, and the perpendicular from the centre 72.426? Ans. 17382.

2. What is the area of a regular decagon whose sides are 46 each, and the perpendicular 70.7867?

16. If only the length and number of sides of a regular polygon be given, the *perpendicular* from the centre may be easily found by trigonometry. The periphery of the circle in which the polygon is inscribed, is divided into as many equal parts as the polygon has sides. (Euc. 16. 4. Schol.) The arc, of which one of the sides is a chord, is therefore known; and of course, the angle at the centre subtended by this arc.

Let AB (Fig. 7.) be one side of a regular polygon, inscribed in the circle ABDG. The perpendicular CP bisects the line AB, and the angle ACB. (Euc. 3. 3.) Therefore BCP is the same part of 360°, which BP is of the perimeter of the polygon. Then, in the right angled triangle BCP, if BP be radius, (Trig. 122.)

R : BP : : Cotan BCP : CP. That is,

As Radius,
To half of one of the sides of the polygon;
So is the cotangent of the opposite angle,
To the perpendicular from the centre.

Ex. 1. If the side of a regular hexagon (Fig. 7.) be 38 inches, what is the area?

The angle BCP$=\frac{1}{12}$ of 360°=30°. Then,

R : 19::Cot 30° : 32.909=CP, the perpendicular.

And the area=19×32.909×6=3751.6.

2. What is the area of a regular decagon whose sides are each 62 feet? Ans. 29576.

17. From the proportion in the preceding article, a *table* of perpendiculars and areas may be easily formed, for a series of polygons, of which each side is a unit. Putting R=1, (Trig. 100.) and n=the number of sides, the proportion becomes

$$1 : \tfrac{1}{2} :: \text{Cot}\frac{360}{2n} : \textit{the perpendicular.}$$

$$\text{So that, } \textit{the perp.} = \tfrac{1}{2}\text{Cot}\frac{360}{2n}$$

And the *area* is equal to half the product of the perpendicular into the number of sides. (Art. 15.)

Thus, in the trigon, or equilateral triangle, the perpendicular$=\frac{1}{2}\text{ Cot}\frac{360°}{6}=\frac{1}{2}\text{ Cot } 60°=0.2886752$.

And the area=0.4330127.

In the tetragon or square, the perpendicular$=\frac{1}{2}\text{ Cot}\frac{360°}{8}$ $=\frac{1}{2}\text{ Cot. } 45°=0.5$. And the area=1.

In this manner, the following table is formed, in which the side of each polygon is supposed to be a unit.

A TABLE OF REGULAR POLYGONS.

Names.	*Sides.*	*Angles.*	*Perpendiculars*	*Areas.*
Trigon	3	60°	0.2886752	0.4330127
Tetragon	4	45°	0.5000000	1.0000000
Pentagon	5	36°	0.6881910	1.7204774
Hexagon	6	30°	0.8660254	2.5980762
Heptagon	7	$25\frac{5}{7}$	1.0382601	3.6339124
Octagon	8	$22\frac{1}{2}$	1.2071069	4.8284271
Nonagon	9	20°	1.3737385	6.1818242
Decagon	10	18°	1.5388418	7.6942088
Undecagon	11	$16\frac{4}{11}$	1.7028439	9.3656399
Dodecagon	12	15°	1.8660252	11.1961524

By this table may be calculated the area of any other regular polygon, of the same number of sides with one of these. For the areas of similar polygons are as the *squares* of their homologous sides. (Euc. 20, 6.)

To find, then, the area of a regular polygon, *multiply the square of one of its sides by the area of a similar polygon of which the side is a unit.*

Ex. 1. What is the area of a regular decagon whose sides are each 102 rods? Ans. 80050.5 rods.

2. What is the area of a regular dodecagon whose sides are each 87 feet?

SECTION II.*

THE QUADRATURE OF THE CIRCLE AND ITS PARTS.

ART. 18. *Definition* I. A *CIRCLE* is a plane bounded by a line which is equally distant in all its parts from a point within called the centre. The bounding line is called the *circumference* or periphery. An *arc* is any portion of the circumference. A semi-circle is half, and a quadrant one-fourth, of a circle.

II. A *Diameter* of a circle is a straight line drawn through the centre, and terminated both ways by the circumference. A *Radius* is a straight line extending from the centre to the circumference. A *Chord* is a straight line which joins the two extremities of an arc.

III. A Circular *Sector* is a space contained between an arc and the two radii drawn from the extremities of the arc. It may be *less* than a semi-circle, as ACBO, (Fig. 9.) or *greater*, as ACBD.

IV. A circular *Segment* is the space contained between an arc and its chord, as ABO or ABD. (Fig. 9) The chord is sometimes called the *base* of the segment. The *height* of a segment is the perpendicular from the middle of the base to the arc, as PO. (Fig. 9.)

V. A Circular *Zone* is the space between two parallel chords, as AGHB. (Fig. 15.) It is called the *middle* zone, when the two chords are equal.

VI. A Circular *Ring* is the space between the peripheries of two concentric circles, as AA′ BB′. (Fig. 13.)

VII. A *Lune* or Crescent is the space between two circular arcs which intersect each other, as ACBD. (Fig. 14.)

19. The *Squaring of the Circle* is a problem which has exercised the ingenuity of distinguished mathematicians for

* Wallis's Algebra, Le Gendre's Geometry, Book IV. and Note IV. Hutton's Mensuration, Horseley's Trigonometry, Book I, Sec. 3; Introduction to Euler's Analysis of Infinites, London Phil. Trans. Vol VI No. 75, LXVI. p. 476, LXXXIV. p. 217, and Hutton's abridgment of do. Vol. II. p. 547.

many centuries. The result of their efforts has been only an *approximation* to the value of the area. This can be carried to a degree of exactness far beyond what is necessary for practical purposes.

20. If the *circumference* of a circle of given diameter were known, its area could be easily found. For the area is equal to the product of half the circumference into half the diameter (Sup. Euc 5, 1.*) But the circumference of a circle has never been exactly determined. The method of approximating to it is by inscribing and circumscribing *polygons*, or by some process of calculation which is, in principle, the same. The perimeters of the polygons can be easily and exactly determined. That which is circumscribed is *greater*, and that which is inscribed is *less*, than the periphery of the circle; and by increasing the number of sides, the difference of the two polygons may be made less than any given quantity. (Sup. Euc. 4, 1.)

21. The side of a *hexagon* inscribed in a circle, as AB, (Fig. 7.) is the chord of an arc of 60°, and therefore equal to the radius. (Trig. 95.) The chord of *half* this arc, as BO, is the side of a polygon of 12 equal sides. By repeatedly bisecting the arc, and finding the chord, we may obtain the side of a polygon of an immense number of sides. Or we may calculate the *sine*, which will be half the chord of double the arc; (Trig. 82, cor.) and the *tangent*, which will be half the side of a similar *circumscribed* polygon. Thus the sine AP (Fig. 7.) is half of AB, a side of the inscribed hexagon; and the tangent NO is half of NT, a side of the circumscribed hexagon. The difference between the sine and the arc AO is less, than the difference between the sine and the tangent. In the section on the computation of the canon, (Trig. 223.) by 12 successive bisections, beginning with 60 degrees, an arc is obtained which is the $\frac{1}{24576}$ of the whole circumference..

The *cosine* of this, if radius be 1, is found to be .99999996732
The *sine* is .00025566346

And the tangent $=\frac{sine}{cosine}$ (Trig. 228.) $=.00025566347$

The diff. between the sine and tangent is only .00000000001
And the difference between the sine and the *arc* is still less.

* In this manner, the *Supplement* to *Playfair's Euclid* is referred to in this work.

Taking then .000255663465 for the length of the arc, multiplying by 24567, and retaining 8 places of decimals, we have 6.28318531 for the whole circumference, the radius being 1. Half of this,

3.14159265

is the circumference of a circle whose radius is $\frac{1}{2}$, and *diameter* 1.

22. If this be multiplied by 7, the product is 21 99+ or 22 nearly. So that,

Diam : Circum : : 7 : 22, nearly.

If 3.14159265 be multiplied by 113, the product is 354.9999+, or 355, very nearly. So that,

Diam : Circum : : 113 : 355, very nearly.

The first of these ratios was demonstrated by Archimedes.

There are various methods, principally by infinite series and fluxions, by which the labour of carrying on the approximation to the periphery of a circle may be very much abridged. The calculation has been extended to nearly 150 places of decimals.* But four or five places are sufficient for most practical purposes.

After determining the ratio between the diameter and the circumference of a circle, the following problems are easily solved.

PROBLEM I.

To find the CIRCUMFERENCE *of a circle from its diameter.*

MULTIPLY THE DIAMETER BY 3.14159.†

Or,

Multiply the diameter by 22 and divide the product by 7.

Or, multiply the diameter by 355, and divide the product by 113. (Art. 22.)

Ex. 1. If the diameter of the earth be 7930 miles, what is the circumference? Ans. 24912 8 miles.

2. How many miles does the earth move, in revolving round the sun; supposing the orbit to bea circle whose diameter is 190 million miles? Ans. 596,992,100.

* See note A.

† In many cases, 3.1416 will be sufficiently accurate.

What is the circumference of a circle whose diameter is 769843 rods?

PROBLEM II.

To find the DIAMETER *of a circle from its circumference.*

24. DIVIDE THE CIRCUMFERENCE BY **3.14159.**

Or,

Multiply the circumference by 7, and divide the product by 22. Or, multiply the circumference by 113, and divide the product by 355. (Art. 22.)

Ex. 1. If the circumference of the sun be 2,800,000 miles, what is his diameter? Ans. 891,267.

2. What is the diameter of a tree which is $5\frac{1}{2}$ feet round?

25. As multiplication is more easily performed than division, there will be an advantage in exchanging the *divisor* 3.14159 for a *multiplier* which will give the same result. In the proportion 3.14159 : 1 : : Circum : Diam.
to find the fourth term, we may divide the second by the first, and multiply the quotient into the third. Now $1 \div 3.14159 = 0.31831$. If then the circumference of a circle be multiplied by .31831, the product will be the diameter.*

Ex. 1. If the circumference of the moon be 6850 miles, what is her diameter? Ans. 2180.

2. If the whole extent of the orbit of Saturn be 5650 million miles, how far is he from the sun?

3. If the periphery of a wheel be 4 feet 7 inches, what is its diameter.

PROBLEM III.

To find the length of an ARC *of a circle.*

26. *As* 360°, *to the number of degrees in the arc;*
So is the circumference of the circle, to the length of the arc.

The circumference of a circle being divided into 360°, (Trig. 73.) it is evident that the length of an arc of any less number of degrees must be a proportional part of the whole.

* See Note B.

Ex. What is the length of an arc of 16°, in a circle whose radius is 50 feet?

The circumference of the circle is 314.159 feet. (Art. 23.)

Then 360 : 16 : : 314.159 : 13.96 feet.

2. If we are 95 millions of miles from the sun, and if the earth revolves round it in $365\frac{1}{4}$ days, how far are we carried in 24 hours? Ans. 1 million 634 thousand miles.

27. The length of an arc may also be found, by multiplying the diameter into the number of degrees in the arc, and this product into .0087266, which is the length of *one* degree, in a circle whose diameter is 1. For $3.14159 \div 360 = 0.0087266$. And in different circles, the circumferences, and of course the degrees, are as the diameters. (Sup. Euc. 8, 1.)

Ex. 1. What is the length of an arc of 10° 15′ in a circle whose radius is 68 rods? Ans. 12.165 rods.

2. If the circumference of the earth be 24913 miles, what is the length of a degree at the equator?

28. The length of an arc is frequently required, when the *number of degrees* is not given. But if the radius of the circle, and either the *chord* or the *height* of the arc, be known; the number of degrees may be easily found.

Let AB (Fig. 9.) be the chord, and PO the height, of the arc AOB. As the angles at P are right angles, and AP is equal to BP; (Art 18. Def. 4.) AO is equal to BO. (Euc. 4, 1.) Then

BP is the *sine*, CP the *cosine*, } of *half* the arc AOB.
OP the *versed sine*, and BO the *chord* }

And in the right angled triangle CBP,

$$CB : R :: \begin{cases} BP : \text{Sin } BCP \text{ or } BO \\ CP : \text{Cos } BCP \text{ or } BO \end{cases}$$

Ex. 1. If the radius CO (Fig. 9.) =25, and the chord AB=43 3; what is the length of the arc AOB?

CB : R : : BP : Sin BCP or BO=60° very nearly.

The circumference of the circle $= 3.14159 \times 50 = 157.08$.

And 360° : 60° :: 157.08 : 26.18=OB. Therefore AOB=52.36.

2. What is the length of an arc whose chord is $216\frac{1}{2}$ in a circle whose radius is 125? Ans. 261.8.

29. If only the *chord* and the *height* of an arc be given, the radius of the circle may be found, and then the length of the arc.

If BA (Fig 9.) be the chord, and PO the height of the arc AOB, then (Euc. 35.3.)

$$DP=\frac{\overline{BP}^2}{OP}. \quad \text{And } DO=OP+DP=OP+\frac{\overline{BP}^2}{OP}.$$

That is, the *diameter* is equal to the height of the arc, + the square of half the chord divided by the height.

The diameter being found, the length of the arc may be calculated by the two preceding articles.

Ex. 1 If the chord of an arc be 173.2, and the height 50, what is the length of the arc?

The diameter $=50+\frac{\overline{86.6}^2}{50}=200$. The arc contains 120°; (Art. 28.) and its length is 209.44. (Art. 26.)

2. What is the length of an arc whose chord is 120, and height 45? Ans. 160.8.*

PROBLEM IV.

To find the AREA *of a* CIRCLE.

30. MULTIPLY THE SQUARE OF THE DIAMETER BY THE DECIMALS .7854.

Or,

MULTIPLY HALF THE DIAMETER INTO HALF THE CIRCUMFERENCE. Or, multiply the whole diameter into the whole circumference, and take $\frac{1}{4}$ of the product.

The area of a circle is equal to the product of half the diameter into half the circumference; (Sup. Euc. 5, 1.) or which is the same thing, $\frac{1}{4}$ the product of the diameter and circumference. If the diameter be 1, the circumference is 3.14159; (Art. 23.) one fourth of which is 0.7854 nearly. But the areas of different circles are to each other, *as the squares of their diameters*. (Sup. Euc 7, 1.†) The area of any circle, therefore, is equal to the product of the square

* See note C. † Euclid 2, 12

of its diameter into 0.7854, which is the area of a circle whose diameter is 1.

Ex. 1. What is the area of a circle whose diameter is 623 feet? Ans. 304836 square feet.

2. How many acres are there in a circular island whose diameter is 124 rods? Ans. 75 acres, and 76 rods.

3. If the diameter of a circle be 113, and the circumference 355, what is the area? Ans. 10029.

4. How many square yards are there in a circle whose diameter is 7 feet?

31. If the *circumference* of a circle be given, the area may be obtained, by first finding the diameter; or, without finding the diameter, by multiplying the square of the circumference by .07958.

For, if the circumference of a circle be 1, the diameter $=1\div3.14159=0.31831$; and $\frac{1}{4}$ the product of this into the circumference is .07958 the area. But the areas of different circles, being as the squares of their diameters, are also as the squares of their *circumferences*. (Sup. Euc. 8. 1.)

Ex. 1. If the circumference of a circle be 136 feet, what is the area? Ans. 1472 feet.

2. What is the surface of a circular fish-pond, which is 10 rods in circumference?

32. If the area of a circle be *given*, the diameter may be found, by dividing the area by .7854, and extracting the square root of the quotient.

This is reversing the rule in art. 30.

Ex. 1. What is the diameter of a circle whose area is 380.1336 feet? Ans. $380.1336\div.7854=484$. And $\sqrt{484}=22$.

2. What is the diameter of a circle whose area is 19.635?

33. The area of a circle, is to the area of the *circumscribed square*; as .7854 to 1; and to that of the *inscribed* square as .7854 to $\frac{1}{2}$.

Let ABDF (Fig. 10.) be the inscribed square and LMNO the circumscribed square, of the circle ABDF. The area of the circle is equal to $\overline{AD}^2\times.7854$. (Art. 30.) But the area of the circumscribed square (Art. 4.) is equal to $\overline{ON}^2=\overline{AD}^2$. And the smaller square is half of the larger one. For the latter contains 8 equal triangles, of which the former contains only 4.

Ex. What is the area of a square inscribed in a circle whose area is 159? Ans. .7854 : $\frac{1}{2}$: : 159 : 101.22.

PROBLEM V.

To find the area of a SECTOR *of a circle.*

34. MULTIPLY THE RADIUS INTO HALF THE LENGTH OF THE ARC.

Or,

As 360, TO THE NUMBER OF DEGREES IN THE ARC;
So IS THE AREA OF THE CIRCLE, TO THE AREA OF THE SECTOR.

It is evident, that the area of the sector has the same ratio to the area of the circle, which the length of the arc has to the length of the whole circumference; or which the number of *degrees* in the arc has to the number of degrees in the circumference.

Ex. 1. If the arc AOB (Fig. 9.) be 120°, and the diameter of the circle 226; what is the area of the sector AOBC?

The area of the whole circle is 40115. (Art. 30.)

And 360° : 120° : : 40115 : 13371$\frac{2}{3}$, the area of the sector.

2. What is the area of a quadrant whose radius is 621?

3. What is the area of a semi-circle whose diameter is 328?

4. What is the area of a sector which is less than a semicircle, if the radius be 15, and the chord of its arc 12?

Half the chord is the sine of 23° 34′$\frac{3}{4}$ nearly. (Art. 28.)
The whole arc, then, is 47° 9′$\frac{1}{2}$
The area of the circle is 706.86
And 360° : 47° 9′$\frac{1}{2}$: : 706.86 : 92.6 the area of the sector.

5. If the arc ADB (Fig. 9.) be 240 degrees, and the radius of the circle 113, what is the area of the sector ADBC?

PROBLEM VI.

To find the area of a SEGMENT *of a circle.*

35. FIND THE AREA OF THE SECTOR WHICH HAS THE SAME ARC, AND ALSO THE AREA OF THE TRIANGLE FORMED BY THE CHORD OF THE SEGMENT AND THE RADII OF THE SECTOR.

THEN, IF THE SEGMENT BE LESS THAN A SEMI-CIRCLE, SUBTRACT THE AREA OF THE TRIANGLE FROM THE AREA OF THE SECTOR. BUT, IF THE SEGMENT BE GREATER THAN A SEMI-CIRCLE, ADD THE AREA OF THE TRIANGLE TO THE AREA OF THE SECTOR.

If the triangle ABC (Fig. 9.) be taken from the sector AOBC, it is evident the difference will be the segment AOBP, less than a semi-circle. And if the same triangle be added to the sector ADBC, the sum will be the segment ADBP, greater than a semi-circle.

The area of the triangle (Art. 8.) is equal to the product of half the chord AB into CP which is the difference between the radius and PO the height of the segment. Or CP is the *cosine* of half the arc BOA. If this cosine, and the chord of the segment are not given, they may be found from the arc and the radius.

Ex. 1. If the arc AOB (Fig. 9.) be 120°, and the radius of the circle be 113 feet, what is the area of the segment AOBP?

In the right angled triangle BCP,

R : BC::Sin BCO : BP=97.86, half the chord. (Art. 28.)

The cosine $PC=\frac{1}{2}CO$ (Trig. 96, Cor.)	=56.5
The area of the sector AOBC (Art. 34.)	= 13371.67
The area of the triangle $ABC=BP\times PC$	= 5528.97
The area of the segment, therefore,	= 7842.7

2. If the base of a segment, less than a semi-circle, be 10 feet, and the radius of the circle 12 feet, what is the area of the segment?

The arc of the segment contains	$49\frac{1}{4}$ degrees.	(Art. 28.)
The area of the sector	=61.89	(Art. 34.)
The area of the triangle	=54.54	
And the area of the segment	= 7.35 square feet.	

3. What is the area of a circular segment, whose height is 19.2 and base 70? Ans. 947.86.

4. What is the area of the segment ADBP, (Fig. 9.) if the base AB be 195.7, and the height PD 169.5?

Ans. 32272.*

36. The area of any figure which is bounded *partly* by arcs of circles, and partly by right lines, may be calculated, by finding the areas of the segments under the arcs, and then the area of the rectilinear space between the chords of the arcs and the other right lines.

Thus the Gothic arch ACB, (Fig. 11.) contains the two segments ACH, BCD, and the plane triangle ABC.

Ex. If AB (Fig. 11.) be 110, each of the lines AC and BC [illegible]00, and the height of each of the segments ACH, BCD 10.435; what is the area of the whole figure?

The areas of the two segments are	1404
The area of the triangle ABC is	4593.4
And the whole figure is	5997.4

PROBLEM VII.

To find the area of a circular ZONE.

37. FROM THE AREA OF THE WHOLE CIRCLE, SUBTRACT THE TWO SEGMENTS ON THE SIDES OF THE ZONE.

If from the whole circle (Fig. 12.) there be taken the two segments ABC and DFH, there will remain the zone ACDH.

Or, the area of the zone may be found, by subtracting the segment ABC from the segment HBD: Or, by adding the two small segments GAH and VDC to the trapezoid ACDH. See art. 36.

The latter method is rather the most expeditious in practice, as the two segments at the end of the zone are *equal*.

Ex. 1. What is the area of the zone ACDH, (Fig. 12.) if AC is 7.75, DH 6.93, and the diameter of the circle 8?

* For the method of finding the areas of segments by a *table*, see note D

The area of the whole circle is		50.26
of the segment ABC	17.32	
of the segment DFH	9.82	
of the zone ACDH		23.12

2. What is the area of a zone, one side of which is 23.25, and the other side 20.8, in a circle whose diameter is 24?

Ans. 208.

38 If the *diameter* of the circle is not given, it may be found from the sides and the breadth of the zone.

Let the centre of the circle be at O. (Fig. 12.) Draw ON perpendicular to AH, NM perpendicular to LR, and HP perpendicular to AL. Then

$AN=\frac{1}{2}AH$, (Euc. 3. 3.) $MN=\frac{1}{2}(LA+RH)$

$LM=\frac{1}{2}LR$, (Euc. 2. 6.) $PA=LA-RH$.

The triangles APH and OMN are similar, because the sides of one are perpendicular to those of the other, each to each. Therefore

$$PH : PA :: MN : MO$$

MO being found, we have $ML-MO=OL$.

And the *radius* $CO=\sqrt{OL^2+CL^2}$. (Euc. 47. 1.)

Ex. If the breadth of the zone ACDH (Fig. 12.) be 6.4, and the sides 6.8 and 6; what is the radius of the circle?

$PA=3.4-3=0.4$. And $MN=\frac{1}{2}(3.4+3)=3.2$.

Then $6.4 : 0.4 :: 3.2 : 0.2=MO$. And $3.2-0.2=3=OL$

And the radius $CO=\sqrt{3^2+(3.4)^2}=4.534$.

PROBLEM VIII.

To find the area of a LUNE *or crescent.*

39. FIND THE DIFFERENCE OF THE TWO SEGMENTS WHICH ARE BETWEEN THE ARCS OF THE CRESCENT AND ITS CHORD.

If the segment ABC (Fig 14.) be taken from the segment ABD; there will remain the lune or crescent ACBD.

Ex. If the chord AB be 88, the height CH 20, and the height DH 40; what is the area of the crescent ACBD?

The area of the segment ABD is	2698
of the segment ABC	1220
of the crescent ACBD	1478

PROBLEM IX.

To find the area of a RING, included between the peripheries of two concentric circles.

40. FIND THE DIFFERENCE OF THE AREAS OF THE TWO CIRCLES.

Or,

Multiply the product of the sum and difference of the two diameters by .7854.

The area of the ring (Fig. 13.) is evidently equal to the difference between the areas of the two circles AB and A'B'.

But the area of each circle is equal to the square of its diameter multiplied into .7854. (Art. 30.) And the *difference* of these squares is equal to the product of the sum and difference of the diameters. (Alg. 235.) Therefore the area of the ring is equal to the product of the sum and difference of the two diameters multiplied by .7854.

Ex. 1. If AB (Fig. 13.) be 221, and A'B' 106, what is the area of the ring?

Ans. $(\overline{221}^2 \times .7854) - (\overline{106}^2 \times .7854) = 29535.$

2. If the diameters of Saturn's larger ring be 205,000 and 190,000 miles, how many square miles are there on one side of the ring?

Ans. $395000 \times 15000 \times .7854 = 4{,}653{,}495{,}000.$

PROMISCUOUS EXAMPLES OF AREAS.

Ex. 1. What is the expense of paving a street 20 rods long, and 2 rods wide, at 5 cents for a square foot?

Ans. $544\frac{1}{2}$ dollars.

2. If an equilateral triangle contains as many square feet as there are inches in one of its sides; what is the area of the triangle?

Let $x=$ the number of square feet in the area.

Then $\frac{x}{12}=$ the number of linear feet in one of the sides.

And (Art. 11.) $x=\frac{1}{4}\left(\frac{x}{12}\right)^2\times\sqrt{3}=\frac{x^2}{576}\times\sqrt{3}$.

Reducing the equation, $x=\frac{476}{\sqrt{3}}=332.55$ the area.

3. What is the side of a square whose area is equal to that of a circle 452 feet in diameter?

Ans. $\sqrt{(452)^2\times.7854}=400.574$. (Art. 30 and 7.)

4. What is the diameter of a circle which is equal to a square whose side is 36 feet?

Ans. $\sqrt{(36)^2\div0.7854}=40.6217$. (Art. 4. and 32.)

5. What is the area of a square inscribed in a circle whose diameter is 132 feet?

Ans. 8712 square feet. (Art. 33.)

6. How much carpeting, a yard wide, will be necessary to cover the floor of a room which is a regular octagon, the sides being 8 feet each? Ans. $34\frac{1}{3}$ yards.

7. If the diagonal of a square be 16 feet, what is the area?

Ans. 128 feet. (Art. 14.)

8. If a carriage wheel four feet in diameter revolve 300 times, in going round a circular green; what is the area of the green?

Ans. $4154\frac{1}{3}$ sq. rods, or 25 acres, 3 qrs. and $34\frac{1}{3}$ rods.

9. What will be the expense of papering the sides of a room, at 10 cents a square yard; if the room be 21 feet long, 18 feet broad, and 12 feet high; and if there be deducted 3 windows, each 5 feet by 3, two doors 8 feet by $4\frac{1}{2}$, and one fire-place 6 feet by $4\frac{1}{2}$? Ans. 8 dollars 80 cents.

10. If a circular pond of water 10 rods in diameter be surrounded by a gravelled walk $8\frac{1}{4}$ feet wide; what is the area of the walk? Ans. $16\frac{1}{2}$ sq. rods. (Art. 40.)

11. If CD (Fig. 17,) the base of the isosceles triangle VCD, be 60 feet, and the area 1200 feet; and if there be cut off, by the line LG parallel to CD, the triangle VLG, whose area is 432 feet; what are the sides of the latter triangle?

Ans. 30, 30, and 36 feet.

12. What is the area of an equilateral triangle inscribed in a circle whose diameter is 52 feet?

Ans. 878.15 sq. feet.

13. If a circular piece of land is enclosed by a fence, in which 10 rails make a rod in length; and if the field contains as many square rods, as there are rails in the fence; what is the value of the land at 120 dollars an acre?

Ans. 942.48 dollars.

14. If the area of the equilateral triangle ABD (Fig. 9.) be 219.5375 feet; what is the area of the circle OBDA, in which the triangle is inscribed?

The sides of the triangle are each 22.5167. (Art. 11.)
And the area of the circle is 530.93

15. If 6 concentric circles are so drawn, that the space
between the least or 1st, and the 2d is 21.2058,
between the 2d and 3d 35.343,
between the 3d and 4th 49.4802,
between the 4th and 5th 63.6174,
between the 5th and 6th 77.7546;
what are the several diameters, supposing the longest to be equal to 6 times the shortest?

Ans. 3, 6, 9, 12, 15, and 18.

SECTION III.

SOLIDS BOUNDED BY PLANE SURFACES.

Art. 41. Definition I. A *PRISM* is a solid bounded by plane figures or faces, two of which are parallel, similar, and equal; and the others are parallelograms.

II. The parallel planes are sometimes called the *bases* or *ends;* and the other figures, the *sides* of the prism. The latter taken together constitute the *lateral surface.*

III. A prism is *right* or *oblique*, according as the sides are perdendicular or oblique to the bases.

IV. The *height* of a prism is the perpendicular distance between the planes of the bases. In a right prism, therefore, the height is equal to the length of one of the sides.

V. A *Parallelopiped* is a prism whose bases are parallelograms.

VI. A *Cube* is a solid bounded by six equal squares. It is a right prism whose sides and bases are all equal.

VII. A *Pyramid* is a solid bounded by a plane figure called the base, and several triangular planes, proceeding from the sides of the base, and all terminating in a single point. These triangles taken together constitute the *lateral surface.*

VIII. A pyramid is *regular*, if its base is a regular polygon, and if a line from the centre of the base to the vertex of the pyramid is *perpendicular* to the base. This line is called the *axis* of the pyramid.

IX. The *height* of a pyramid is the perpendicular distance from the summit to the plane of the base. In a *regular* pyramid, it is the length of the *axis.*

X. The *slant-height* of a regular pyramid, is the distance from the summit to the middle of one of the sides of the base.

XI. A *frustum* or *trunk* of a pyramid is a portion of the solid next the base, cut off by a plane parallel to the base. The *height* of the frustum is the perpendicular distance of the two parallel planes. The *slant-height* of a frustum of a *regular* pyramid, is the distance from the middle of one of the sides of the base, to the middle of the corresponding side in the plane above. It is a line passing on the surface of the frustum, through the middle of one of its sides.

XII. A *Wedge* is a solid of five sides, viz. a rectangular base, two rhomboidal sides meeting in an edge, and two triangular ends; as ABHG. (Fig 20.) The base is ABCD, the sides are ABHG and DCHG, meeting in the edge GH, and the ends are BCH and ADG. The *height* of the wedge is a perpendicular drawn from any point in the edge, to the plane of the base, as GP.

XIII. A *Prismoid* is a solid whose ends or bases are parallel, but not similar, and whose sides are quadrilateral. It differs from a prism or a frustum of a pyramid, in having its ends dissimilar. It is a *rectangular* prismoid, when its ends are right parallelograms.

XIV. A *linear side* or *edge* of a solid is the line of intersection of two of the planes which form the surface.

42. The common *measuring unit* of solids is a *cube*, whose sides are squares of the same name. The sides of a cubic inch are square inches; of a cubic foot, square feet, &c. Finding the *capacity*, *solidity*,* or *solid contents* of a body, is finding the number of cubic measures, of some given denomination contained in the body.

In solid measure.

1728	cubic inches	=1	cubic foot,
27	cubic feet	=1	cubic yard,
4492⅛	cubic feet	=1	cubic rod,
32768000	cubic rods	=1	cubic mile,
282	cubic inches	=1	ale gallon,
231	cubic inches	=1	wine gallon,
2150.42	cubic inches	=1	bushel,
1	cubic foot of pure water weighs 1000 Avoirdupois ounces, or 62½ pounds.		

* See note E.

PROBLEM I.

To find the SOLIDITY *of a* PRISM.

43. MULTIPLY THE AREA OF THE BASE BY THE HEIGHT.

This is a general rule, applicable to parallelopipeds whether right or oblique, cubes. triangular prisms, &c.

As *surfaces* are measured, by comparing them with a right *parallelogram* (Art. 3.); so *solids* are measured, by comparing them with a right *parallelopiped.*

If ABCD (Fig. 1.) be the base of a right parallelopiped, as a stick of timber standing erect, it is evident that the number of *cubic* feet contained in *one* foot of the height, is equal to the number of *square* feet in the area of the base. And if the solid be of any other height, instead of one foot, the contents must have the same ratio. For parallelopipeds of the same base are to each other as their heights. (Sup. Euc. 9. 3.) The solidity of a right parallelopiped, therefore, is equal to the *product of its length, breadth, and thickness.* See Alg. 523.

And an *oblique* parallelopiped being equal to a right one of the same base and altitude, (Sup. Euc. 7, 3.) is equal to the area of the base multiplied into the perpendicular height. This is true also of *prisms*, whatever be the form of their bases. (Sup. Euc. 2. Cor. to 8. 3.)

44. As the sides of a *cube* are all *equal*, the solidity is found by *cubing one of its edges.* On the other hand, if the solid contents be given, the length of the edges may be found, by *extracting the cube root.*

45. When solid measure is cast by *Duodecimals*, it is to be observed that *inches* are not *primes* of feet, but *thirds.* If the unit is a cubic foot, a solid which is an inch thick and a foot square is a prime; a parallelopiped a foot long, an inch broad, and an inch thick is a second, or the twelfth part of a prime; and a cubic inch is a third, or a twelfth part of a second. A linear inch is $\frac{1}{12}$ of a foot, a square inch $\frac{1}{144}$ of a foot, and a cubic inch $\frac{1}{1728}$ of a foot.

Ex. 1. What are the solid contents of a stick of timber which is 31 feet long, 1 foot 3 inches broad, and 9 inches thick? Ans. 29 feet 9″, or 29 feet 108 inches.

2. What is the solidity of a wall which is 22 feet long, 12 feet high, and 2 feet 6 inches thick?

Ans. 660 cubic feet.

3. What is the capacity of a cubical vessel which is 2 feet 3 inches deep?

Ans. 11F. 4′ 8″ 3‴, or 11 feet 675 inches.

4. If the base of a prism be 108 square inches, and the height 36 feet, what are the solid contents?

Ans. 27 cubic feet.

5, If the height of a square prism be $2\frac{1}{4}$ feet, and each side of the base $10\frac{1}{3}$ feet, what is the solidity?

The area of the base $=10\frac{1}{3}\times10\frac{1}{3}\times106\frac{7}{9}$ sq. feet.

And the solid contents $=106\frac{7}{9}\times2\frac{1}{4}=240\frac{1}{4}$ cubic feet.

6. If the height of a prism be 23 feet, and its base a regular pentagon, whose perimeter is 18 feet, what is the solidity?

Ans. 512.84 cubic feet.

46. The number of *gallons* or *bushels* which a vessel will contain may be found, by calculating the capacity in *inches*, and then dividing by the number of inches in 1 gallon or bushel.

The *weight of water* in a vessel of given dimensions is easily calculated; as it is found by experiment, that a cubic foot of pure water weighs 1000 ounces Avoirdupois. For the weight in ounces, then, multiply the cubic feet by 1000; or for the weight in pounds, multiply by $62\frac{1}{2}$.

Ex. 1. How many ale gallons are there in a cistern which is 11 feet 9 inches deep, and whose base is 4 feet 2 inches square?

The cistern contains 352500 cubic inches;

And $352500\div282=1250$.

2. How many wine gallons will fill a ditch 3 feet 11 inches wide, 3 feet deep, and 462 feet long? Ans. 40608.

3. What weight of water can be put into a cubical vessel 4 feet deep? Ans. 4000 lbs.

PROBLEM II.

To find the LATERAL SURFACE *of a* RIGHT PRISM.

47. MULTIPLY THE LENGTH INTO THE PERIMETER OF THE BASE.

Each of the sides of the prism is a right parallelogram, whose area is the product of its length and breadth. But the breadth is one side of the base; and therefore, the sum of the breadths is equal to the perimeter of the base.

Ex. 1. If the base of a right prism be a regular hexagon whose sides are each 2 feet 3 inches, and if the height be 16 feet, what is the lateral surface?

Ans. 216 square feet.

If the areas of the two ends be added to the lateral surface, the sum will be the whole surface of the prism. And the superficies of any solid bounded by planes, is evidently equal to the areas of all its sides.

Ex. 2. If the base of a prism be an equilateral triangle whose perimeter is 6 feet, and if the height be 17 feet, what is the surface?

The area of the triangle is 1.732. (Art. 11.)
And the whole surface is 105.464.

PROBLEM III.

To find the SOLIDITY *of a* PYRAMID.

48. MULTIPLY THE AREA OF THE BASE INTO $\frac{1}{3}$ OF THE HEIGHT.

The solidity of a *prism* is equal to the product of the area of the base into the height. (Art. 43.) And a pyramid is $\frac{1}{3}$ of a prism of the same base and altitude. (Sup. Euc. 15. 3. Cor. 1.) Therefore the solidity of a pyramid whether right or oblique, is equal to the product of the base into $\frac{1}{3}$ of the perpendicular height.

Ex. 1. What is the solidity of a triangular pyramid, whose height is 60, and each side of whose base is 4?

The area of the base is 6.928
And the solidity is 138.56.

2. Let ABC (Fig. 16.) be one side of an oblique pyramid whose base is 6 feet square; let BC be 20 feet, and make an angle of 70 degrees with the plane of the base; and let CP be perpendicular to this plane. What is the solidity of the pyramid?

In the right angled triangle BCP, (Trig. 134.)

$R : BC :: \text{Sin}\, B : PC = 18.79.$

And the solidity of the pyramid is 225.48 feet.

3. What is the solidity of a pyramid whose perpendicular height is 72, and the sides of whose base are 67, 54, and 40?

Ans. 25920.

PROBLEM IV.

To find the LATERAL SURFACE *of a* REGULAR PYRAMID.

49. MULTIPLY HALF THE SLANT-HEIGHT INTO THE PERIMETER OF THE BASE.

Let the triangle ABC (Fig. 18.) be one of the sides of a regular pyramid. As the sides AC and BC are equal, the angles A and B are equal. Therefore a line drawn from the vertex C to the middle of AB is *perpendicular* to AB. The area of the triangle is equal to the product of half this perpendicular into AB. (Art. 8.) The perimeter of the base is the sum of its sides, each of which is equal to AB. And the areas of all the equal triangles which constitute the lateral surface of the pyramid, are together equal to the product of the perimeter into half the slant-height CP.

The *slant-height* is the hypothenuse of a right angled triangle, whose legs are the axis of the pyramid, and the distance from the centre of the base to the middle of one of the sides. See Def. 10.

Ex. 1. What is the lateral surface of a regular hexagonal pyramid, whose axis is 20 feet, and the sides of whose base are each 8 feet?

The square of the distance from the centre of the base to one of the sides (Art. 16.) $=48$.

The slant-height (Euc. 47. 1.) $=\sqrt{48+20^2}=21.16$.

And the lateral surface $=21.16\times4\times6=507.84$ sq. feet.

2. What is the whole surface of a regular triangular pyramid whose axis is 8, and the sides of whose base are each 20.78?

The lateral surface is	312
The area of the base is	187
And the whole surface is	499

3. What is the lateral surface of a regular pyramid whose axis is 12 feet, and whose base is 18 feet square?

Ans. 540 square feet.

The lateral surface of an *oblique* pyramid may be found, by taking the sum of the areas of the unequal triangles which form its sides.

PROBLEM V.

To find the SOLIDITY *of a* FRUSTUM *of a pyramid.*

50. ADD TOGETHER THE AREAS OF THE TWO ENDS, AND THE SQUARE ROOT OF THE PRODUCT OF THESE AREAS; AND MULTIPLY THE SUM BY $\frac{1}{3}$ OF THE PERPENDICULAR HEIGHT OF THE SOLID.

Let CDGL (Fig. 17.) be a vertical section, through the middle of a frustum of a right pyramid CDV whose base is a *square*.

Let $CD=a$, $LG=b$, $RN=h$

By similar triangles, $LG : CD :: RV : NV$.

Subtracting the antecedents, (Alg. 389.)

$LG : CD-LG :: RV : NV-RV=RN$.

$$\text{Therefore } RV=\frac{RN\times LG}{CD-LG}=\frac{hb}{a-b}$$

The square of CD is the base of the pyramid CDV;
And the square of LG is the base of the small pyramid LGV.
Therefore, the solidity of the larger pyramid (Art. 48.) is

$$\overline{CD}^2\times\tfrac{1}{3}(RN+RV)=a^2\times\tfrac{1}{3}\left(h+\frac{hb}{a-b}\right)=\frac{ha^3}{3a-3b}$$

And the solidity of the smaller pyramid is equal to

$$\overline{LG}^2\times\tfrac{1}{3}RV=b^2\times\frac{hb}{3a-3b}=\frac{hb^3}{3a-3b}.$$

If the smaller pyramid be taken from the larger, there will remain the frustum CDLG, whose solidity is equal to

$$\frac{ha^3-hb^3}{3a-3b}=\tfrac{1}{3}h\times\frac{a^3-b^3}{a-b}=\tfrac{1}{3}h\times(a^2+ab+b^2)\quad\text{(Alg. 466.)}$$

Or, because $\sqrt{a^2b^2}=ab$, (Alg. 259.)

$$\tfrac{1}{3}h\times(a^2+b^2+\sqrt{a^2b^2})$$

Here h, the height of the frustum, is multiplied into a^2 and b^2, the areas of the two ends, and into $\sqrt{a^2b^2}$ the square root of the products of these areas.

In this demonstration, the pyramid is supposed to be *square*. But the rule is equally applicable to a pyramid of any other form. For the solid contents of pyramids are equal, when they have equal heights and bases, whatever be the *figure* of their bases. (Sup. Euc. 14. 3.) And the sections parallel to the bases, and at equal distances, are equal to one another. (Sup. Euc. 12. 3. Cor. 2.)*

Ex. 1. If one end of the frustum of a pyramid be 9 feet square, the other end 6 feet square, and the height 36 feet, what is the solidity?

The areas of the two ends are 81 and 36.
The square root of their product is 54.
And the solidity of the frustum $=(81+36+54)\times12=2052$.

2. If the height of a frustum of a pyramid be 24, and the areas of the two ends 441 and 121; what is the solidity?
Ans. 6344.

3. If the height of a frustum of a hexagonal pyramid be 48, each side of one end 26, and each side of the other end 16; what is the solidity?
Ans. 56034.

PROBLEM VI.

To find the LATERAL SURFACE *of a* FRUSTUM *of a regular pyramid.*

51. MULTIPLY HALF THE SLANT-HEIGHT BY THE SUM OF THE PERIMETERS OF THE TWO ENDS.

Each side of a frustum of a regular pyramid is a *trapezoid*, as ABCD. (Fig. 19.) The slant-height HP, (Def. 11.) though it is oblique to the base of the solid, is perpendicular to the line AB. The area of the trapezoid is equal to the product of half this perpendicular into the sum of the parallel sides AB and DC. (Art. 12) Therefore the area of all the equal trapezoids which form the lateral surface of

* See note F.

the frustum, is equal to the product of half the slant-height into the sum of the perimeters of the ends.

Ex. If the slant-height of a frustum of a regular octagonal pyramid be 42 feet, the sides of one end 5 feet each, and the sides of the other end 3 feet each; what is the lateral surface? Ans. 1344 square feet.

52. If the slant-height be not given, it may be obtained from the perpendicular height, and the dimensions of the two ends. Let GD (Fig. 17.) be the slant-height of the frustum CDGL, RN or GP the perpendicular height, ND and RG the radii of the circles inscribed in the perimeters of the two ends. Then PD is the difference of the two radii:

And the slant-height $GD=\sqrt{GH^2+PD^2}$

Ex. If the perpendicular height of a frustum of a regular hexagonal pyramid be 24, the sides of one end 13 each, and the sides of the other end 8 each; what is the whole surface?

$\sqrt{BC^2-BP^2}=CP$, (Fig. 7.) that is, $\sqrt{13^2-6.5^2}=11.258$

And $\sqrt{8^2-4^2}=$ 6.928

The difference of the two radii is, therefore, 4.33

The slant-height $=\sqrt{24^2+4.33^2}$ $=24.3875$
The lateral surface is 1536.4
And the whole surface, 2141.75

53. The height of the *whole pyramid* may be calculated from the dimensions of the frustum. Let VN (Fig. 17.) be the height of the pyramid, RN or GP the height of the frustum, ND and RG the radii of the circles inscribed in the perimeters of the ends of the frustum.

Then, in the similar triangles GPD and VND,

$DP : GP :: DN : VN.$

The height of the frustum subtracted from VN, gives VR the height of the small pyramid VLG. The *solidity* and *lateral surface* of the frustum may then be found, by subtracting from the whole pyramid, the part which is above the cutting plane. This method may serve to verify the calculations which are made by the rules in arts. 50 and 51.

Ex. If oue end of the frustum CDGL (Fig. 17.) be 90 feet square, the other end 60 feet square, and the height RN 36 feet; what is the height of the whole pyramid VCD: and what are the solidity and lateral surface of the frustum?

DP=DN−GR=45−30=15. And GP=RN=36.

Then 15 : 36 :: 45 : 108=VN, the height of the whole pyramid.

And 108−36=72=VR, the height of the part VLG.

The solidity of the large pyramid is	291600	(Art. 48.)
of the small pyramid	86400	
of the frustum CDGL	205200	
The lateral surface of the large pyramid is	21060	(Art. 49.)
of the small pyramid	9360	
of the frustum	11700	

PROBLEM VII.

To find the SOLIDITY *of a* WEDGE.

54. ADD THE LENGTH OF THE EDGE TO TWICE THE LENGTH OF THE BASE, AND MULTIPLY THE SUM BY $\frac{1}{6}$ OF THE PRODUCT OF THE HEIGHT OF THE WEDGE AND THE BREADTH OF THE BASE.

Let L=AB the length of the base. (Fig. 20.)
l=GH the length of the edge.
b=BC the breadth of the base.
h=PG the height of the wedge.
Then L−l=AB−GH=AM.

If the length of the base and the edge be *equal*, as BM and GH, (Fig. 20.) the wedge MBHG is half a parallelopiped of the same base and height. And the solidity (Art. 43.) is equal to half the product of the height, into the length and breadth of the base; that is to $\frac{1}{2}$ bhl.

If the length of the base be *greater* than that of the edge, as ABGH; let a section be made by the plane GMN, parallel to HBC. . This will divide the whole wedge into two parts MBHG and AMG. The latter is a pyramid, whose solidity (Art. 48.) is $\frac{1}{3}bh \times (L-l)$

The solidity of the parts together, is, therefore,

$$\tfrac{1}{2}bhl+\tfrac{1}{3}bh \times (L-l)=\tfrac{1}{6}bh3l+\tfrac{1}{6}bh2L-\tfrac{1}{6}bh2l=\tfrac{1}{6}bh \times (2L+l)$$

If the length of the base be *less* than that of the edge, it is evident that the pyramid is to be *subtracted* from half a parallelopiped, which is equal in height and breadth to the wedge, and equal in length to the edge.

The solidity of the wedge is, therefore,

$$\tfrac{1}{2}bhl-\tfrac{1}{3}bh \times (l-L)=\tfrac{1}{6}bh3l-\tfrac{1}{6}bh2l+\tfrac{1}{6}bh2L=\tfrac{1}{6}bh \times (2L+l)$$

Ex. 1. If the base of a wedge be 35 by 15, the edge 55, and the perpendicular height 12.4; what is the solidity?

$$\text{Ans. } (70+55) \times \frac{15 \times 12.4}{6}=3875.$$

2. If the base of a wedge be 27 by 8, the edge 36, and the perpendicular height 42; what is the solidity?

Ans. 5040.

PROBLEM VIII.

To find the SOLIDITY *of a rectangular* PRISMOID.

55. TO THE AREAS OF THE TWO ENDS, ADD FOUR TIMES THE AREA OF A PARALLEL SECTION EQUALLY DISTANT FROM THE ENDS, AND MULTIPLY THE SUM BY $\frac{1}{6}$ OF THE HEIGHT.

Let L and B (Fig. 21.) be the length and breadth of one end,
l and b the length and breadth of the other end,
M and m the length and breadth of the section in the middle,
And h the height of the prismoid.

The solid may be divided into two wedges, whose bases are the ends of the prismoid, and whose edges are L and l. The solidity of the whole by the preceding article, is

$$\tfrac{1}{6}Bh \times (2L+l)+\tfrac{1}{6}bh \times (2l+L)=\tfrac{1}{6}h(2BL+Bl+2bl+bL)$$

As M is equally distant from L and l,

$$2M=L+l, 2m=B+b, \text{and} 4Mm=(L+l)(B+b)=BL+Bl+bL+lb.$$

Substituting $4Mm$ for its value in the preceding expression for the solidity, we have

$$\tfrac{1}{6}h(BL+bl+4Mm)$$

That is, the solidity of the prismoid is equal to $\frac{1}{6}$ of the height, multiplied into the areas of the two ends, and 4 times the area of the section in the middle.

This rule may be applied to prismoids of other forms. For, whatever be the figure of the two ends, there may be drawn in each, such a number of small rectangles, that the sum of them shall differ less, than by any given quantity, from the figure in which they are contained. And the solids between these rectangles will be rectangular prismoids.

Ex. 1. If one end of a rectangular prismoid be 44 feet by 23, the other end 36 by 21, and the perpendicular height 72; what is the solidity?

The area of the larger end $=44\times23=1012$
of the smaller end $=36\times21=\ 756$
of the middle section $=40\times22=\ 880$
And the solidity $=(1012+756+4\times880)\times12=63456$ feet.

2. What is the solidity of a stick of hewn timber, whose ends are 30 inches by 27, and 24 by 18, and whose length is 48 feet? Ans. 204 feet.

Other solids not treated of in this section, if they be bounded by plane surfaces, may be measured by supposing them to be divided into prisms, pyramids, and wedges. And, indeed, every such solid may be considered as made up of triangular pyramids.

THE FIVE REGULAR SOLIDS.

56. A SOLID IS SAID TO BE REGULAR, WHEN ALL ITS SOLID ANGLES ARE EQUAL, AND ALL ITS SIDES ARE EQUAL AND REGULAR POLYGONS.

The following figures are of this description;

1. The *Tetraedron*	whose sides are	four triangles;
2. The *Hexaedron or cube*		six squares;
3. The *Octaedron*		eight triangles;
4. The *Dodecaedron*		twelve pentagons;
5. The *Icosaedron*		twenty triangles.*

Besides these five, there can be no other regular solids. The only plane figures which can form such solids, are triangles, squares, and pentagons. For the plane angles which contain any solid angle, are together less than four right angles or 360°. (Sup. Euc. 21. 2.) And the least number which can form a solid angle is three. (Sup. Euc. Def. 8. 2.) If they are angles of equilateral *triangles*, each is 60°. The sum of *three* of them is 180°, of *four* 240°, of *five* 300°, and of *six* 360°. The latter number is too great for a solid angle.

The angles of *squares* are 90° each. The sum of *three* of these is 270°, of four 360°, and of any other greater number still more.

The angles of regular *pentagons* are 108° each. The sum of *three* of them is 324°; of four, or any other greater number, more than 360°. The angles of all other regular polygons are still greater.

In a regular solid, then, each solid angle must be contained by three, four, or five equilateral triangles, by three squares, or by three regular pentagons.

57. As the sides of a regular solid are similar and equal, and the angles are also alike; it is evident that the sides are all equally distant from a central point in the solid. If then, planes be supposed to proceed from the several edges to the centre, they will divide the solid into as many equal *pyramids*, as it has sides. The base of each pyramid will be one of the sides; their common vertex will be the central point; and their height will be a perpendicular from the centre to one of the sides.

* For the geometrical construction of these solids, see Legendre's Geometry; Appendix to Books VI and VII.

PROBLEM IX.

To find the SURFACE *of a* REGULAR SOLID.

58. MULTIPLY THE AREA OF ONE OF THE SIDES BY THE NUMBER OF SIDES.

Or,

MULTIPLY THE SQUARE OF ONE OF THE EDGES, BY THE SURFACE OF A SIMILAR SOLID WHOSE EDGES ARE 1.

As all the sides are *equal*, it is evident that the area of one of them multiplied by the number of sides, will give the area of the whole.

Or, if a *table* is prepared, containing the surfaces of the several regular solids whose linear edges are *unity;* this may be used for other regular solids, upon the principle, that the areas of similar polygons are as the squares of their homologous sides. (Euc. 20. 6.) Such a table is easily formed, by multiplying the area of one of the sides, as given in art. 17, by the number of sides. Thus the area of an equilateral triangle whose side is 1, is 0.4330127. Therefore the surface

Of a regular tetraedron $=.4330127\times4=1.7320508$.
Of a regular octaedron $=.4330127\times8=3.4641016$.
Of a regular icosaedron $=.4330127\times20=8.6602540$.

See the table in the following article.

Ex. 1. What is the surface of a regular dodecaedron whose edges are each 25 inches?

The area of one of the sides is 1075.3.
And the surface of the whole solid $=1075.3\times12=12903.6$.

2. What is the surface of a regular icosaedron whose edges are each 102? Ans. 90101.3.

PROBLEM X.

To find the SOLIDITY *of a* REGULAR SOLID.

59. MULTIPLY THE SURFACE BY $\frac{1}{3}$ OF THE PERPENDICULAR DISTANCE FROM THE CENTRE TO ONE OF THE SIDES.

Or,

MULTIPLY THE CUBE OF ONE OF THE EDGES, BY THE SOLIDITY OF A SIMILAR SOLID WHOSE EDGES ARE 1.

As the solid is made up of a number of equal pyramids, whose bases are the sides, and whose height is the perpendicular distance of the sides from the centre; (Art. 57.) the solidity of the whole must be equal to the areas of all the sides, multiplied into $\frac{1}{3}$ of this perpendicular. (Art. 48.)

If the contents of the several regular solids whose edges are 1, be inserted in a *table*, this may be used to measure other similar solids. For two similar regular solids contain the same number of similar pyramids; and these are to each other as the *cubes* of their linear sides or edges. (Sup. Euc. 15. 3. Cor. 3.)

A TABLE OF REGULAR SOLIDS WHOSE EDGES ARE 1.

Names.	*No. of sides.*	*Surfaces.*	*Solidities.*
Tetraedron	4	1.7320508	0.1178513
Hexaedron	6	6.0000000	1.0000000
Octaedron	8	3.4641016	0.4714045
Dodecaedron	12	20.6457288	7.6631189
Icosaedron	20	8.6602540	2.1816950

For the method of calculating the last column of this table, see Hutton's Mensuration, Part III. Sec. 2.

Ex. What is the solidity of a regular octaedron whose edges are each 32 inches? Ans. 15447 inches.

SECTION IV*.

THE CYLINDER, CONE, AND SPHERE.

ART. 61. DEFINITION I. A *RIGHT CYLINDER* is a solid described by the revolution of a rectangle about one of its sides. The *ends or bases* are evidently equal and parallel circles. And the *axis*, which is a line passing through the middle of the cylinder, is perpendicular to the bases.

The ends of an *oblique* cylinder are also equal and parallel circles; but they are not perpendicular to the axis. The *height* of a cylinder is the perpendicular distance from one base to the plane of the other. In a right cylinder, it is the length of the axis.

II. A *right cone* is a solid described by the revolution of a right angled triangle about one of the sides which contain the right angle. The *base* is a circle, and is perpendicular to the *axis*, which proceeds from the middle of the base to the vertex.

The base of an *oblique* cone is also a circle, but is not perpendicular to the axis. The *height* of a cone is the perpendicular distance from the vertex to the plane of the base. In a right cone, it is the length of the axis. The *slant-height* of a right cone is the distance from the vertex to the circumference of the base.

III. A *frustum* of a cone is a portion cut off, by a plane parallel to the base. The *height* of the frustum is the perpendicular distance of the two ends. The *slant-height* of a frustum of a right cone, is the distance between the peripheries of the two ends, measured on the outside of the solid; as AD. (Fig. 23.)

IV. A *sphere* or *globe* is a solid which has a centre equally distant from every part of the surface. It may be described by the revolution of a semicircle about a diameter. A *radius* of the sphere is a line drawn from the centre to any

* Hutton's Mensuration, West's Mathematics, Legendre's, Clairaut's, and Camus's Geometry.

part of the surface. A *diameter* is a line passing through the centre, and terminated at both ends by the surface. The *circumference* is the same as the circumference of a circle whose plane passes through the centre of the sphere. Such a circle is called a *great circle*.

V. A *segment* of a sphere is a part cut off by any plane. The *height* of the segment is a perpendicular from the middle of the base to the convex surface, as LB. (Fig. 12.)

VI. A *spherical zone* or frustum is a part of the sphere included between two parallel planes. It is called the *middle zone*, if the planes are equally distant from the centre. The *height* of a zone is the distance of the two planes, as LR. (Fig. 12.*)

VII. A *spherical sector* is a solid produced by a *circular* sector, revolving in the same manner as the semicircle which describes the whole sphere. Thus a spherical sector is described by the circular sector ACP (Fig. 15.) or GCE revolving on the axis CP.

VIII. A solid described by the revolution of any figure about a fixed axis, is called a *solid of revolution*.

PROBLEM I.

To find the CONVEX SURFACE *of a* RIGHT CYLINDER

62. MULTIPLY THE LENGTH INTO THE CIRCUMFERENCE OF THE BASE.

If a right cylinder be covered with a thin substance like paper, which can be spread out into a plane; it is evident that the plane will be a *parallelogram*, whose length and breadth will be equal to the length and circumference of the cylinder. The area must, therefore, be equal to the length multiplied into the circumference. (Art. 4.)

Ex. 1. What is the convex surface of a right cylinder which is 42 feet long, and 15 inches in diameter?

Ans. $42 \times 1.25 \times 3.14159 = 164.938$ sq. feet.

* According to some writers, a spherical *segment* is either a solid which is cut off from a sphere by a single plane, or one which is included between two planes: and a *zone* the *surface* of either of these. In this sense, the term zone is commonly used in geography.

2. What is the whole surface of a right cylinder, which is 2 feet in diameter and 36 feet long?

The convex surface is	226.1945
The area of the two ends (Art. 30.) is	6.2832
The whole surface is	232.4777

3. What is the whole surface of a right cylinder whose axis is 82, and circumference 71? Ans. 6624.32.

63. It will be observed that the rules for the *prism* and *pyramid* in the preceding section, are substantially the same, as the rules for the *cylinder* and *cone* in this. There may be some advantage, however, in considering the latter by themselves.

In the base of a *cylinder*, there may be inscribed a polygon, which shall differ from it less than by any given space. (Sup. Euc. 6. 1. Cor.) If the polygon be the base of a *prism*, of the same height as the cylinder, the two solids may differ less than by any given quantity. In the same manner, the base of a *pyramid* may be a polygon of so many sides, as to differ less than by any given quantity, from the base of a *cone* in which it is inscribed. A cylinder is therefore considered, by many writers, as a prism of an infinite number of sides; and a cone, as a pyramid of an infinite number of sides. For the meaning of the term "infinite," when used in the mathematical sense, see Alg. Sec. xv.

PROBLEM II.

To find the SOLIDITY *of a* CYLINDER.

64. MULTIPLY THE AREA OF THE BASE BY THE HEIGHT.

The solidity of a *parallelopiped* is equal to the product of the base into the perpendicular altitude. (Art. 43.) And a parallelopiped and a cylinder which have equal bases and altitudes are equal to each other. (Sup. Euc. 17. 3.)

Ex. 1. What is the solidity of a cylinder, whose height is 121, and diameter 45.2?

Ans. $\overline{45.2}^2 \times .7854 \times 121 = 194156.6$.

2. What is the solidity of a cylinder whose height is 424, and circumference 213? Ans. 1530837.

3. If the side AC of an oblique cylinder (Fig. 22.) be 27, and the area of the base 32.61, and if the side make an angle of 62° 44′ with the base, what is the solidity?

R : AC : : Sin A : BC=24 the perpendicular height.

And the solidity is 782.64.

4. The Winchester bushel is a hollow cylinder, $18\frac{1}{2}$ inches in diameter, and 8 inches deep. What is its capacity?

The area of the base$=(18.5)^2\times.7853982=268.8025$.

And the capacity is 2150.42 cubic inches. See the table in art. 42.

PROBLEM III.

To find the CONVEX SURFACE *of a* RIGHT CONE.

65. MULTIPLY HALF THE SLANT-HEIGHT INTO THE CIRCUMFERENCE OF THE BASE.

If the convex surface of a right cone be spread out into a plane, it will evidently form a *sector* of a circle whose radius is equal to the slant-height of the cone. But the area of the sector is equal to the product of half the radius into the length of the arc. (Art. 34.) Or if the cone be considered as a pyramid of an infinite number of sides, its lateral surface is equal to the product of half the slant-height into the perimeter of the base. (Art. 49.)

Ex. 1. If the slant-height of a right cone be 82 feet, and the diameter of the base 24, what is the convex surface?

Ans. $41\times24\times3.14159=3091.3$ square feet.

2. If the axis of a right cone be 48, and the diameter of the base 72, what is the whole surface?

The slant-height $=\sqrt{\overline{36}^2+\overline{48}^2}=60$. (Euc. 47. 1.)

The convex surface is	6786
The area of the base	4071.6
And the whole surface	10857.6

3. If the axis of a right cone be 16, and the circumference of the base 75.4; what is the whole surface?

Ans. 1206.4.

PROBLEM IV.

To find the SOLIDITY *of a* CONE.

66. MULTIPLY THE AREA OF THE BASE INTO $\frac{1}{3}$ OF THE HEIGHT.

The solidity of a *cylinder* is equal to the product of the base into the perpendicular height. (Art. 64.) And if a cone and a cylinder have the same base and altitude, the cone is $\frac{1}{3}$ of the cylinder. (Sup. Euc. 18. 3.) Or if a cone be considered as a pyramid of an infinite number of sides, the solidity is equal to the product of the base into $\frac{1}{3}$ of the height, by art. 48.

Ex. 1. What is the solidity of a right cone whose height is 663, and the diameter of whose base is 101?

Ans. $\overline{101}^2 \times .7854 \times 221 = 1770622$.

2. If the axis of an oblique cone be 738, and make an angle of 30° with the plane of the base; and if the circumference of the base be 355, what is the solidity?

Ans. 1233536.

PROBLEM V.

To find the CONVEX SURFACE *of a* FRUSTUM *of a right cone.*

67. MULTIPLY HALF THE SLANT-HEIGHT BY THE SUM OF THE PERIPHERIES OF THE TWO ENDS.

This is the rule for a frustum of a *pyramid*; (Art. 51.) and is equally applicable to a frustum of a *cone*, if a cone be considered as a pyramid of an infinite number of sides. (Art. 63.)

Or thus,

Let the sector ABV (Fig. 23.) represent the convex surface of a right cone, (Art. 65.) and DCV the surface of a portion of the cone, cut off by a plane parallel to the base. Then will ABCD be the surface of the frustum.

Let $AB=a$, $DC=b$, $VD=d$, $AD=h$.

Then the area $ABV=\frac{1}{2}a\times(h+d)=\frac{1}{2}ah+\frac{1}{2}ad$. (Art. 34,)

And the area $DCV=\frac{1}{2}bd$.

Subtracting the one from the other,

The area $ABDC=\frac{1}{2}ah+\frac{1}{2}ad-\frac{1}{2}bd$.

But $d:d+h::b:a$. (Sup.Euc.8.1) Therefore $\frac{1}{2}ad-\frac{1}{2}bd=\frac{1}{2}bh$.

The surface of the frustum, then, is equal to

$\frac{1}{2}ah+\frac{1}{2}bh$. or $\frac{1}{2}h\times(a+b)$

Cor. The surface of the frustum is equal to the product of the slant-height into the circumference of a circle which is *equally distant* from the two ends. Thus the surface ABCD (Fig. 23.) is equal to the product of AD into MN. For MN is equal to half the sum of AB and DC.

Ex. 1. What is the convex surface of a frustum of a right cone, if the diameters of the two ends be 44 and 33, and the slant-height 84? Ans. 10159.8.

2. If the perpendicular height of a frustum of a right cone be 24, and the diameters of the two ends 80 and 44, what is the whole surface?

Half the difference of the diameters is 18.

And $\sqrt{18^2+24^2}=30$, the slant-height, (Art. 52.)

The convex surface of the frustum is	5843
The sum of the areas of the two ends is	6547
And the whole surface is	12390

PROBLEM VI.

To find the SOLIDITY *of a* FRUSTUM *of a cone.*

68. ADD TOGETHER THE AREAS OF THE TWO ENDS, AND THE SQUARE ROOT OF THE PRODUCT OF THESE AREAS; AND MULTIPLY THE SUM BY $\frac{1}{3}$ OF THE PERPENDICULAR HEIGHT.

This rule, which was given for the frustum of a *pyramid*, (Art, 50.) is equally applicable to the frustum of a cone; because a cone and a pyramid which have equal bases and altitudes are equal to each other.

Ex. 1. What is the solidity of a mast which is 72 feet long, 2 feet in diameter at one end, and 18 inches at the other? Ans. 174.36 cubic feet.

2. What is the capacity of a conical cistern which is 9 feet deep, 4 feet in diameter at the bottom, and 3 feet at the top? Ans. 87.18 cubic feet =652.15 wine gallons.

3. How many gallons of ale can be put into a vat in the form of a conic frustum, if the larger diameter be 7 feet, the smaller diameter 6 feet, and the depth 8 feet?

PROBLEM VII.

To find the SURFACE *of a* SPHERE.

69. MULTIPLY THE DIAMETER BY THE CIRCUMFERENCE.

Let a hemisphere be described by the quadrant CPD, (Fig. 25.) revolving on the line CD. Let AB be a side of a regular polygon inscribed in the circle of which DBP is an arc. Draw AO and BN perpendicular to CD, and BH perpendicular to AO. Extend AB till it meet CD continued. The triangle AOV, revolving on OV as an axis, will describe a right *cone*. (Defin. 2.) AB will be the slant-height of a *frustum* of this cone extending from AO to BN. From G the middle of AB, draw GM parallel to AO. The surface of the frustum described by AB, (Art. 67. Cor.) is equal to

$$AB \times circ\ GM.^*$$

From the centre C draw CG, which will be perpendicular to AB, (Euc. 3. 3.) and the radius of a circle inscribed in the polygon. The triangles ABH and CGM are similar, because the sides are perpendicular, each to each. Therefore,

$$HB \text{ or } ON : AB :: GM : GC :: circ\ GM : circ\ GC.$$

So that $ON \times circ\ GC = AB \times circ\ GM$, that is, the surface of the frustum is equal to the product of ON the perpendicular height, into *circ* GC, the perpendicular distance from the centre of the polygon to one of the sides.

* By *circ* GM is meant the circumference of a circle the radius of which is GM.

In the same manner it may be proved, that the surfaces produced by the revolution of the lines BD and AP about the axis DC, are equal to

$$ND \times circ\ GC, \quad \text{and} \quad CO \times circ\ GC.$$

The surface of the whole solid, therefore, (Euc. 1.2.) is equal to

$$CD \times circ\ GC.$$

The demonstration is applicable to a solid produced by the revolution of a polygon of *any* number of sides. But a polygon may be supposed which shall differ less than by any given quantity from the circle in which it is inscribed; (Sup. Euc. 4. 1.) and in which the perpendicular GC shall differ less than by any given quantity from the radius of the circle. Therefore the surface of a *hemisphere* is equal to the product of its radius into the circumference of its base; and *the surface of a sphere is equal to the product of its diameter into its circumference.*

Cor. 1. From this demonstration it follows, that the surface of any *segment* or *zone* of a sphere is equal to the product of the height of the segment or zone into the circumference of the sphere. The surface of the zone produced by the revolution of the arc AB about ON, is equal to $ON \times circ\ CP$. And the surface of the segment produced by the revolution of BD about DN is equal to $DN \times circ\ CP$.

Cor. 2. The surface of a sphere is equal to four times the area of a circle of the same diameter; and therefore, the convex surface of a hemisphere is equal to twice the area of its base. For the area of a circle is equal to the product of half the diameter into half the circumference; (Art. 30.) that is, to $\frac{1}{4}$ the product of the diameter and circumference.

Cor. 3. The surface of a sphere, or the convex surface of any spherical segment or zone, is equal to that of the circumscribing cylinder. A hemisphere described by the revolution of the arc DBP, is circumscribed by a cylinder produced by the revolution of the parallelogram DdCP. The convex surface of the cylinder is equal to its height multiplied by its circumference. (Art. 62.) And this is also the surface of the hemisphere.

So the surface produced by the revolution of AB is equal to that produced by the revolution of *ab*. And the surface produced by BD is equal to that produced by *bd*.

Ex. 1. Considering the earth as a sphere 7930 miles in diameter, how many square miles are there on its surface?

Ans. 197,558,500.

2. If the circumference of the sun be 2,800,000 miles, what is his surface? Ans. 2,495,547,600,000 sq. miles.

3. How many square feet of lead will it require, to cover a hemispherical dome whose base is 13 feet across?

Ans. 265½.

PROBLEM VIII.

To find the SOLIDITY *of a* SPHERE.

70. 1. MULTIPLY THE CUBE OF THE DIAMETER BY .5236.

Or,

2. MULTIPLY THE SQUARE OF THE DIAMETER BY $\frac{1}{6}$ OF THE CIRCUMFERENCE.

Or,

3. MULTIPLY THE SURFACE BY $\frac{1}{6}$ OF THE DIAMETER.

1. A sphere is *two thirds* of its circumscribing cylinder. (Sup. Euc. 21. 3.) The height and diameter of the cylinder are each equal to the diameter of the sphere. The solidity of the cylinder is equal to its height multiplied into the area of its base, (Art. 64.) that is putting D for the diameter,

$$D \times D^2 \times .7854 \quad \text{or} \quad D^3 \times .7854.$$

And the solidity of the *sphere*, being $\frac{2}{3}$ of this, is

$$D^3 \times .5236.$$

2. The base of the circumscribing cylinder is equal to hal the circumference multiplied into half the diameter, (Art. 30.) that is, if C be put for the circumference,

$$\tfrac{1}{4}C \times D, \text{ and the solidity is } \tfrac{1}{4}C \times D^2.$$

Therefore the solidity of the sphere is

$$\tfrac{2}{3} \text{ of } \tfrac{1}{4}C \times D^2 = D^2 \times \tfrac{1}{6}C.$$

3. In the last expression, which is the same as $C \times D \times \frac{1}{6}D$, we may substitute S, the surface, for $C \times D$. (Art. 69.) We then have the solidity of the sphere equal to

$$S \times \tfrac{1}{6}D.$$

Or, the sphere may be supposed to be filled with small *pyramids*, standing on the surface of the sphere, and having their common vertex in the centre. The number of these may be such, that the difference between their sum and the sphere shall be less than any given quantity. The solidity of each pyramid is equal to the product of its base into $\frac{1}{3}$ of its height. (Art. 48.) The solidity of the whole, therefore, is equal to the product of the surface of the sphere into $\frac{1}{3}$ of its radius, or $\frac{1}{6}$ of its diameter.

71. The numbers 3.14159, .7854, .5236, should be made perfectly familiar. The first expresses the ratio of the *circumference* of a circle to the *diameter*; (Art. 23.) the second, the ratio of the *area* of a circle to the square of the diameter (Art. 30.); and the third, the ratio of the *solidity* of a sphere to the *cube* of the diameter. The second is $\frac{1}{4}$ of the first, and the third is $\frac{1}{6}$ of the first.

As these numbers are frequently occurring in mathematical investigations, it is common to represent the first of them by the Greek letter Π. According to this notation,

$$\pi = 3.14159, \quad \tfrac{1}{4}\pi = .7854. \quad \tfrac{1}{6}\pi = .5236.$$

If D = the *diameter*, and R = the *radius* of any circle or sphere;

Then $D = 2R \quad D^2 = 4R^2 \quad D^3 = 8R^3$.

And $\left.\begin{matrix}\pi D \\ \text{Or } 2\pi R\end{matrix}\right\}$ = the *periph.* $\left.\begin{matrix}\frac{1}{4}\pi D^2 \\ \text{or } \pi R^2\end{matrix}\right\}$ = the *area* of the circ. $\left.\begin{matrix}\frac{1}{6}\pi D^3 \\ \text{or } \frac{4}{3}\pi R^3\end{matrix}\right\}$ = the *solidity* of the sphere.

Ex. 1. What is the solidity of the earth, if it be a sphere 7930 miles in diameter?

Ans. 261,107,000,000 cubic miles.

2. How many wine gallons will fill a hollow sphere 4 feet in diameter?

Ans. The capacity is 33.5104 feet = $250\frac{2}{3}$ gallons.

3. If the diameter of the moon be 2180 miles, what is its solidity? Ans. 5,424,600,000 miles.

72. If the solidity of a sphere be *given*, the diameter may be found by reversing the first rule in the preceding article; that is, *dividing by* .5236 and *extracting the cube root of the quotient.*

Ex. 1. What is the diameter of a sphere whose solidity is 65.45 cubic feet? Ans. 5 feet.

2. What must be the diameter of a globe to contain 16755 pounds of water? Ans. 8 feet.

PROBLEM IX.

To find the CONVEX SURFACE *of a* SEGMENT *or* ZONE *of a sphere.*

73. MULTIPLY THE HEIGHT OF THE SEGMENT OR ZONE INTO THE CIRCUMFERENCE OF THE SPHERE.

For the demonstration of this rule, see art. 69.

Ex. 1. If the earth be considered a perfect sphere 7930 miles in diameter, and if the polar circle be 23° 28′ from the pole, how many square miles are there in one of the frigid zones?

If PQOE (Fig. 15.) be a meridian on the earth, ADB one of the polar circles, and P the pole; then the frigid zone is a spherical segment described by the revolution of the arc APB about PD. The angle ACD subtended by the arc AP is 23° 28′. And in the right angled triangle ACD,

$$R : AC :: \text{Cos } ACD : CD = 3637.$$

Then $CP - CD = 3965 - 3637 = 328 = PD$ the height of the segment.

And $328 \times 7930 \times 3.14159 = 8171400$ the surface.

2. If the diameter of the earth be 7930 miles, what is the surface of the torrid zone, extending 23° 28′ on each side of the equator?

If EQ (Fig. 15.) be the equator, and GH one of the tropics, then the angle ECG is 23° 28′. And in the right angled triangle GCM,

R : CG : : Sin ECG : GM=CN=1578.9 the height of half the zone.

The surface of the whole zone is 78669700.

3. What is the surface of each of the temperate zones?

The height DN=CP−CN−PD=2058.1

And the surface of the zone is 51273000.

The surface of the two temperate zones is	102,546,000
of the two frigid zones	16,342,800
of the torrid zone	78,669,700
of the whole globe	197,558,500

PROBLEM X.

To find the SOLIDITY *of a spherical* SECTOR.

74. MULTIPLY THE SPHERICAL SURFACE BY $\frac{1}{3}$ OF THE RADIUS OF THE SPHERE.

The spherical sector, (Fig. 24.) produced by the revolution of ACBD about CD, may be supposed to be filled with *small pyramids*, standing on the spherical surface ADB, and terminating in the point C. Their number may be so great, that the height of each shall differ less than by any given length from the radius CD, and the sum of their bases shall differ less than by any given quantity from the surface ABD. The solidity of each is equal to the product of its base into $\frac{1}{3}$ of the radius CD. (Art. 48.) Therefore, the solidity of all of them, that is, of the sector ADBC, is equal to the product of the spherical surface into $\frac{1}{3}$ of the radius.

Ex. Supposing the earth to be a sphere 7930 miles in diameter, and the polar circle ADB (Fig. 15.) to be 23° 28′ from the pole; what is the solidity of the spherical sector ACBP?

Ans. 10,799,867,000 miles

PROBLEM XI.

To find the SOLIDITY *of a spherical* SEGMENT.

75. MULTIPLY HALF THE HEIGHT OF THE SEGMENT INTO THE AREA OF THE BASE, AND THE CUBE OF THE HEIGHT INTO .5236; AND ADD THE TWO PRODUCTS.

As the *circular* sector AOBC (Fig. 9.) consists of two parts, the segment AOBP and the triangle ABC; (Art. 35.) so the *spherical* sector produced by the revolution of AOC about OC consists of two parts, the *segment* produced by the revolution of AOP, and the *cone* produced by the revolution of ACP. If then the cone be subtracted from the sector, the remainder will be the segment.

Let $CO=R$ the radius of the sphere,
$PB=r$ the radius of the base of the segment,
$PO=h$ the height of the segment,
Then $PC=R-h$ the axis of the cone.

The sector$=2\pi R\times h\times\frac{1}{3}R$ (Arts. 71, 73, 74.)$=\frac{2}{3}\pi hR^2$
The cone$=\pi r^2\times\frac{1}{3}(R-h)$ (Arts. 71, 66.)$=\frac{1}{3}\pi r^2R-\frac{1}{3}\pi hr^2$.

Subtracting the one from the other,

The segment $=\frac{2}{3}\pi hR^2-\frac{1}{3}\pi r^2R+\frac{1}{3}\pi hr^2$.

But $DO\times PO=\overline{BO}^2$ (Trig. 97.*)$=\overline{PO}^2+\overline{PB}^2$ (Euc. 47. 1.)

That is, $2Rh=h^2+r^2$. So that, $R=\dfrac{h^2+r^2}{2h}$

And $R^2=\left(\dfrac{h^2+r^2}{2h}\right)^2=\dfrac{h^4+2h^2r^2+r^4}{4h^2}$

Substituting then, for R and R^2, their values, and multiplying the factors,

$$\text{The segment}=\tfrac{1}{6}\pi h^3+\tfrac{1}{3}\pi hr^2+\tfrac{1}{6}\,\frac{\pi r^4}{h}-\tfrac{1}{6}\pi hr^2-\tfrac{1}{6}\,\frac{\pi r^4}{h}+\tfrac{1}{3}\,hr^2$$

which, by uniting the terms, becomes
$\frac{1}{2}\pi hr^2+\frac{1}{6}\pi h^3$.

* Euclid 31, 3, and 8, 6. Cor.

The first term here is $\frac{1}{2}h\times\pi r^2$, half the height of the segment multiplied into the area of the base; (Art. 71.) and the other $h^3\times\frac{1}{6}\pi$, the cube of the height multiplied into .5236.

If the segment be *greater* than a hemisphere, as ABD; (Fig. 9.) the cone ABC must be *added* to the sector ACBD.

Let $PD=h$ the height of the segment,
Then $PC=h-R$ the axis of the cone.

$$\text{The sector ACBD}=\tfrac{2}{3}\pi hR^2$$

$$\text{The cone}=\pi r^2\times\tfrac{1}{3}(h-R)=\tfrac{1}{3}\pi hr^2-\tfrac{1}{3}\pi r^2R$$

Adding them together, we have as before,

$$\text{The segment}=\tfrac{2}{3}\pi hR^2-\tfrac{1}{3}\pi r^2R+\tfrac{1}{3}\pi hr^2.$$

Cor. The solidity of a spherical segment is equal to half a cylinder of the same base and height + a sphere whose diameter is the height of the segment. For a cylinder is equal to its height multiplied into the area of its base; and a sphere is equal to the cube of its diameter multiplied by .5236.

Thus if Oy (Fig. 15.) be half Ox, the spherical segment produced by the revolution of Oxt is equal to the cylinder produced by $tvyx$ + the sphere produced by $Oyxz$; supposing each to revolve on the line Ox.

Ex. 1. If the height of a spherical segment be 8 feet, and the diameter of its base 25 feet; what is the solidity?

Ans. $25^2\times.7854\times4+8^3\times.5236=2231.58$ feet.

2. If the earth be a sphere 7930 miles in diameter, and the polar circle 23° 28′ from the pole, what is the solidity of one of the frigid zones?

Ans. 1,303,000,000 miles.

PROBLEM XII.

To find the SOLIDITY *of a spherical* ZONE *or frustum.*

76. FROM THE SOLIDITY OF THE WHOLE SPHERE, SUBTRACT THE TWO SEGMENTS ON THE SIDES OF THE ZONE.

Or,

Add together the squares of the radii of the two ends, and $\frac{1}{3}$ the square of their distance; and multiply the sum by three times this distance, and the product by .5236.

If from the whole sphere, (Fig. 15.) there be taken the two segments ABP and GHO, there will remain the zone or frustum ABGH.

Or, the zone ABGH is equal to the difference between the segments GHP and ABP.

Let $\left.\begin{array}{r}NP=H\\ DP=h,\end{array}\right\}$ the *heights* of the two segments.

$\left.\begin{array}{r}GN=R\\ AD=r\end{array}\right\}$ the *radii* of their bases.

$DN=d=H-h$ the *distance* of the two bases, or the height of the zone.

Then the larger segment $=\frac{1}{2}\pi HR^2+\frac{1}{6}\pi H^3$
And the smaller segment $=\frac{1}{2}\pi hr^2+\frac{1}{6}\pi h^3$ } (Art. 75.)

Therefore the zone $ABGH=\frac{1}{6}\pi(3HR^2+H^3-3hr^2-h^3)$

By the properties of the circle, (Euc. 35, 3.)

$ON\times H=R^2$. Therefore $(ON+H)\times H=R^2+H^2$.

$$\text{Or } OP=\frac{R^2+H^2}{H}$$

In the same manner, $OP=\dfrac{r^2+h^2}{h}$

Therefore $3H\times(r^2+h^2)=3h\times(R^2+H^2)$.

Or $3Hr^2+3Hh^2-3hR^2-3H^2h=0$. (Alg. 178.)

To reduce the expression for the solidity of the zone to the required form, without altering its value, let these terms be added to it: and it will become

$$\tfrac{1}{6}\pi(3HR^2+3Hr^2-3hR^2-3hr^2+H^3-3H^2h+3Hh^2-h^3)$$

Which is equal to

$$\tfrac{1}{6}\pi\times3(H-h)\times(R^2+r^2+\tfrac{1}{3}(H-h)^2).$$

Or, as $\frac{1}{6}\pi$ equals .5236 (Art. 71.) and $H-h$ equals d,

The zone$=.5236\times3d\times(R^2+r^2+\frac{1}{3}d^2)$

Ex. 1. If the diameter of one end of a spherical zone is 24 feet, the diameter of the other end 20 feet, and the distance of the two ends, or the height of the zone 4 feet; what is the solidity? Ans. 1566.6 feet.

2. If the earth be a sphere 7930 miles in diameter, and the obliquity of the ecliptic 23° 28′; what is the solidity of one of the temperate zones?

Ans. 55,390,500,000 miles.

3. What is the solidity of the torrid zone?

Ans. 147,720,000,000 miles.

The solidity of the two temperate zones is	110,781,000,000
of the two frigid zones	2,606,000,000
of the torrid zone	147,720,000,000
of the whole globe	261,107,000,000

PROMISCUOUS EXAMPLES OF SOLIDS.

Ex. 1. How much water can be put into a cubical vessel three feet deep, which has been previously filled with cannon balls of the same size, 2, 4, 6, or 9 inches in diameter, regularly arranged in tiers, one directly above another?

Ans. $96\frac{1}{5}$ wine gallons.

2. *If* a cone or pyramid, whose height is three feet, be divided into three equal portions, by sections parallel to the base; what will be the heights of the several parts?

Ans. 24.961, 6.488, and 4.551 inches.

3. What is the solidity of the greatest square prism which can be cut from a cylindrical stick of timber, 2 feet 6 inches in diameter and 56 feet long?*

Ans. 175 cubic feet.

4. How many such globes as the earth are equal in bulk to the sun; if the former is 7930 miles in diameter, and the latter 890,000? Ans. 1,413,678.

5. How many cubic feet of wall are there in a conical tower 66 feet high, if the diameter of the base be 20 feet from outside to outside, and the diameter of the top 8 feet; the thickness of the wall being 4 feet at the bottom, and decreasing regularly, so as to be only 2 feet at the top?

Ans. 7188.

* The common rule for measuring *round timber* is to multiply the square of the *quarter-girt* by the length. The quarter girt is one fourth of the circumference. This method does not give the whole solidity. It makes an allowance of about one-fifth, for waste in hewing, bark, &c. The solidity of a cylinder is equal to the product of the length into the area of the base.

If C=the circumference, and π=3.14159, then (Art. 31.)

$$\text{The area of the base} = \frac{C^2}{4\pi} = \left(\frac{C}{\sqrt{4\pi}}\right)^2 = \left(\frac{C}{3.545}\right)^2$$

If then the circumference were divided by 3.545, instead of 4, and the quotient squared, the area of the base would be correctly found. See note G.

6. If a metallic globe is filled with wine, which cost as much at 5 dollars a gallon, as the globe itself at 20 cents for every square inch of its surface; what is the diameter of the globe? Ans. 55.44 inches.

7. If the circumference of the earth be 25,000 miles, what must be the diameter of a metallic globe, which, when drawn into a wire $\frac{1}{20}$ of an inch in diameter, would reach round the earth? Ans. 15 feet and 1 inch.

8. If a conical cistern be 3 feet deep, $7\frac{1}{2}$ feet in diameter at the bottom, and 5 feet at the top; what will be the depth of a fluid occupying half its capacity?

Ans. 14.535 inches.

9. If a globe 20 inches in diameter be perforated by a cylinder 16 inches in diameter, the axis of the latter passing through the centre of the former; what part of the solidity, and the surface of the globe will be cut away by the cylinder?

Ans. 3284 inches of the solidity, and 502,655 of the surface.

10. What is the solidity of the greatest cube which can be cut from a sphere three feet in diameter?

Ans. $5\frac{1}{5}$ feet

SECTION V.

ISOPERIMETRY.*

ART. 77. IT is often necessary to compare a number of different figures or solids, for the purpose of ascertaining which has the *greatest area*, within a given perimeter, or the *greatest capacity* under a given surface. We may have occasion to determine, for instance, what must be the form of a fort, to contain a given number of troops, with the least extent of wall; or what the shape of a metallic pipe to convey a given portion of water, or of a cistern, to hold a given quantity of liquor, with the least expense of materials.

78. Figures which have equal perimeters are called *Isoperimeters.* When a quantity is *greater* than any other of the same class, it is called a *maximum.* A multitude of straight lines, of different lengths, may be drawn within a circle. But among them all, the *diameter* is a *maximum.* Of all *sines* of angles, which can be drawn in a circle, the sine of 90° is a *maximum.*

When a quantity is *less* than any other of the same class, it is called a *minimum.* Thus, of all straight lines drawn from a given point to a given straight line, that which is *perpendicular* to the given line is a *minimum.* Of all straight lines drawn from a given point in a circle, to the circumference, the *maximum* and *minimum* are the two parts of the diameter which pass through that point. (Euc, 7, 3.)

In isoperimetry, the object is to determine, on the one hand, in what cases the area is a *maximum*, within a given perimeter; or the capacity a *maximum*, within a given surface: and on the other hand, in what cases the perimeter is a *minimum* for a given area, or the surface a *minimum*, for a given capacity.

* Emerson's, Simpson's, and Legendre's Geometry, Lhuillier. Fontenelle, Hutton's Mathematics, and Lond. Phil. Trans. Vol. 75.

PROPOSITION I.

79. *An* Isosceles Triangle *has a greater area than any scalene triangle, of equal base and perimeter.*

If ABC (Fig. 26.) be an isosceles triangle whose equal sides are AC and BC; and if ABC′ be a scalene triangle on the same base AB, and having AC′+BC=AC+BC; then the area of ABC is greater than that of ABC′.

Let perpendiculars be raised from each end of the base, extend AC to D, make C′D′ equal to AC′, join BD, and draw CH and C′H′ parallel to AB.

As the angle CAB=ABC, (Euc. 5, 1.) and ABD is a right angle, ABC+CBD=CAB+CDB=ABC+CDB. Therefore CBD=CDB, so that CD=CB; and by construction, C′D′=AC′. The perpendiculars of the equal right angled triangles CHD and CHB are equal; therefore BH=$\frac{1}{2}$BD. In the same manner, AH′=$\frac{1}{2}$AD′. The line AD=AC+BC =AC′+BC′=D′C′+BC′. But D′C′+BC′>BD′. (Euc. 20, 1.) Therefore, AD>BD′; BD>AD′, (Euc. 47, 1.) and $\frac{1}{2}$BD>$\frac{1}{2}$AD′. But $\frac{1}{2}$BD, or BH, is the height of the isosceles triangle; (Art. 1.) and $\frac{1}{2}$AD′ or AH′, the height of the scalene triangle; and the areas of two triangles which have the same base are as their heights. (Art. 8.) Therefore the area of ABC is greater than that of ABC′. Among all triangles, then, of a given perimeter, and upon a given base, the isosceles triangle is a *maximum.*

Cor. The isosceles triangle has a *less perimeter* than any scalene triangle of the same base and area. The triangle ABC′ being less than ABC, it is evident the perimeter of the former must be enlarged, to make its area equal to the area of the latter.

PROPOSITION II.

80. *A triangle in which two given sides make a* right angle, *has a greater area than any triangle in which the same sides make an oblique angle.*

If BC, BC′, and BC″ (Fig. 27.) be equal, and if BC be perpendicular to AB; then the right angled triangle ABC, has a greater area than the acute angled triangle ABC′, or the oblique angled triangle ABC″.

Let P′C′ and PC″ be perpendicular to AP. Then, as the three triangles have the same base AB, their areas are as their heights; that is, as the perpendiculars BC, P′C′, and PC″. But BC is equal to BC′, and therefore greater than P′C′. (Euc. 47, 1.) BC is also equal to BC″, and therefore greater than PC″.

PROPOSITION III.

81. *If all the sides* EXCEPT ONE *of a polygon be given, the area will be the greatest, when the given sides are so disposed, that the figure may be* INSCRIBED IN A SEMICIRCLE, *of which the undetermined side is the diameter.*

If the sides AB, BC, CD, DE, (Fig. 28.) be given, and if their position be such that the area, included between these and another side whose length is not determined, is a *maximum;* the figure may be inscribed in a semicircle, of which the undetermined side AE is the diameter.

Draw the lines AD, AC, EB, EC. By varying the angle at D, the triangle ADE may be enlarged or diminished, without affecting the area of the other parts of the figure. The whole area, therefore, cannot be a *maximum*, unless this triangle be a *maximum*, while the sides AD and ED are given. But if the triangle ADE be a *maximum*, under these conditions, the angle ADE is a right angle; (Art. 80.) and therefore the point D is in the circumference of a circle, of which AE is the diameter. (Euc. 31, 3.) In the same manner it may be proved, that the angles ACE and ABE are right angles, and therefore that the points C and B are in the circumference of the same circle.

The term *polygon* is used in this section to include *triangles*, and *four-sided* figures, as well as other right-lined figures.

82. The area of a polygon, inscribed in a semi-circle, in the manner stated above, will not be altered by varying the *order* of the given sides.

The sides AB, BC, CD, DE, (Fig. 28.) are the *chords* of so many arcs. The sum of these arcs, in whatever order they are arranged, will evidently be equal to the semicircumference. And the *segments* between the given sides and the arcs will be the same, in whatever part of the circle they are situated. But the area of the polygon is equal to the area of the semicircle, diminished by the sum of these segments.

83. If a polygon, of which all the sides except one are given, be inscribed in a semicircle whose diameter is the undetermined side; a polygon having the same given sides, cannot be inscribed in any *other* semicircle which is either greater or less than this, and whose diameter is the undetermined side.

The given sides AB, BC, CD, DE, (Fig. 28.) are the chords of arcs whose sum is 180 degrees. But in a larger circle, each would be the chord of a less number of degrees, and therefore the sum of the arcs would be less than 180° : and in a smaller circle, each would be the chord of a greater number of degrees, and the sum of the arcs would be greater than 180°.

PROPOSITION IV.

84. *A polygon* INSCRIBED IN A CIRCLE *has a greater area, than any polygon of equal perimeter, and the same number of sides, which cannot be inscribed in a circle.*

If in the circle ACHF, (Fig. 30.) there be inscribed a polygon ABCDEFG ; and if another polygon *abcdefg* (Fig. 31.) be formed of sides which are the same in number and length, but which are so disposed, that the figure cannot be inscribed in a circle ; the area of the former polygon is greater than that of the latter.

Draw the diameter AH, and the chords DH and EH. Upon *de* make the triangle *deh* equal and similar to DEH, and join *ah*. The line *ah* divides the figure *abcdhefg* into two parts, of which *one at least* cannot, by supposition, be inscribed in a semicircle of which the diameter is AH, nor in any other semicircle of which the diameter is the undetermined side. (Art. 83.) It is therefore less than the corresponding part of the figure ABCDHEFG. (Art. 81.) And the other part of *abcdhefg* is not greater than the corresponding part of ABCDHEFG. Therefore the whole figure ABCDHEFG is greater than the whole figure *abcdhefg*. If from these there be taken the equal triangles DEH and *deh*, there will remain the polygon ABCDEFG greater than the polygon *abcdefg*.

85. A polygon of which all the sides are given in number and length, can not be inscribed in circles of different diameters. (Art. 83.) And the area of the polygon will not be altered, by changing the *order* of the sides. (Art. 82.)

PROPOSITION V.

86. *When a polygon has a greater area than any other, of the same number of sides, and of equal perimeter, the sides are* EQUAL.

The polygon ABCDF (Fig. 29.) cannot be a *maximum*, among all polygons of the same number of sides, and of equal perimeters unless it be equilateral. For if any two of the sides, as CD and FD, are unequal, let CH and FH be equal, and their sum the same as the sum of CD and FD. The isosceles triangle CHF is greater than the scalene triangle CDF (Art. 79.); and therefore the polygon ABCHF is greater than the polygon ABCDF; so that the latter is not a *maximum*.

PROPOSITION VI.

87. *A* REGULAR POLYGON *has a greater area than any other polygon of equal perimeter, and of the same number of sides.*

For, by the preceding article, the polygon which is a *maximum* among others of equal perimeters, and the same number of sides is *equilateral*, and by art. 84, it may be *inscribed in a circle*. But if a polygon inscribed in a circle is equilateral, as ABDFGH (Fig. 7.) it is also *equiangular*. For the sides of the polygon are the bases of so many isosceles triangles, whose common vertex is the centre C. The angles at these bases are all equal; and two of them, as AHC and GHC, are equal to AHG one of the angles of the polygon. The polygon, then, being equiangular, as well as equilateral, is a *regular* polygon. (Art. 1. Def. 2.)

Thus an *equilateral triangle* has a greater area, than any other triangle of equal perimeter. And a *square* has a greater area, than any other four-sided figure of equal perimeter.

Cor. A regular polygon has a *less perimeter* than any other polygon of equal area, and the same number of sides.

For if, with a given perimeter, the regular polygon is greater than one which is not regular; it is evident the perimeter of the former must be diminished, to make its area equal to that of the latter.

PROPOSITION VII.

88. *If a polygon be* DESCRIBED ABOUT A CIRCLE, *the areas of the two figures are as their perimeters.*

Let ST (Fig. 32.) be one of the sides of a polygon, either regular or not, which is described about the circle LNR. Join OS and OT, and to the point of contact M draw the radius OM, which will be perpendicular to ST. (Euc. 18, 3.) The triangle OST is equal to half the base ST multiplied into the radius OM. (Art. 8.) And if lines be drawn, in the same manner, from the centre of the circle, to the extremities of the several sides of the circumscribed polygon, each of the triangles thus formed will be equal to half its base multiplied into the radius of the circle. Therefore the area of the whole polygon is equal to half its perimeter multiplied into the radius: and the area of the circle is equal to half its circumference multiplied into the radius. (Art. 30.) So that the two areas are to each other as their perimeters.

Cor. 1. If different polygons are described about the same circle, their areas are to each other as their perimeters. For the area of each is equal to half its perimeter, multiplied into the radius of the inscribed circle.

Cor. 2. The *tangent* of an arc is always greater than the arc itself. The triangle OMT (Fig. 32.) is to OMN, as MT to MN. But OMT is greater than OMN, because the former includes the latter. Therefore the tangent MT is greater than the arc MN.

PROPOSITION VIII.

89. *A* CIRCLE *has a greater area than any polygon of equal perimeter.*

If a circle and a regular polygon have the same centre, and equal perimeters; each of the sides of the polygon must fall partly *within* the circle. For the area of a *circumscribing* polygon is greater than the area of the circle, as the one includes the other: and therefore, by the preceding article, the *perimeter* of the former is greater than that of the latter.

Let AD then (Fig. 32.) be one side of a regular polygon, whose perimeter is equal to the circumference of the circle RLN. As this falls partly within the circle, the perpendicular OP is less than the radius OR. But the area of the polygon is equal to half its perimeter multiplied into this perpendicular (Art. 15.); and the area of the circle is equal to half its circumference multiplied into the radius. (Art. 30.) The circle then is greater than the given regular polygon; and therefore greater than any other polygon of equal perimeter. (Art. 87.)

Cor. 1. A circle has a *less perimeter*, than any polygon of equal area.

Cor. 2 Among regular polygons of a given perimeter, that which has the *greatest number of sides*, has also the *greatest area*. For the greater the number of sides, the more nearly does the perimeter of the polygon approach to a coincidence with the circumference of a circle.*

PROPOSITION IX.

90. *A right* PRISM *whose bases are* REGULAR POLYGONS, *has a less surface than any other right prism of the same solidity, the same altitude, and the same number of sides.*

If the altitude of a prism is given, the area of the base is as the solidity (Art. 43.); and if the number of sides is also given, the perimeter is a *minimum* when the base is a regular

* For a rigorous demonstration of this, see Legendre's Geometry, Appendix to Book iv.

polygon. (Art. 87. Cor.) But the lateral surface is as the perimeter (Art. 47.) Of two right prisms, then, which have the same altitude, the same solidity, and the same number of sides. that whose bases are regular polygons has the least *lateral* surface, while the areas of the ends are equal.

Cor. A right prism whose bases are regular polygons has a *greater solidity*, than any other right prism of the same surface, the same altitude, and the same number of sides.

PROPOSITION X.

91. *A right* CYLINDER *has a less surface, than any right prism of the same altitude and solidity.*

For if the prism and cylinder have the same altitude and solidity, the areas of their bases are equal. (Art. 64.) But the *perimeter* of the cylinder is less, than that of the prism (Art. 89. Cor. 1.); and therefore its lateral surface is less, while the areas of the ends are equal.

Cor. A right cylinder has a *greater solidity*, than any right prism of the same altitude and surface.

PROPOSITION XI.

92. *A* CUBE *has a less surface than any other right parallelopiped of the same solidity.*

A parallelopiped is a prism, any one of whose faces may be considered a base. (Art. 41. Def. I. and V.) If these are not all *squares*, let one which is not a square be taken for a base. The perimeter of this may be diminished, without altering its area (Art. 87. Cor.); and therefore the surface of the solid may be diminished, without altering its altitude or solidity. (Art. 43, 47.) The same may be proved of each of the other faces which are not squares. The surface is therefore a *minimum*, when *all* the faces are squares, that is, when the solid is a *cube*.

Cor. A cube has a *greater solidity* than any other right parallelopiped of the same surface.

PROPOSITION XII.

93. *A* CUBE *has a greater solidity, than any other right parallelopiped, the sum of whose length, breadth, and depth is equal to the sum of the corresponding dimensions of the cube.*

The solidity is equal to the product of the length, breadth, and depth. If the length and breadth are unequal, the solidity may be increased, without altering the sum of the three dimensions. For the product of two factors whose sum is given, is the greatest when the factors are equal. (Euc. 27. 6.) In the same manner, if the breadth and depth are unequal, the solidity may be increased, without altering the sum of the three dimensions. Therefore, the solid can not be a *maximum,* unless its length, breadth, and depth are equal.

PROPOSITION XIII.

94, *If a* PRISM BE DESCRIBED ABOUT A CYLINDER, *the capacities of the two solids are as their surfaces.*

The capacities of the solids are as the *areas* of their bases, that is, as the *perimeters* of their bases. (Art. 88.) But the lateral surfaces are also as the perimeters of the bases. Therefore the *whole* surfaces are as the solidities.

Cor. The capacities of different prisms, described about the same right cylinder, are to each other as their surfaces.

PROPOSITION XIV.

95. *A right cylinder* WHOSE HEIGHT IS EQUAL TO THE DIAMETER OF ITS BASE *has a greater solidity than any other right cylinder of equal surface.*

Let C be a right cylinder whose height is equal to the diameter of its base; and C′ another right cylinder having the same surface, but a different altitude. If a square prism P be described about the former, it will be a *cube.* But a square prism P′ described about the latter will not be a cube.

Then the surfaces of C and P are as their bases (Arts. 47 and 88.); which are as the bases of C′ and P′ (Sup. Euc. 7, 1.); so that,

$surf C : surf P :: base C : base P :: base C' : base P' :: surf C' : surf P'$

But the surface of C is, by supposition, equal to the surface of C′. Therefore, (Alg. 395.) the surface of P is equal to the surface of P′. And by the preceeding article,

$solid P : solid C :: surf P : surf C :: surf P' : surf C' :: solid P' : solid C'$

But the solidity of P is greater than that of P′. (Art. 92. Cor.) Therefore the solidity of C is greater than that of C′.

Schol. A right cylinder whose height is equal to the diameter of its base, is that which *circumscribes a sphere.* It is also called *Archimedes' cylinder;* as he discovered the ratio of a sphere to its circumscribing cylinder; and these are the figures which were put upon his tomb.

Cor. Archimedes' cylinder has a *less surface,* than any other right cylinder of the same capacity.

PROPOSITION XV.

96. *If a* SPHERE BE CIRCUMSCRIBED *by a solid bounded by plane surfaces; the capacities of the two solids are as their surfaces.*

If planes be supposed to be drawn from the centre of the sphere, to each of the edges of the circumscribing solid, they will divide it into as many pyramids as the solid has faces. The base of each pyramid will be one of the faces; and the height will be the radius of the sphere. The capacity of the pyramid will be equal, therefore, to its base multiplied into $\frac{1}{3}$ of the radius (Art. 48.); and the capacity of the whole circumscribing solid, must be equal to its whole surface multiplied into $\frac{1}{3}$ of the radius. But the capacity of the sphere is also equal to its surface multiplied into $\frac{1}{3}$ of its radius. (Art. 70.)

Cor. The capacities of different solids circumscribing the same sphere, are as their surfaces.

PROPOSITION XVI.

97. *A* SPHERE *has a greater solidity, than any regular polyedron of equal surface.*

If a sphere and a regular polyedron have the same centre, and equal surfaces; each of the faces of the polyedron must fall partly *within* the sphere. For the solidity of a *circumscribing* solid is greater than the solidity of the sphere, as the one includes the other: and therefore, by the preceding article, the *surface* of the former is greater than that of the latter. But if the faces of the polyedron fall partly within the sphere, their perpendicular distance from the centre must be less than the radius. And therefore, if the surface of the polyedron be only equal to that of the sphere, its solidity must be less. For the solidity of the polyedron is equal to its surface multiplied into $\frac{1}{3}$ of the distance from the centre. (Art. 59.) And the solidity of the sphere is equal to its surface multiplied into $\frac{1}{3}$ of the radius.

Cor. A sphere has a *less surface*, than any regular polyedron of the same capacity.

For other cases of Isoperimetry, see Fluxions.

APPENDIX.—PART I.

*Containing rules, without demonstrations, for the mensuration of the Conic Sections, and other figures not treated of in the Elements of Euclid.**

PROBLEM I.

To find the area of an ELLIPSE.

101. Multiply the product of the transverse and conjugate axes into .7854.

Ex. What is the area of an ellipse whose transverse axis is 36 feet, and conjugate 28? Ans. 791.68 feet.

PROBLEM II.

To find the area of a SEGMENT *of an ellipse,* cut off by a line perpendicular to either axis.

102. If either axis of an ellipse be made the diameter of a circle; and if a line perpendicular to this axis cut off a segment from the ellipse, and from the circle;

The diameter of the circle is, to the other axis of the ellipse;
As the circular segment, to the elliptic segment.

* For demonstrations of these rules, see Conic Sections, Spherical Trigonometry, and Fluxions, or Hutton's Mensuration.

Ex. What is the area of a segment cut off from an ellipse whose transverse axis is 415 feet, and conjugate 332; if the height of the segment is 96 feet, and its base is perpendicular to the transverse axis?

The circular segment is 23680 feet.
And the elliptic segment 18944.

PROBLEM III.

To find the area of a conic PARABOLA.

103. Multiply the base by $\frac{2}{3}$ of the height.

Ex. If the base of a parabola is 26 inches, and the height 9 feet; what is the area? Ans. 13 feet.

PROBLEM IV.

To find the area of a FRUSTUM *of a parabola*, cut off by a line parallel to the base.

104. Divide the difference of the cubes of the diameters of the two ends, by the difference of their squares; and multiply the quotient by $\frac{2}{3}$ of the perpendicular height.

Ex. What is the area of a parabolic frustum, whose height is 12 feet, and the diameters of its ends 20 and 12 feet? Ans. 196 feet.

PROBLEM V.

To find the area of a conic HYPERBOLA.

105. Multiply the base by $\frac{2}{3}$ of the height; and correct the product by subtracting from it the series

$$2bh \times \left(\frac{z}{1.3.5}+\frac{z^2}{3.5.7}+\frac{z^3}{5.7.9}+\frac{z^4}{7.9.11}+\&c.\right)$$

In which $\begin{cases} b = \text{the base or double ordinate,} \\ h = \text{the height or abscissa,} \\ z = \text{the height divided by the sum of the height and transverse axis.} \end{cases}$

The series converges so rapidly, that a few of the first terms will generally give the correction with sufficient exactness. This correction is the difference between the hyperbola, and a parabola of the same base and height.

Ex. If the base of a hyperbola be 24 feet, the height 10 and the transverse axis 30; what is the area?

The base $\times \frac{2}{3}$ the height is	160.
The first term of the series is	0.016666
The second	0.000592
The third	0.000049
The fourth	0.000006
Their sum	0.017313
This into $2bh$ is	8.31
And the area corrected is	151.69

PROBLEM VI.

To find the area of a SPHERICAL TRIANGLE formed by three arcs of great circles of a sphere.

106. As 8 right angles or 720°,
To the excess of the 3 given angles above 180°;
So is the whole surface of the sphere,
To the area of the spherical triangle.

Ex. What is the area of a spherical triangle, on a sphere whose diameter is 30 feet, if the angles are 130°, 102°, and 68°? Ans. 471.24 feet.

PROBLEM VII.

To find the area of a SPHERICAL POLYGON formed by arcs of great circles.

107. As 8 right angles, or 720°,
To the excess of all the given angles above the product of the number of angles − 2 into 180°;
So is the whole surface of the sphere,
To the area of the spherical polygon.

Ex. What is the area of a spherical polygon of seven sides, on a sphere whose diameter is 17 inches; if the sum of all the angles is 1080°? Ans. 227 inches.

PROBLEM VIII.

To find the lunar surface included between two great circles of a sphere.

108. As 360°, to the angle made by the given circles;
So is the whole surface of the sphere, to the surface between the circles.

Or,

The lunar surface is equal to the breadth of the middle part of it, multiplied into the diameter of the sphere.

Ex. If the earth be 7930 miles in diameter, what is the surface of that part of it which is included between the 65th and 83d degree of longitude?

Ans. 9,878,000 square miles.

PROBLEM IX.

To find the solidity of a SPHEROID, formed by the revolution of an ellipse about either axis.

109. Multiply the product of the fixed axis and the square of the revolving axis, into .5236.

Ex. 1. What is the solidity of an oblong spheroid, whose longest and shortest diameters are 40 and 30 feet?

Ans. $40 \times \overline{30}^2 \times .5236 = 18850$ feet.

2. If the earth be an oblate spheroid, whose polar and equatorial diameters are 7930 and 7960 miles; what is its solidity? Ans. 263,000,000,000 miles.

PROBLEM X.

To find the solidity of the MIDDLE FRUSTUM *of a spheroid,* included between two planes which are perpendicular to the axis, and equally distant from the centre.

110. Add together the square of the diameter of one end, and twice the square of the middle diameter; multiply the sum by $\frac{1}{3}$ of the height, and the product by .7854.

If D and d = the two diameters, and h = the height;
The solidity $=(2D^2+d^2)\times\frac{1}{3}h\times.7854$.

Ex. If the diameter of one end of a middle frustum of a spheroid be 8 inches, the middle diameter 10, and the height 30, what is the solidity?

Ans. 2073.4 inches.

Cor. *Half* the middle frustum is equal to a frustum of which one of the ends passes through the centre.

If then D and d=the diam'rs of the two ends, and h=the hei't,
The solidity $=(2D^2+d^2)\times\frac{1}{3}h\times.7854$.

PROBLEM XI.

To find the solidity of a PARABOLOID.

111. Multiply the area of the base by half the height.

Ex. If the diameter of the base of a paraboloid be 12 feet, and the height 22 feet, what is the solidity?

Ans. 1243 feet.

PROBLEM XII.

To find the solidity of a FRUSTUM *of a paraboloid.*

112. Multiply the sum of the areas of the two ends by half their distance.

Ex. If the diameter of one end of a frustum of a paraboloid be 8 feet, the diameter of the other end 6 feet, and the length 24 feet; what is the solidity?

Ans. $942\frac{1}{2}$ feet.

Cor. If a cask be in the form of *two equal* frustums of a paraboloid; and

If D=the middle diam. d=the end diam. and h=the length;

$$\text{The solidity}=(D^2+d^2)\times\tfrac{1}{2}h\times.7854.$$

PROBLEM XIII.

To find the solidity of a HYPERBOLOID, produced by the revolution of a hyperbola on its axis.

113. Add together the square of the radius of the base, and the square of the diameter of a section which is equally distant from the base and the vertex; multiply the sum by the height, and the product by .5236.

If R=the radius of the base, D=the middle diameter, and h=the height;

$$\text{The solidity}=(R^2+D^2)\times h\times.5236.$$

Ex. If the diameter of the base of a hyperboloid be 24, the square of the middle diameter 252, and the height 10, what is the solidity? Ans. 2073.4.

PROBLEM XIV.

Ta find the solidity of a FRUSTUM *of a hyperboloid.*

114. Add together the squares of the radii of the two ends, and the square of the middle diameter; multiply the sum by the height, and the product by .5236.

If R and r=the two radii, D=the middle diameter, and h=the height;

$$\text{The solidity}=(R^2+r^2+D^2)\times h\times.5236.$$

Ex. If the diameter of one end of a frustum of a hyperboloid be 32, the diameter of the other end 24, the square of the middle diameter $793\frac{3}{5}$, and the length 20, what is the solidity? Ans. 12409.3.

PROBLEM XV.

To find the solidity of a CIRCULAR SPINDLE, produced by the revolution of a circular segment about its base or chord as an axis.

115. From $\frac{1}{3}$ of the cube of half the axis, subtract the product of the central distance into half the revolving circular segment, and multiply the remainder by four times 3.14159.

If a=the area of the revolving circular segment,
l=half the length or axis of the spindle,
c=the distance of this axis from the centre of the circle to which the revolving segment belongs;
The solidity$=(\frac{1}{3}l^3-\frac{1}{2}ac)\times4\times3.14159$.

Ex. Let a circular spindle be produced by the revolution of the segment ABO (Fig. 9.) about AB. If the axis AB be 140, and OP half the middle diameter of the spindle be 38.4; what is the solidity?

The area of the revolving segment is	3791
The central distance PC	44.6
The solidity of the spindle	374402

PROBLEM XVI.

To find the solidity of the MIDDLE FRUSTUM *of a circular spindle.*

116. From the square of half the axis of the whole spindle, subtract $\frac{1}{3}$ of the square of half the length of the frustum; multiply the remainder by this half length; from the product subtract the product of the revolving area into the central distance; and multiply the remainder by twice 3.14159.

If L=half the length or axis of the whole spindle,
l=half the length of the middle frustum,
c=the distance of the axis from the centre of the circle,
a=the area of the figure which, by revolving, produces the frustum;
The solidity$=(\overline{L^2-\frac{1}{3}l^2}\times l-ac)\times2\times3.14159$.

Ex. If the diameter of each end of a frustum of a circular spindle be 21.6, the middle diameter 60, and the length 70; what is the solidity?

The length of the whole spindle is	79.75
The central distance	11.5
The revolving area	1703.8
The solidity	136751.5

PROBLEM XVII.

To find the solidity of a PARABOLIC SPINDLE, produced by the revolution of a parabola about a double ordinate or base.

117. Multiply the square of the middle diameter by $\frac{3}{15}$ of the axis, and the product by .7854.

Ex. If the axis of a parabolic spindle be 30, and the middle diameter 17, what is the solidity?

Ans. 3631.7.

PROBLEM XVIII.

To find the solidity of the MIDDLE FRUSTUM *of a parabolic spindle.*

118. Add together the square of the end diameter, and twice the square of the middle diameter, from the sum subtract $\frac{2}{5}$ of the square of the difference of the diameters; and multiply the remainder by $\frac{1}{3}$ of the length, and the product by .7854.

If D and d = the two diameters, and l = the length;
The solidity $=(2D^2+d^2-\frac{2}{5}(D-d)^2)\times\frac{1}{3}l\times.7854.$

Ex. If the end diameters of a frustum of a parabolic spindle be each 12 inches, the middle diameter 16, and the length 30; what is the solidity? Ans. 5102 inches.

APPENDIX.—PART II.

GAUGING OF CASKS.

Art. 119. GAUGING is a practical art, which does not admit of being treated in a very scientific manner. Casks are not commonly constructed in exact conformity with any regular mathematical figure. By most writers on the subject, however, they are considered as nearly coinciding with one of the following forms;

1. } The middle frustum { of a spheroid,
2. } The middle frustum { of a parabolic spindle.
3. } Two equal frustums { of a paraboloid,
4. } Two equal frustums { of a cone.

The *second* of these varieties agrees more nearly, than any of the others, with the forms of casks, as they are commonly made. The first is too much curved, the third too little, and the fourth not at all, from the head to the hung.

120. Rules have already been given, for finding the capacity of each of the four varieties of casks. (Arts. 68, 110, 112, 118.) As the dimensions are taken in *inches*, these rules will give the contents in cubic inches. To abridge the computation, and adapt it to the particular measures used in gauging, the factor .7854 is divided by 282 or 231; and the quotient is used instead of .7854, for finding the capacity in ale gallons or wine gallons.

$$\text{Now } \frac{.7854}{282} = .002785, \text{ or } .0028 \text{ nearly;}$$

$$\text{And } \frac{.7854}{231} = .0034.$$

If then .0028 and .0034 be substituted for .7854, in the rules referred to above; the contents of the cask will be given in ale gallons and wine gallons. These numbers are to each other nearly as 9 to 11.

PROBLEM I.

To calculate the contents of a cask, in the form of the middle frustum of a SPHEROID.

121. Add together the square of the head diameter, and twice the square of the bung diameter; multiply the sum by $\frac{1}{3}$ of the length, and the product by .0028 for ale gallons, or by .0034 for wine gallons.

If D and d=the two diameters, and l=the length;
The capacity in inches$=(2D^2+d^2)\times\frac{1}{3}l\times.7854$. (Art. 110.)

And by substituting .0028 or .0034 for .7854, we have the capacity in ale gallons or wine gallons.

Ex. What is the capacity of a cask of the first form, whose length is 30 inches, its head diameter 18, and its bung diameter 24?

Ans. 41.3 ale gallons, or 52.2 wine gallons.

PROBLEM II.

To calculate the contents of a cask, in the form of the middle frustum of a PARABOLIC SPINDLE.

122. Add together the square of the head diameter, and twice the square of the bung diameter, and from the sum subtract $\frac{2}{5}$ of the square of the difference of the diameters; multiply the remainder by $\frac{1}{3}$ of the length, and the product by .0028 for ale gallons, or .0034 for wine gallons.

The capacity in inches $=(2D^2+d^2-\frac{2}{5}(D-d)^2)\times\frac{1}{3}l\times .7854$. (Art. 118.)

Ex. What is the capacity of a cask of the second form, whose length is 30 inches, its head diameter 18, and its bung diameter 24?

Ans. 40.9 ale gallons, or 49.7 wine gallons.

PROBLEM III.

To calculate the contents of a cask, in the form of two equal frustums of a PARABOLOID.

123. Add together the square of the head diameter, and the square of the bung diameter; multiply the sum by half the length, and the product by .0028 for ale gallons, or .0034 for wine gallons.

The capacity in inches $=(D^2+d^2)\times\frac{1}{2}l\times.7854$. (Art. 112. Cor.)

Ex. What is the capacity of a cask of the third form, whose dimensions are, as before, 30, 18, 24?

Ans. 37.8 ale gallons, or 45.9 wine gallons.

PROBLEM IV.

To calculate the contents of a cask, in the form of two equal frustums of a CONE.

124. Add together the square of the head diameter, the square of the bung diameter. and the product of the two diameters; multiply the sum by $\frac{1}{3}$ of the length, and the product by .0028 for ale gallons, or 0034 for wine gallons.

The capacity in inches $=(D^2+d^2+Dd)\times\frac{1}{3}l\times.7854$ (Art. 68.)

Ex. What is the capacity of a cask of the fourth form, whose length is 30, and its diameters 18 and 24?

Ans. 37.3 ale gallons, or 45.3 wine gallons.

125. The preceding rules, though correct in theory, are not very well adapted to practice, as they suppose the form of the cask to be *known*. The two following rules, taken from Hutton's Mensuration, may be used for casks of the usual forms. For the first, *three* dimensions are required; the length, the head diameter, and the bung diameter. It is evident that no allowance is made by this, for different degrees of curvature from the head to the bung. If the cask is more or less curved than usual, the following rule is to be preferred, for which *four* dimensions are required, the head and

bung diameters, and a third diameter taken in the middle between the bung and the head. For the demonstration of these rules, see Hutton's Mensuration, Part v. Sec. 2. Ch. 5. and 7.

PROBLEM V.

To calculate the contents of any common cask from THREE *dimensions.*

126. Add together
25 times the square of the head diameter,
39 times the square of the bung diameter, and
26 times the product of the two diameters;
Multiply the sum by the length, divide the product by 90, and multiply the quotient by .0028 for ale gallons, or .0034 for wine gallons.

$$\text{The capacity in inches} = (39D^2 + 25d^2 + 26Dd) \times \frac{l}{90} \times .7854.$$

Ex. What is the capacity of a cask whose length is 30 inches, the head diameter 18, and the bung diameter 24?
Ans. 39 ale gallons, or $47\frac{1}{2}$ wine gallons.

PROBLEM VI.

To calculate the contents of a cask from FOUR *dimensions*, the length, the head and bung diameters, and a diameter taken in the middle between the head and the bung.

127. Add together the squares of the head diameter, of the bung diameter, and of double the middle diameter; multiply the sum by $\frac{1}{6}$ of the length, and the product by .0028 for ale gallons, or .0034 for wine gallons.

If D=the bung diameter, d=the head diameter, m=the middle diameter, and l=the length;

$$\text{The capacity in inches} = (D^2 + d^2 + \overline{2m}^2) \times \tfrac{1}{6}l \times .7854.$$

Ex. What is the capacity of a cask, whose length is 30 inches, the head diameter 18, the bung diameter 24, and the middle diameter $22\frac{1}{2}$?
Ans. 41 ale gallons, or $49\frac{2}{3}$ wine gallons.

128. In making the calculations in gauging, according to the preceding rules, the multiplications and divisions are frequently performed by means of a *Sliding Rule*, on which are placed a number of logarithmic lines, similar to those on Gunter's Scale. See Trigonom. Sec. vi. and Note I. p. 122.

Another instrument commonly used in gauging is the *Diagonal Rod*. By this, the capacity of a cask is very expeditiously found, from a single dimension, the distance from the bung to the intersection of the opposite stave with the head. The measure is taken by extending the rod through the cask, from the bung to the most distant part of the head. The number of gallons corresponding to the length of the line thus found, is marked on the rod. The *logarithmic* lines on the gauging rod are to be used in the same manner, as on the sliding rule.

ULLAGE OF CASKS.

129. When a cask is *partly* filled, the whole capacity is divided, by the surface of the liquor into two portions; the *least* of which, whether full or empty, is called the *ullage*. In finding the ullage, the cask is supposed to be in one of two positions; either *standing*, with its axis perpendicular to the horizon; or *lying*, with its axis parallel to the horizon. The rules for ullage which are *exact*, particularly those for lying casks, are too complicated for common use. The following are considered as sufficiently near approximations. See Hutton's Mensuration.

PROBLEM VII.

To calculate the ullage of a STANDING *cask.*

130. Add together the squares of the diameter at the surface of the liquor, of the diameter of the nearest end, and of double the diameter in the middle between the other two; multiply the sum by $\frac{1}{6}$ of the distance between the surface and the nearest end, and the product by .0028 for ale gallons, or .0034 for wine gallons.

If D=the diameter of the surface of the liquor,
d=the diameter of the nearest end,
m=the middle diameter, and
l= the distance between the surface and the nearest end;
The ullage in inches$=(D^2+d^2+\overline{2m}^2)\times\frac{1}{6}l\times.7854$.

Ex. If the diameter at the surface of the liquor, in a standing cask, be 32 inches, the diameter of the nearest end 24, the middle diameter 29, and the distance between the surface of the liquor and the nearest end 12; what is the ullage?

Ans. $27\frac{4}{5}$ ale gallons, or $33\frac{3}{4}$ wine gallons.

PROBLEM VIII.

To calculate the ullage of a LYING *cask.*

Divide the distance from the bung to the surface of the liquor, by the whole bung diameter, find the quotient in the column of heights or versed sines in a table of circular segments, take out the corresponding segment, and multiply it by the whole capacity of the cask, and the product by $1\frac{1}{4}$ for the part which is empty.

If the cask be not half full, divide the depth of the liquor by the whole bung diameter, take out the segment, multiply, &c. for the contents of the part which is full.

Ex. If the whole capacity of a lying cask be 41 ale gallons, or $49\frac{2}{3}$ wine gallons, the bung diameter 24 inches and the distance from the bung to the surface of the liquor 6 inches; what is the ullage?

Ans. $7\frac{3}{4}$ ale gallons, or $9\frac{1}{2}$ wine gallons.

NOTES.

Note A. p. 16.

One of the earliest approximations to the ratio of the circumference of a circle to its diameter, was that of *Archimedes.* He demonstrated that the ratio of the perimeter of a regular inscribed polygon of 96 sides, to the diameter of the circle, is greater than $3\frac{10}{71} : 1$; and that the ratio of the perimeter of a circumscribed polygon of 192 sides, to the diameter, is less than $3\frac{10}{70} : 1$, that is, than $22 : 7$.

Metius gave the ratio of 355 : 115, which is more accurate than any other expressed in small numbers. This was confirmed by *Vieta*, who by inscribed and circumscribed polygons of 393216 sides, carried the approximation to ten places of figures, viz.

3.141592653.

Van Ceulen of Leyden afterwards extended it, by the laborious process of repeated bisections of an arc, to 36 places. This calculation was deemed of so much consequence at the time, that the numbers are said to have been put upon his tomb.

But since the invention of *fluxions*, methods much more expeditious have been devised, for approximating to the required ratio. These principally consist in finding the sum of a series, in which the length of an arc is expressed in terms of its *tangent.*

If t=the tangent of an arc, the radius being 1,

The arc $= t - \frac{t^3}{3} + \frac{t^5}{5} - \frac{t^7}{7} + \frac{t^9}{9} -$ &c. See Fluxions.

This series is in itself very simple. Nothing more is necessary to make it answer the purpose in practice, than that the arc be *small*, so as to render the series sufficiently converging, and that the tangent be expressed in such simple numbers, as can easily be raised in the several powers. The given series will be expressed in the most simple numbers, when the arc is 45°, whose tangent is equal to radius. If the radius be 1,

The arc of $45° = 1 - \frac{1}{3} + \frac{1}{5} - \frac{1}{7} + \frac{1}{9} -$ &c. And this multiplied by 8 gives the length of the whole circumference.

But a series in which the tangent is smaller, though it be less simple than this, is to be preferred, for the rapidity with which it converges. As the tangent of $30° = \sqrt{\frac{1}{3}}$, if the radius be 1,

The arc of $30° = \sqrt{\frac{1}{3}} \times (1 - \frac{1}{3.3} + \frac{1}{5.3^2} - \frac{1}{7.3^3} + \frac{1}{9.3^4} -$ &c.

And this multiplied into 12 will give the whole circumference.
This was the series used by Dr. Halley. By this also, Mr *Abraham Sharp* of Yorkshire computed the circumference to 72 places of figures, Mr. *John Machin*, Professor of Astronomy in Gresham college, to 100 places, and M. De Lagny to 128 places. Several expedients have been devised, by Machin, Euler, Dr. Hutton, and others, to reduce the labour of summing the terms of the series. See Euler's Analysis of Infinites, Hutton's Mensuration, Appendix to Maseres on the Negative Sign, and Lond. Phil. Trans. for 1776. For a demonstration that the diameter and the circumference of a circle are incommensurable, see Legendre's Geomety, Note. IV.

The circumference of a circle whose diameter is 1, is
3.1415926535, 8979323846, 2643383279,
5028841971, 6939937510, 5820974944,
5923078164, 0628620899, 8628034825,
3421170679, 8214808651, 3272306647,
0938446 + or 7 –.

Note B. p. 17.

The following multipliers may frequently be useful;

The diam'r of a circle $\times .8862$ = the side of an equal square.
$\times .707$ = the side of an ins'bed sq're.
$\times .866$ = the side of an inscribed [equilateral triangle.

The circumf. $\times .2821$ = the side of an equal square.
$\times .2251$ = the side of an inscribed square.
$\times .2756$ = the side of an ins'bed eq'lat. trian.

The side of a sq. $\times 1.128$ = the diam. of an equ'l circle.
$\times 3.545$ = the circ. of an equal circle.
$\times 1.414$ = the dia. of the circumsc. circle.
$\times 4.443$ = the cir. of the circumsc. circle.

Note C. p. 19.

The following approximating rules may be used for finding the arc of a circle.

1. The arc of a circle is nearly equal to $\frac{1}{3}$ of the difference between the chord of the whole arc, and 8 times the chord of half the arc.

2. If h = the *height* of an arc, and d = the diameter of the circle;

$$\text{The arc} = 2d\sqrt{\frac{3h}{3d-h}} \qquad \text{Or,}$$

3. $$\text{The arc} = 2\sqrt{dh} \times \left(1 + \frac{h}{2.3d} + \frac{3h^2}{2.4.5d^2} + \frac{3\ 5h^3}{2.4.6.7d^3} \text{ \&c.}\right) \text{Or,}$$

4. $$\text{The arc} = (5d\tfrac{1}{5}\sqrt{\frac{5h}{5d-3h}} + 4\sqrt{dh}) \text{ very nearly.}$$

5. If s = the *sine* of an arc, and r = the radius of the circle;

$$\text{The arc} = s \times \left(1 + \frac{s^2}{2.3r^2} + \frac{3s^4}{5.2.4r^4} + \frac{3.5s^6}{7.2.4.6r^6} \text{ \&c.}\right)$$

See Hutton's Mensuration.

Note D. p. 23.

To expedite the calculation of the areas of circular segments, a *table* is provided, which contains the areas of segments in a circle whose diameter is one. See the table at the end of the book, in which the diameter is supposed to be divided into 1000 equal parts. By this may be found the areas of segments of other circles. For the heights of similar segments of different circles are as the diameters. If then the height of any given segment be divided by the diameter of the circle, the quotient will be the height of a similar segment in a circle whose diameter is 1. The area of the latter is found in the table; and from the properties of similar figures, the two segments are to each other, as the squares of the diameters of the circles. We have then the following rule:

To find the area of a circular SEGMENT *by the* TABLE.

Divide the height of the segment by the diameter of the circle; look for the quotient in the column of heights in the table; take out the corresponding number in the column of areas; and multiply it by the square of the diameter.

It is to be observed, that the figures in each of the columns in the table are *decimals*.

If accuracy is required, and the quotient of the height divided by the diameter, is *between* two numbers in the column of heights: allowance may be made for a *proportional part* of the difference of the corresponding numbers in the column of areas; in the same manner, as in taking out logarithms.

Segments *greater than a semicircle* are not contained in the table. If the area of such a segment is required, as ABD (Fig. 9.), find the area of the segment ABO, and subtract his from the area of the whole circle,

Or,

Divide the height of the given segment by the diameter, subtract the quotient from 1, find the remainder in the column of heights, subtract the corresponding area from .7854 and multiply this remainder by the square of the diameter.

Ex. 1. What is the area of a segment whose height is 16, the diameter of the circle being 48? Ans. 528.

2. What is the area of a segment whose height is 32, the diameter being 48? Ans. 1281.55.

The following rules may also be used for a circular segment.

1. To the chord of the whole arc, add $\frac{4}{3}$ of the chord of half the arc, and multiply the sum by $\frac{2}{5}$ of the height.

If C and c=the two chords, and h=the height;

The segment $=(C+\frac{4}{3}c)\frac{2}{5}h$ nearly.

2. If h= the height of the segment, and d=the diameter of the circle;

$$\text{The segment}=2h\sqrt{dh}\times\left(\tfrac{2}{3}-\frac{h}{5d}-\frac{h^2}{28d^2}-\frac{h^3}{72d^3}\ \&c.\right)$$

NOTE E. p. 29.

The term *solidity* is used here in the customary sense, to express the magnitude of any geometrical quantity of three dimensions, length, breadth, and thickness; whether it be a solid body, or a fluid, or even a portion of empty space. This use of the word, however, is not altogether free from objection. The same term is applied to one of the general properties of matter; and also to that peculiar quality by which certain substances are distinguished from *fluids*. There seems to be an impropriety in speaking of the *solidity* of a body of *water*, or of a vessel which is *empty*. Some writers have therefore substituted the word *volume* for solidity. But the latter term, if it be properly defined, may be retained without danger of leading to mistake.

NOTE F. p. 35.

The *geometrical* demonstration of the rule for finding the solidity of a frustum of a pyramid, depends on the following proposition:

A frustum of a triangular pyramid is equal to three pyramids ; the greatest and least of which are equal in height to the frustum, and have the two ends of the frustum for their bases ; and the third is a mean proportional between the other two.

Let ABCDFG (Fig. 34.) be a frustum of a triangular pyramid. If a plane be supposed to pass through the points AFC, it will cut off the pyramid ABCF. The height of this is evidently equal to the height of the frustum, and its base is ACB, the greater end of the frustum.

Let another plane pass through the points AFD. This will divide the remaining part of the figure into two triangular pyramids AFDG and AFDC. The height of the former is equal to the height of the frustum, and its base is DFG, the smaller end of the frustum.

To find the magnitude of the third pyramid AFCD, let F be now considered as the vertex of this, and of the second pyramid AFDG. Their bases will then be the triangles ADC and ADG. As these are in the same plane, the two pyramids have the same altitude, and are to each other as their bases. But these triangular bases, being between the same parallels, are as the lines AC and DG. Therefore the pyramid AFDC is to the pyramid AFDG as AC to DG ; and $\overline{AFCD}^2 : \overline{AFDG}^2 :: \overline{AC}^2 : \overline{DG}^2$. (Alg. 391.) But the pyramids ABCF and AFDG, having the same altitude, are as their bases ABC and DFG, that is, as $\overline{AC}^2$ and $\overline{DG}^2$. (Euc. 19, 6.) We have then

$$\left.\begin{array}{l}\overline{AFDC}^2 : \overline{AFDG}^2 :: \overline{AC}^2 : \overline{DG}^2 \\ ABCF : AFDG :: \overline{AC}^2 : \overline{DG}^2\end{array}\right\}$$

$$\text{Therefore } \overline{AFDC}^2 : \overline{AFDG}^2 :: ABCF : AFDG.$$

$$\text{And } \overline{AFDC}^2 = AFDG \times ABCF.$$

That is, the pyramid AFDC is a mean proportional between AFDG and ABCF.

Hence, the solidity of a frustum of a triangular pyramid is equal to $\frac{1}{3}$ of the height, multiplied into the sum of the areas of the two ends and the square root of the product of these areas. This is true also of a frustum of any other pyramid. (Sup. Euc. 12, 3. Cor. 2.)

If the smaller end of a frustum of a pyramid be enlarged, till it is made equal to the other end ; the frustum will become a *prism*, which may be divided into three *equal* pyramids. (Sup. Euc. 15, 3.)

NOTE G. p. 59.

The following simple rule for the solidity of round timber, or of any cylinder, is nearly exact :

Multiply the length into twice the square of ⅕ of the circumference.

If C=the circumference of a cylinder;

The area of the base $=\frac{C^2}{4\pi}=\frac{C^2}{12.566}$ But $2\left(\frac{C}{5}\right)^2=\frac{C^2}{12.2}$

It is common to measure *hewn* timber, by multiplying the length into the square of the *quarter-girt*. This gives exactly the solidity of a parallelopiped, if the ends are *squares*. But if the ends are parallelograms, the area of each is *less* than the square of the quarter-girt. (Euc. 27, 6.)

Timber which is *tapering* may be exactly measured by the rule for the frustum of a pyramid or cone (Art. 50, 68.); or, if the ends are not similar figures, by the rule for a prismoid. (Art. 55.) But for common purposes, it will be sufficient to multiply the length by the area of a section *in the middle* between the two ends.

A TABLE

OF THE SEGMENTS OF A CIRCLE, WHOSE DIAMETER IS 1, AND IS SUPPOSED TO BE DIVIDED INTO 1000 EQUAL PARTS.

Height	Area Seg.	Height.	Area Seg.	Height	Area Seg.
.001	.000042	.034	.008273	.067	.022652
002	000119	035	008638	068	023154
003	000219	036	009008	069	023659
004	000337	037	009383	070	024168
005	000471	038	009763	071	024680
006	000618	039	010148	072	025195
007	000779	040	010537	073	025714
008	000952	041	010932	074	026236
009	001135	042	011331	075	026761
010	001329	043	011734	076	027289
011	001533	044	012142	077	027821
012	001746	045	012554	078	028356
013	001968	046	012971	079	028894
014	002199	047	013392	080	029435
015	002438	048	013818	081	029979
016	002685	049	014247	082	030526
017	002940	050	014681	083	031076
018	003202	051	015119	084	031629
019	003472	052	015561	085	032186
020	003748	053	016007	086	032745
021	004032	054	016457	087	033307
022	004322	055	016911	088	033872
023	004618	056	017369	089	034441
024	004921	057	017831	090	035011
025	005231	058	018296	091	035585
026	005546	059	018766	092	036162
027	005867	060	019239	093	036741
028	006194	061	019716	094	037323
029	006527	062	020206	095	037909
030	006865	063	020690	096	038496
031	007209	064	021178	097	039087
032	007558	065	021659	098	039680
.033	.007913	.066	.022154	.099	.040276

Height	Area Seg.	Height.	Area Seg.	Height.	Area Seg.
.100	.040875	.144	.069625	.188	.102334
101	041476	145	070328	189	103116
102	042080	146	071033	190	103900
103	042687	147	071741	191	104685
104	043296	148	072450	192	105472
105	043908	149	073161	193	106261
106	044522	150	073874	194	107051
107	045139	151	074589	195	107842
108	045759	152	075306	196	108636
109	046381	153	076026	197	109430
110	047005	154	076747	198	110226
111	047632	155	077469	199	111024
112	048262	156	078194	200	111823
113	048894	157	078921	201	112624
114	049528	158	079649	202	113426
115	050165	159	080380	203	114230
116	050804	160	081112	204	115035
117	051446	161	081846	205	115842
118	052090	162	082582	206	116650
119	052736	163	083320	207	117460
120	053385	164	084059	208	118271
121	054036	165	084801	209	119083
122	054689	166	085544	210	119897
123	055345	167	086289	211	120712
124	056003	168	087036	212	121529
125	056663	169	087785	213	122347
126	057326	170	088535	214	123167
127	057991	171	089287	215	123988
128	058658	172	090041	216	124810
129	059327	173	090797	217	125634
130	059999	174	091554	218	126459
.131	060672	175	092313	219	127285
132	061348	176	093074	220	128113
133	062026	177	093836	221	128942
134	062707	178	094601	222	129773
135	063389	179	095366	223	130605
136	064074	180	096134	224	131438
137	064760	181	096903	225	132272
138	065449	182	097674	226	133108
139	066140	183	098447	227	133945
140	066833	184	099221	228	134784
141	067528	185	099997	229	135624
142	068225	186	100774	230	136465
.143	.068924	.187	.101553	.231	.137307

Height	Area Seg.	Height.	Area Seg	Height.	Aaea. Seg.
.232	.138150	.277	.177330	.322	.218533
233	138995	278	178225	323	219468
234	139841	279	179122	324	220404
235	140688	280	180019	325	221340
236	141537	281	180918	326	222277
237	142387	282	181817	327	223215
238	143238	283	182718	328	224154
239	144091	284	183619	329	225093
240	144944	285	184521	330	226033
241	145799	286	185425	331	226974
242	146655	287	186329	332	227915
243	147512	288	187234	333	228858
244	148371	289	188140	334	229801
245	149230	290	189047	335	230745
246	150091	291	189955	336	231689
247	150953	292	190864	337	232634
248	151816	293	191775	338	233580
249	152680	294	192684	339	234526
250	153546	295	193596	340	235473
251	154412	296	194509	341	236421
252	155280	297	195422	342	237369
253	156149	298	196337	343	238318
254	157019	299	197252	344	239268
255	157890	300	198168	345	240218
256	158762	301	199085	346	241169
257	159636	302	200003	347	242121
258	160510	303	200922	348	243074
259	161386	304	201841	349	244026
260	162263	305	202761	350	244980
261	163140	306	203683	351	245934
262	164019	307	204605	352	246889
263	164899	308	205527	353	247845
264	165780	309	206451	354	248801
265	166663	310	207376	355	249757
266	167546	311	208301	356	250715
267	168430	312	209227	357	251673
268	169315	313	210154	358	252631
269	170202	314	211082	359	253590
270	171089	315	212011	360	254550
271	171978	316	212940	361	255510
272	172867	317	213871	362	256471
273	173758	318	214802	363	257433
274	174649	319	215733	364	258395
275	175542	320	216666	365	259357
.276	.176435	.321	.217599	.366	.260320

Height.	Area Seg.	Height.	Area Seg	Height.	Area Seg.
.367	.261284	.412	.305155	.457	.349752
368	262248	413	306140	458	350748
369	263213	414	307125	459	351745
370	264178	415	308110	460	352742
371	265144	416	309095	461	353739
372	266111	417	310081	462	354736
373	267078	418	311068	463	355732
374	268045	419	312054	464	356730
375	269013	420	313041	465	357727
376	269982	421	314029	466	358725
377	270951	422	315016	467	359723
378	271920	423	316004	468	360721
379	272890	424	316992	469	361719
380	273861	425	317981	470	362717
381	274832	426	318970	471	363715
282	275803	427	319959	472	364713
383	276775	428	320948	473	365712
384	277748	429	321938	474	366710
385	278721	430	322928	475	367709
386	279694	431	323918	476	368708
387	280668	432	324909	477	369707
388	281642	433	325900	478	370706
389	282617	434	326892	479	371705
390	283592	435	327882	480	372704
391	284568	436	328874	481	373703
392	285544	437	329866	482	374702
393	286521	438	330858	483	375702
394	287498	439	331850	484	376702
395	288476	440	332843	485	377701
396	289454	441	333836	486	378701
397	290432	442	334829	487	379700
398	291411	443	335822	488	380700
399	292390	444	336816	489	381699
400	293369	445	337810	490	382699
401	294349	446	338804	491	383699
402	295330	447	339798	492	384699
403	296311	448	340793	493	385699
404	297292	449	341787	494	386699
405	298273	450	342782	495	387699
406	299255	451	343777	496	388699
407	300238	452	344772	497	389699
408	301220	453	345768	498	390699
409	302203	454	346764	499	391699
410	303187	455	347759	.500	.392699
.411	.304171	.456	.348755		

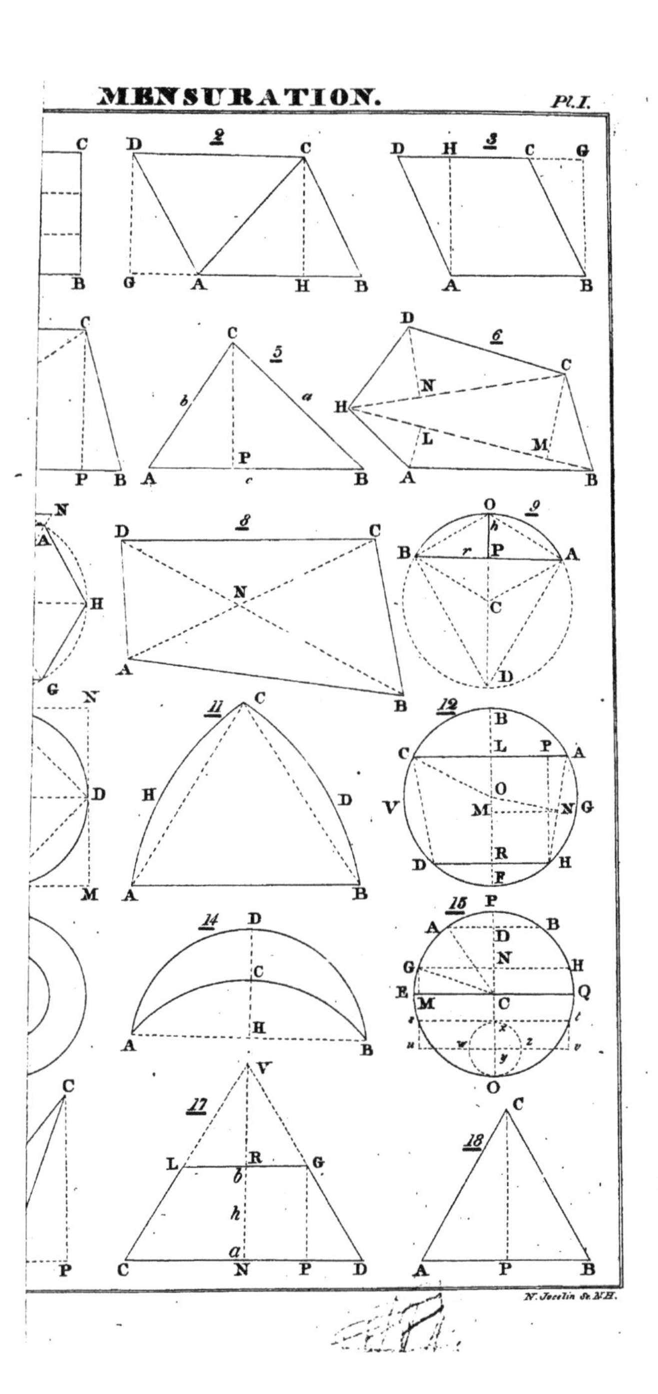

2
3
5
6
8
9
11
12
14
15
17
18
N. Jocelin Sc. N.H.

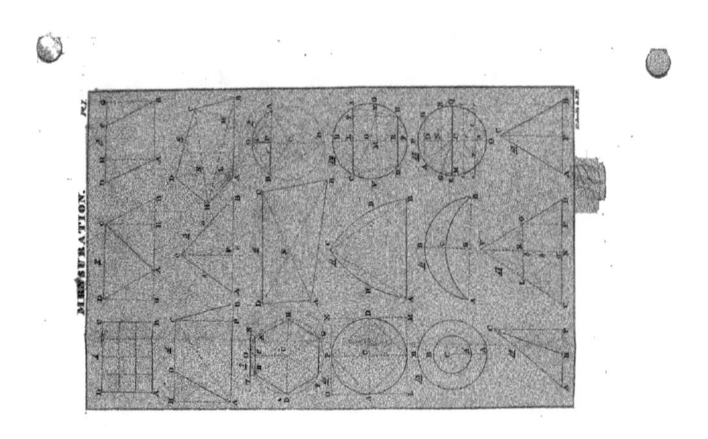
MENSURATION.

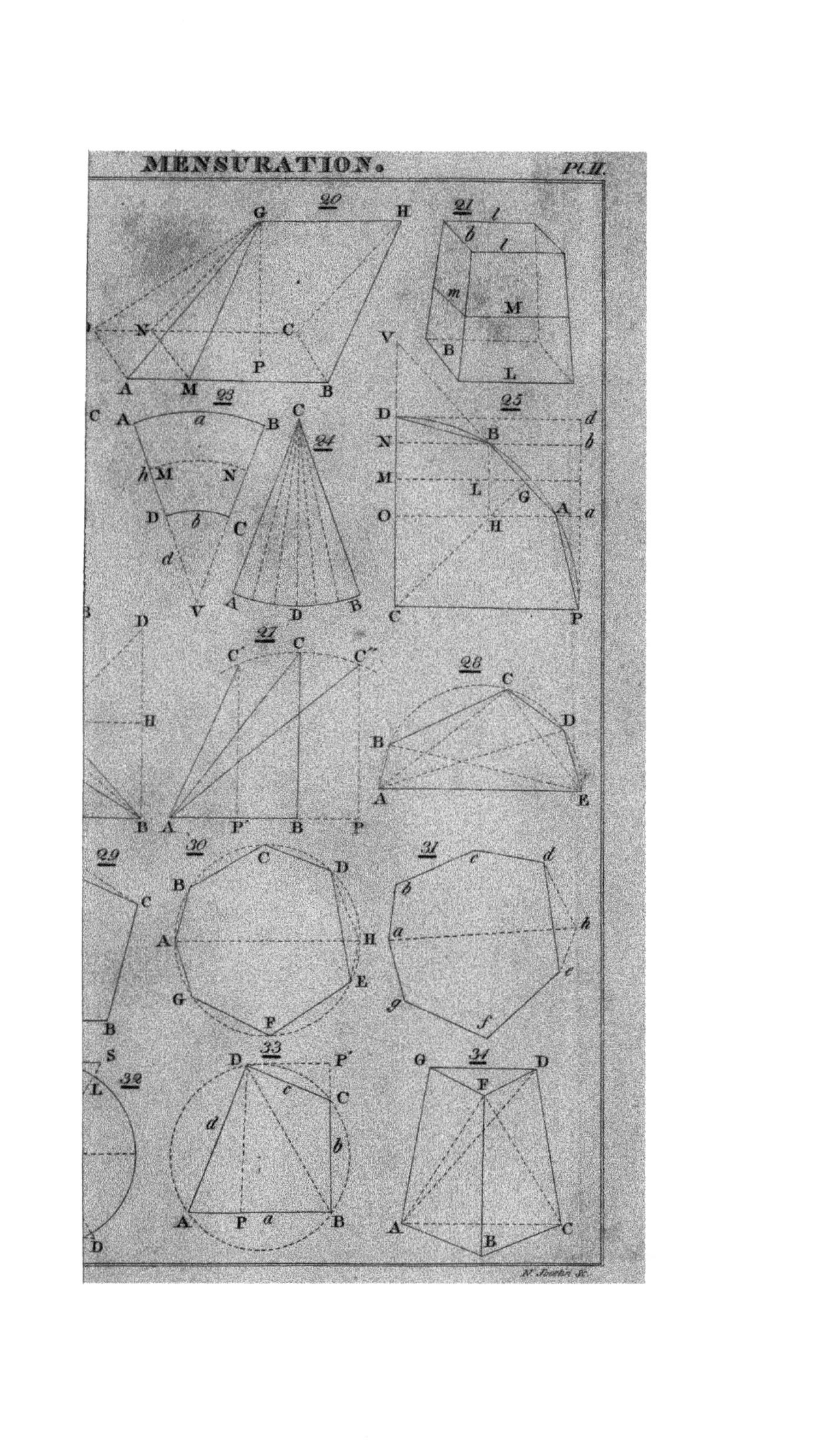

N. Jocelyn Sc.

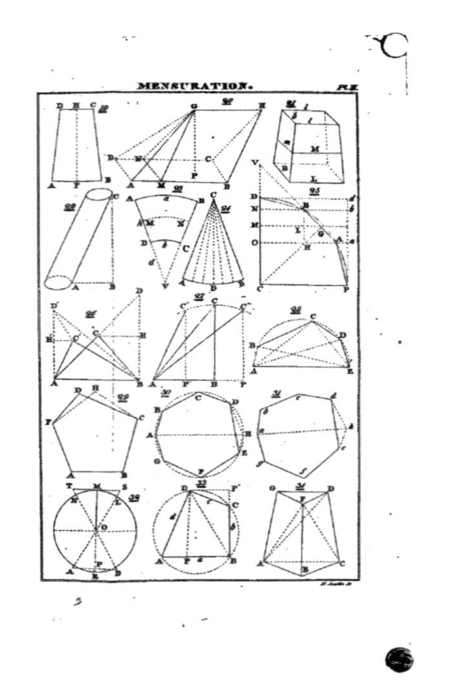
MENSURATION.

THE

MATHEMATICAL PRINCIPLES

OF

VIGATION AND SURVEYING,

WITH THE

MENSURATION

OF

HEIGHTS AND DISTANCES.

BEING

THE FOURTH PART

OF

COURSE OF MATHEMATICS.

ADAPTED TO THE METHOD OF INSTRUCTION IN

THE AMERICAN COLLEGES.

BY JEREMIAH DAY. D.D. LL.D.
President of Yale College.

THE SECOND EDITION.

NEW-HAVEN:

PRINTED AND PUBLISHED BY S. CONVERSE.

1824.

AS the following treatise has been prepared for the use of a class in College, it does not contain all the details which would be requisite for a practical navigator or surveyor. The object of a scientific education is rather to teach *principles*, than the minute rules which are called for in professional practice. The principles should indeed be accompanied with such illustrations and examples as will render it easy for the student to make the applications for himself whenever occasion shall require. But a collection of rules merely, would be learned only to be forgotten, except by a few who might have use for them in the course of their business. There are many things belonging to the art of navigation, which are not comprehended in the mathematical part of the subject. Seamen will of course make use of the valuable system of Mackay, or the still more complete work of Bowditch.

The student is supposed to be familiar with the principles of Geometry and Trigonometry, before he enters upon the present number, which contains little more than the application of those principles to some of the most simple problems in heights and distances, navigation, and surveying.

CONTENTS.

HEIGHTS AND DISTANCES.

ART. 1. THE most direct and obvious method of determining the distance or height of any object, is to apply to it some known measure of length, as a foot, a yard, or a rod. In this manner, the height of a room is found, by a joiner's rule; or the side of a field by a surveyors chain. But in many instances, the object, or a part, at least of the line which is to be measured is *inaccessible*. We may wish to determine the breadth of a river, the height of a cloud, or the distances of the heavenly bodies. In such cases it is necessary to measure some *other* line; from which the required line may be obtained, by geometrical construction, or more exactly, by trigonometrical calculation. The line first measured is frequently called a *base* line.

2. In measuring *angles*, some instrument is used which contains a portion of a graduated circle divided into degrees and minutes. For the proper measure of an angle is an arc of a circle, whose centre is the angular point. (Trig. 74.) The instruments used for this purpose are made in different forms, and with various appendages. The essential parts are a graduated circle, and an index with sight-holes, for taking the directions of the lines which include the angles.

3. Angles of *elevation*, and of *depression*. are in a plane perpendicular to the horizon, which is called a *vertical plane*. An angle of *elevation* is contained between a parallel to the horizon, and an ascending line, as BAC. (Fig. 2.) An angle of *depression* is contained between a parallel to the horizon, and a descending line, as DCA. The *complement* of this is the angle ACB.

4. The instrument by which angles of elevation, and of depression, are commonly measured, is called a *Quadrant*. In its most simple form, it is a portion of a circular board

ABC, (Fig. 1.) on which is a graduated arc of 90 degrees, AB, a plumb line CP, suspended from the central point C, and two sight-holes D and E, for taking the direction of the object.

To measure an angle of *elevation* with this, hold the plane of the instrument perpendicular to the horizon, bring the centre C to the angular point, and direct the edge AC in such a manner, that the object G may be seen through the two sight-holes. Then the arc BO measures the angle BCO, which is equal to the angle of elevation FCG. For as the plumb-line is perpendicular to the horizon, the angle FCO is a right angle, and therefore equal to BCG. Taking from these the common angle BCF, there will remain the angle BCO=FCG.

In taking an angle of *depression*, as HCL (Fig. 1.) the eye is placed at C, so as to view the object at L, through the sight-holes D and E.

5. In treating of the mensuration of heights and distances, no *new principles* are to be brought into view. We have only to make an application of the rules for the solution of triangles, to the particular circumstances in which the observer may be placed, with respect to the line to be measured. These are so numerous, that the subject may be divided into a great number of distinct cases. But as they are all solved upon the same general principles, it will not be necessary to give examples under each. The following problems may serve as a specimen of those which most frequently occur in practice.

PROBLEM I.

TO FIND THE PERPENDICULAR HEIGHT OF AN ACCESSIBLE OBJECT STANDING ON A HORIZONTAL PLANE.

6. MEASURE FROM THE OBJECT TO A CONVENIENT STATION, AND THERE TAKE THE ANGLE OF ELEVATION SUBTENDED BY THE OBJECT.

If the distance AB (Fig. 2.) be measured, and the angle of elevation BAC; there will be given in the right angled triangle ABC, the base and the angles, to find the perpendicular. (Trig. 137.)

As the instrument by which the angle at A is measured, is commonly raised a few feet above the ground; a point B must be taken in the object, so that AB shall be parallel to the horizon. The part BP, may afterwards be added to the height BC, found by trigonometrical calculation.

Ex. 1. What is the height of a tower BC, (Fig. 2.) if the distance AB, on a horizontal plane, be 98 feet; and the angle BAC $35\frac{1}{3}$ degrees?

Making the hypothenuse radius, (Trig. 121.)
Cos BAC : AB::Sin BAC : BC=69.9 feet.

For the *geometrical construction* of the problem, see Trig. 169.

2. What is the height of the perpendicular sheet of water at the falls of Niagara, if it subtends an angle of 40 degrees, at the distance of 163 feet from the bottom, measured on a horizontal plane? Ans. $136\frac{3}{4}$ feet.

7. If the height of the object be *known*, its *distance* may be found by the angle of elevation. In this case the angles, and the perpendicular of the triangle are given, to find the base.

Ex. A person on shore, taking an observation of a ship's mast which is known to be 99 feet high, finds the angle of elevation $3\frac{1}{2}$ degrees. What is the distance of the ship from the observer? Ans. 98 rods.

8. If the observer be stationed at the *top* of the perpendicular BC, (Fig. 2.) whose height is known; he may find the length of the base line AB, by measuring the angle of *depression* ACD, which is equal to BAC

Ex. A seaman at the top of a mast 66 feet high, looking at another ship, finds the angle of depression 10 degrees. What is the distance of the two vessels from each other?

Ans. $22\frac{2}{3}$ rods.

We may find the distance between *two objects* which are in the same vertical plane with the perpendicular, by calculating the distance of each from the perpendicular. Thus AG (Fig. 2.) is equal to the difference between AB and GB.

Problem. II.

To find the height of an accessible object standing on an inclined plane.

9. Measure the distance from the object to a convenient station, and take the angles which this base makes with lines drawn from its two ends to the top of the object.

If the base AB (Fig. 3.) be measured, and the angles BAC and ABC; there will be given, in the oblique angled triangle ABC, the side AB and the angles, to find BC. (Trig. 150.)

Or the height BC may be found by measuring the distances BA, AD, and taking the angles BAC and BDC. There will then be given in the triangle ADC, the angles and the side AD, to find AC; and consequently, in the triangle ABC, the sides AB and AC with the angle BAC, to find BC.

Ex. If AB (Fig. 3.) be 76 feet, the angle B 101° 25′ and the angle A 44° 42′; what is the height of the tree BC?

$$\text{Sin C} : \text{AB} :: \text{Sin A} : \text{BC} = 95.9 \text{ feet.}$$

For the *geometrical construction* of the problem, see Trig. 169.

10. The following are some of the methods by which the height of an object may be found, without measuring the angle of elevation.

1. *By shadows.* Let the staff *bc* (Fig. 4.) be parallel to an object BC whose height is required. If the shadow of BC extend to A, and that of *bc* to *a*; the rays of light CA and *ca* coming from the sun may be considered parallel; and therefore the triangles ABC and *abc* are similar; so that

$$ab : bc :: AB : BC.$$

Ex. If *ab* be 3 feet, *bc* 5 feet, and AB 69 feet, what is the height of BC? Ans. 115 feet.

2. *By parallel rods.* If two poles *am* and *cn* (Fig. 5.) be placed parallel to the object BC, and at such distances as to bring the points C, *c*, *a* in a line, and if *ab* be made parallel to AB; the triangles ABC, and *abc* will be similar; and we shall have

$$ab : bc :: AB : BC.$$

One pole will be sufficient, if the observer can place his eye at the point A, so as to bring A, *a*, and C in a line.

3. *By a mirror.* Let the smooth surface of a body of water at A, (Fig. 6.) or any plane mirror parallel to the horizon, be so situated, that the eye of the observer at *c* may view the top of the object C reflected from the mirror. By a law of Optics, the angle BAC is equal to *b*A*c*; and if *bc*

be made parallel to BC, the triangle bAc will be similar to BAC; so that

$$Ab : bc :: AB : BC.$$

Problem III.

To find the height of an inaccessible object above a horizontal plane.

11. Take two stations in a vertical plane passing through the top of the object, measure the distance from one station to the other, and the angle of elevation at each.

If the base AB (Fig. 7.) be measured with the angles CBP and CAB; as ABC is the supplement of CBP, there will be given, in the oblique angled triangle ABC, the side AB and the angles, to find BC; and then, in the right angled triangle BCP, the hypothenuse and the angles, to find the perpendicular CP.

Ex. 1. If C (Fig. 7.) be the top of a spire, the horizontal base line AB 100 feet, the angle of elevation BAC 40°, and the angle PBC 60°; what is the perpendicular height of the spire?

The difference between the angles PBC and BAC is equal to ACB. (Euc. 32. 1)

Then Sin ACB : AB :: Sin BAC : BC=187.9
And R : BC :: Sin PBC : CP=$162\frac{3}{4}$ feet.

2. If two persons 120 rods from each other, are standing on a horizontal plane, and also in a vertical plane passing through a *cloud*, both being on the same side of the cloud: and if they find the angles of elevation at the two stations to be 68° and 76°; what is the height of the cloud?

Ans. 2 miles 135.7 rods.

12. The preceding problems are useful in particular cases. But the following is a *general* rule, which may be used for finding the height of any object whatever, within moderate distances.

Problem IV.

To find the height of any object, by observations at two stations.

13. Measure the base line between the two stations, the angles between this base and lines drawn from each of the stations to each end of the object, and the angle subtended by the object, at one of the stations.

If BC (Fig. 8.) be the object whose height is required, and if the distance between the stations A and D be measured, with the angles ADC, DAC, ADB, DAB, and BAC; there will be given, in the triangle ADC, the side AD and the angles, to find AC; in the triangle ADB, the side AD and the angles, to find AB; and then, in the triangle BAC, the sides AB and AC with the included angle, to find the required height BC.

If the two stations A and D be in the *same plane* with BC, the angle BAC will be equal to the difference between BAD and CAD. In this case it will not be necessary to measure BAC.

Ex. If AD=83 feet, (Fig. 8.) $\begin{cases} ADC=51^\circ \\ DAC=95^\circ \end{cases}$ $\begin{cases} ADB=33^\circ, \\ DAB=121^\circ \end{cases}$ $BAC=26^\circ$,

What is the height of the object BC?

Sin ACD : AD::ADC : AC=115.3

Sin ABD : AD::ADB : AB=103.1

$(AC+AB) : (AC-AB)::\text{Tan}\ \frac{1}{2}(ABC+ACB) : \text{Tan}\ \frac{1}{2}(ABC-ACB)=13^\circ\ 38'$

Sin ACB : AB::Sin BAC : BC=50.57 feet.

If the object BC be perpendicular to the horizon, its height, after obtaining AB and BC as before, may be found by taking the *angles* of *elevation* BAP and CAP. The difference of the perpendiculars in the right angled triangles ABP and ACP, will be the height required.

Problem V.

To find the distance of an inaccessible object.

14. Measure a base line between two stations, and the angles between this and lines drawn from each of the stations to the object.

If C (Fig, 9.) be the object, and if the distance between the stations A and B be measured. with the angles at B and A; there will be given, in the oblique angled triangle ABC, the side AB and the angles, to find AC and BC, the distances of the object from the two stations.

For the geometrical construction, see Trig. 169.

Ex. 1. What are the distances of the two stations A and B (Fig. 9.) from the house C, on the opposite side of a river; if AB be 26.6 rods, B 92° 46′, and A 38° 40′?

The angle C=180−(A+B)=48° 34′. Then

$$\text{Sin } C : AB :: \begin{cases} \text{Sin } A : BC = 22.17 \\ \text{Sin } B : AC = 35.44 \end{cases}$$

2. Two ships in a harbour, wishing to ascertain how far they are from a fort on shore, find that their mutual distance is 90 rods, and that the angles formed between a line from one to the other, and lines drawn from each to the fort are 45° and 56° 15′. What are their respective distances from the fort? Ans. 76.3 and 64.9 rods.

15. The *perpendicular* distance of the object from the line joining the two stations may be easily found, after the distance from one of the stations is obtained. The perpendicular distance PC (Fig. 9.) is one of the sides of the right angled triangle BCP. Therefore

$$R : BC :: \text{Sin } B : PC$$

Problem VI.

To find the distance between two objects, when the passage from one to the other, in a straight line, is obstructed.

16. Measure the right lines from one station to each of the objects, and the angle included between these lines.

If A and B (Fig. 10.) be the two objects, and if the distances BC and AC be measured, with the angle at C; there will be given, in the oblique angled triangle ABC, two sides and the included angle to find the other two angles, and the remaining side. (Trig. 153.)

Ex. The passage between the two objects A and B (Fig 10.) being obstructed by a morass, the line BC was measured and found to be 109 rods, the line AC 76 rods, and the angle at C 101° 30′. What is the distance AB?

Ans. 144.7 rods.

Problem VII.

To find the distance between two inaccessible objects.

17. Measure a base line between two stations and the angles between this base and lines drawn from each of the stations, to each of the objects.

If A and B (Fig. 11.) be the two objects, and if the distance between the stations C and D be measured, with the angles BDC, BCD, ADC, and ACD ; the lines AC and BC may be found as in Problem V, and then the distance AB as in Problem VI.

This rule is substantially the same as that in art. 13. The two stations are supposed to be in the *same plane* with the objects. If they are not, it will be necessary to measure the angle ACB.

18. The same process by which we obtain the distance of *two* objects from each other, will enable us to find the distance between one of these and a third, between that and a fourth, and so on, till a connection is formed between a great number of remote points. This is the plan of the great *Trigonometrical Surveys* which have been lately carried on, with surprising exactness, particularly in England and France. See Surveying, Section II.

19. In the preceding problems for determining altitudes, the objects are supposed to be at such moderate distances, that the observations are not sensibly affected by the *spherical figure of the earth*. The height of an object is measured from an *horizontal plane*, passing through the station at which the angle of elevation is taken. But in an extent of several miles, the figure of the earth ought to be taken into account.

Let AB (Fig. 12) be a portion of the earth's surface, H an object above it, and AT a tangent at the point A, or a horizontal line passing through A. Then HT, the oblique height of the object above the horizon of A, is only a *part* of the height above the surface of the earth, or the level of the ocean. To obtain the true altitude, it is necessary to add BT to the height HT found by observation. The height BT may be calculated, if the diameter of the earth and the distance AT be previously known. Or if the height BT be first determined from observation, with the distance AT ; the diameter of the earth may be thence deduced.

PROBLEM VIII.

TO FIND THE DIAMETER OF THE EARTH, FROM THE KNOWN HEIGHT OF A DISTANT MOUNTAIN, WHOSE SUMMIT IS JUST VISIBLE IN THE HORIZON.

20. FROM THE SQUARE OF THE DISTANCE DIVIDED BY THE HEIGHT, SUBTRACT THE HEIGHT.

If BT (Fig. 12) be a mountain whose height is known, with the distance AT; and if the summit T be just visible in the horizon at A; then AT is a *tangent* at the point A.

Let 2BC=D, the diameter of the earth,
AT=d, the distance of the mountain,
BT=h, its height.

Then considering AT as a straight line, and the earth as a sphere, we have (Euc. 36. 3.)

$$(2BC+BT)\times BT=\overline{AT}^2 \text{ ; that is, } (D+h)\times h=d^2,$$

and reducing the equation,

$$D=\frac{d^2}{h}-h$$

Ex. The highest point of the Andes is about 4 miles above the level of the ocean. If a straight line from this touch the surface of the water at the distance of 178¼ miles; what is the diameter of the earth? Ans. 7940 miles.

21. If the distance AT (Fig. 12.) be *unknown*, it may be found by measuring with a quadrant the angle ATC. Draw BG perpendicular to BC; and join CG. The triangles ACG and BCG are equal, because each has a right angle, the sides AC and BC are equal, and the hypothenuse CG is common. Therefore BG and AG are equal. In the right angled triangle BGT, the angle BTG is given, and the perpendicular BT. From these may be found BG and TG, whose sum is equal to AT, the distance required.*

22. In the common measurement of angles, the light is supposed to come from the object to the eye in a *straight line*. But this is not strictly true. The direction of the light is affected by the *refraction of the atmosphere*. If the object be near, the deviation is very inconsiderable. But in an ex-

* This method of determining the diameter of the earth is not as accurate as that by measuring a degree of Latitude. See Surveying, Sec. II.

tent of several miles, and particularly in such nice observations as determining the height of distant mountains, and the diameter of the earth, it is necessary to make allowance for the refraction.*

PROBLEM IX.

TO FIND THE GREATEST DISTANCE AT WHICH A GIVEN OBJECT CAN BE SEEN ON THE SURFACE OF THE EARTH.

23. TO THE PRODUCT OF THE HEIGHT OF THE OBJECT INTO THE DIAMETER OF THE EARTH, ADD THE SQUARE OF THE HEIGHT; AND EXTRACT THE SQUARE ROOT OF THE SUM.

Let $2BC=D$, the diameter of the earth, (Fig. 12.)
$BT=h$, the height of the object,
$AT=d$, the distance required.
Then $(D+h)\times h=d^2$. And $d=\sqrt{Dh+h^2}$.

Ex. If the diameter of the earth be 7940 miles, and Mount Ætna 2 miles high; how far can its summit be seen at sea?
Ans. 126 miles.

The actual distance at which an object can be seen, is increased by the refraction of the air.*

24. In this problem, the eye is supposed to be placed at the level of the ocean. But if the observer be elevated above the surface, as on the deck of a ship, he can see to a greater distance. If BT (Fig. 13.) be the height of the object, and B′T′ the height of the eye above the level of the ocean; the distance at which the object can be seen, is evidently equal to the *sum* of the tangents AT and AT′.

Ex. The top of a ship's mast 132 feet high is just visible in the horizon, to an observer whose eye is 33 feet above the surface of the water. What is the distance of the ship?
Ans. $21\frac{1}{8}$ miles.

25. The distance to which a person can see the smooth surface of the ocean, if no allowance be made for refraction, is equal to a tangent to the earth drawn from his eye, as T′A (Fig. 13.)

Ex. If a man standing on the level of the ocean, has his eye raised $5\frac{1}{2}$ feet above the water; to what distance can he see the surface?
Ans. $2\frac{7}{8}$ miles.

* See Note A.

26. If the distance AT, (Fig. 12.) with the diameter of the earth be given, and the *height* BT be required; the equation in art. 23 gives

$$h=\sqrt{\tfrac{1}{4}D^2+d^2}-\tfrac{1}{2}D$$

See Surveying, Section IV. on *Levelling.*

27. When the diameter of the earth is ascertained, this may be made a *base line* for determining the distances of the *heavenly bodies.* A right angled triangle may be formed, the perpendicular sides of which shall be the distance required, and the semi-diameter of the earth. If then one of the angles be found by observation, the required side may be easily calculated.

Let AC (Fig. 14.) be the semi-diameter of the earth, AH the sensible horizon at A, and CM the rational horizon parallel to AH, passing through the moon M. The angle HAM may be found by astronomical observation. This angle, which is called the *Horizontal Parallax,* is equal to AMC, the angle at the moon subtended by the semi-diameter of the earth. (Euc. 29. 1.)

PROBLEM X.

TO FIND THE DISTANCE OF ANY HEAVENLY BODY WHOSE HORIZONTAL PARALLAX IS KNOWN.

28. AS RADIUS, TO THE SEMI-DIAMETER OF THE EARTH;
SO IS THE COTANGENT OF THE HORIZONTAL PARALLAX, TO THE DISTANCE.

In the right angled triangle ACM, (Fig. 14.) if AC be made radius;

$$R : AC :: \text{Cot } AMC : CM$$

Ex. If the horizontal parallax of the moon be 0° 57′, and the diameter of the earth 7940 miles; what is the distance of the moon from the centre of the earth?

Ans. 239,414 miles.

29. The *fixed stars* are too far distant to have any sensible horizontal parallax. But from late observations it would seem, that some of them are near enough, to suffer a small apparent change of place, from the revolution of the earth round the sun. The distance of the sun, then, which is the semi-diameter of the earth's orbit, may be taken as a *base line,* for finding the distance of these stars.

We thus proceed by degrees, from measuring a line on the surface of the earth, to calculate the distances of the heavenly bodies. From a base line on a plane, is determined the height of a mountain; from the height of the mountain, the diameter of the earth; from the diameter of the earth, the distance of the sun, and from the distance of the sun, the distance of the stars.

30. After finding the distance of a heavenly body, its *magnitude* is easily ascertained; if it have an apparent diameter, sufficiently large to be measured by the instruments which are used for taking angles.

Let AEB (Fig. 15.) be the angle which a heavenly body subtends at the eye. Half this angle, if C be the centre of the body, is AEC; the line EA is a tangent to the surface, and therefore EAC is a right angle. Then making the distance EC radius,

$$R : EC :: \text{Sin } AEC : AC$$

That is, radius is to the distance, as the sine of half the angle which the body subtends, to its semi-diameter.

Ex. If the sun subtends an angle of 32′ 2″, and if his distance from the earth be 95 million miles; what is his diameter? Ans. 885 thousand miles.

Promiscuous Examples.

1. On the bank of a river, the angle of elevation of a tree on the opposite side is found to be 46°; and at another station 100 feet directly back on the same level, 31°. What is the height of the tree?

Ans. 143 feet.

2. On a horizontal plane, observations were taken of a tower standing on the top of a hill. At one station, the angle of elevation of the top of the tower was found to be 50°; that of the bottom 39°; and at another station 150 feet directly back, the angle of elevation of the top of the tower was 32°. What are the heights of the hill and the tower?

Ans. The hill is 134 feet high; the tower 63.

3. What is the altitude of the sun, when the shadow of a tree, cast on a horizontal plane, is to the height of the tree as 4 to 3? Ans. 36° 52′ 12″

4. If a straight line from the top of the White Mountains in New-Hampshire touch the ocean at the distance of $103\frac{1}{3}$ miles; what is the height of the mountains?

Ans. 7100 feet.

5. From the top of a perpendicular rock 55 yards high, the angle of depression of the nearest bank of a river is found to be 55° 54′, that of the opposite bank 33° 20′. Required the breadth of the river, and the distance of its nearest bank from the bottom of the rock.

The breadth of the river is 46.4 yards;
Its distance from the rock 37.2

6. If the moon subtend an angle of 31′ 14″, when her distance is 240,000 miles; what is her diameter?

Ans. 2180 miles.

7. Observations are made on the altitude of a balloon, by two persons standing on the same side of the balloon, and in a vertical plane passing through it. The distance of the stations is half a mile. At one, the angle of elevation is 30° 58′, at the other 36° 52′. What is the height of the balloon above the ground? Ans. $1\frac{1}{3}$ mile.

8. The shadow of the top of a mountain, when the altitude of the sun on the meridian is 32°, strikes a certain point on a level plain below; but when the meridian altitude of the sun is 67°, the shadow strikes half a mile farther south, on the same plain. What is the height of the mountain above the plain. Ans. 2245 feet.

NAVIGATION.

SECTION I.

Plane Sailing.

Art. 33. NAVIGATION is the art of conducting a ship on the ocean. The most accurate method of ascertaining the situation of a vessel at sea is to find, by astronomical observations, her *latitude* and *longitude*. But this requires a view of the heavenly bodies ; and these are often obscured by intervening clouds. The mariner must therefore have recourse to other means for determining the progress which he has made, and the particular part of the ocean through which he is at any time making his way. The common method is to measure the rate of the ship's going by a *log-line*, and to find the direction in which she sails by a *mariner's compass*. From these data, the difference of latitude, the departure, and the difference of longitude, may be calculated. The two first may be found by plane sailing ; the last by middle latitude sailing, or more correctly by Mercator's sailing. See Sec. II. and III.

34. The *log-line* is a cord which is wound round a reel, one end being attached to a piece of wood called a *log*. It is used to determine the distance which a ship runs in an hour, by measuring the distance which she runs in half a minute. The *log* is commonly a small piece of board, in the form of a quadrant of a circle. The arc is loaded with a quantity of lead sufficient to give the board a perpendicular position, when thrown upon the water. This will prevent it from moving forward toward the vessel, while the line is running off the reel. So that the length of line drawn off

by the log in half a minute, is equal to the distance which the vessel moves through the water in that time.

The log-line, which is a hundred fathoms or more, is divided into equal portions called *knots.* Each of these has the same ratio to a nautical mile, which half a minute has to an hour. That is, a knot is the 120th part of a mile. If therefore the motion of the ship is uniform, she sails as many miles in an hour, as she does knots in half a minute.

The *time* is measured by a *half-minute glass*, constructed like an hour glass. This is turned when the log is thrown upon the water; and the knots drawn from the reel, while the sands are running, give the rate of the ship. The log is thrown either every hour, or once in two hours.

35. The *Mariner's compass* is a circular card, attached to a magnetic needle, which is balanced on an upright pin, so as to move freely in any direction. The ends of the needle turn towards the northern and southern points of the horizon. It places itself in the magnetic meridian, which nearly coincides with the astronomical meridian, or a north and south line.* Directly over the needle, a line is drawn on a card, one end of which is marked N, and the other S. The whole circumference is divided into equal parts by 32 *points.* Four of these, the N, S, E, and W, are called *cardinal* points. The interval between two adjacent points is 11° 15′, which is the quotient of 360° divided by 32. The card and the needle are inclosed in a circular box, on the inside of which a black *mark* is drawn perpendicular to the horizon. When the compass is placed in the vessel, a line passing from this mark through the centre of the card should be parallel to the keel. The part of the circumference which coincides with the mark will then shew the point of compass to which the keel is directed. To prevent the needle from being affected by the motion of the vessel, the box, and a brass ring by which it is surrounded, have four points of suspension so contrived as to keep the card nearly parallel to the horizon.

* For the *Variation* of the needle, see SURVEYING, Sec. V.

The following is a table of the number of degrees and minutes corresponding to each point and quarter point of the compass. See Fig. 16.

North-East Quadrant.	South-East Quadrant	Points	D	M.	S	South-West Quadrant	North-West Quadrant.
North.	*South.*	0 0	0	0	0	*South.*	*North.*
N¼E	S¼E	0 ¼	2	48	45	S¼W	N¼W
N½E	S½E	0 ½	5	37	30	S½W	N½W
N¾E	S¾E	0 ¾	8	26	15	S¾W	N¾W
N*b*E	S*b*E	1 0	11	15	0	S*b*W	N*b*W
N*b*E¼E	S*b*E¼E	1 ¼	14	3	45	S*b*W¼W	N*b*W¼W
N*b*E½E	S*b*E½E	1 ½	16	52	30	S*b*W½W	N*b*W½W
N*b*E¾E	S*b*E¾E	1 ¾	19	41	15	S*b*W¾W	N*b*W¾W
NNE	SSE	2 0	22	30	0	SSW	NNW
NNE¼E	SSE¼E	2 ¼	25	18	45	SSW¼W	NNW¼W
NNE½E	SSE½E	2 ½	28	7	30	SSW½W	NNW½W
NNE¾E	SSE¾E	2 ¾	30	56	15	SSW¾W	NNW¾W
NE*b*N	SE*b*S	3 0	33	45	0	SW*b*S	NW*b*N
NE¾N	SE¾S	3 ¼	36	33	45	SW¾S	NW¾N
NE½N	SE½S	3 ½	39	22	30	SW½S	NW½N
NE¼N	SE¼S	3 ¾	42	11	15	SW¼S	NW¼N
NE	SE	4 0	45	0	0	SW	NW
NE¼E	SE¼E	4 ¼	47	48	45	SW¼W	NW¼W
NE½E	SE½E	4 ½	50	37	30	SW½W	NW½W
NE¾E	SE¾E	4 ¾	53	26	15	SW¾W	NW¾W
NE*b*E	SE*b*E	5 0	56	15	0	SW*b*W	NW*b*W
NE*b*E¼E	SE*b*E¼E	5 ¼	59	3	45	SW*b*W¼W	NW*b*W¼W
NE*b*E½E	SE*b*E½E	5 ½	61	52	30	SW*b*W½W	NW*b*W½W
NE*b*E¾E	SE*b*E¾E	5 ¾	64	41	15	SW*b*W¾W	NW*b*W¾W
ENE	ESE	6 0	67	30	0	WSW	WNW
E*b*N¾N	E*b*S¾S	6 ¼	70	18	45	W*b*S¾S	W*b*N¾N
E*b*N½N	E*b*S½S	6 ½	73	7	30	W*b*S½S	W*b*N½N
E*b*N¼N	E*b*S¼S	6 ¾	75	56	15	W*b*S¼S	W*b*N¼N
E*b*N	E*b*S	7 0	78	45	0	W*b*S	W*b*N
E¾N	E¾S	7 ¼	81	33	45	W¾S	W¾N
E½N	E½S	7 ½	84	22	30	W½S	W½N
E¼N	E¼S	7 ¾	87	11	15	W¼S	W¼N
East.	*East.*	8 0	90	0	0	*West.*	*West.*

36. Plane Sailing is the method of calculating the situation and progress of a ship by means of a plane triangle. Though the surface of the ocean, conforming to the general figure of the earth, is nearly *spherical* ;* yet the quantities which are the objects of inquiry in plane sailing, have the same relations to each other, as the sides and angles of a rectilinear triangle. The particulars which are either given or required are *four*, viz.

1. The Course,
2. The Distance,
3. The Difference of Latitude,
4. The Departure.

37. The *Course* is the angle between a meridian line passing through the ship, and the direction in which she sails. It is described by saying that it is so many points or degrees east or west from a north or south line. Thus if the vessel steers NE by E, the course is said to be N 5 points E, or N 56° 15′ E : if SSW, it is said to be S 2 points W, or S $22\frac{1}{2}$° W.

A ship is said to continue on the *same course*, when she cuts every meridian which she crosses at the *same angle*. She is steered in any required direction, by causing the keel to make a constant angle with the needle. The line thus described is not a straight line, nor an arc of a circle, but a peculiar kind of curve called the *Loxodromic*† *spiral* or *Rhumb-line*.

38. The *Distance* is the length of the line which the vessel describes in the given time.

39. *Difference of Latitude* is the distance between two parallels of latitude, measured on a meridian. It is also called *Northing* or *Southing*.

40. *Departure is the deviation of a ship east or west from a meridian.* If she sails on a parallel of latitude, her departure is the length of that portion of the parallel over which she passes. But if her course is *oblique*, she is continually chan-

* The true figure of the earth is nearer a *spheroid* than a sphere. But the difference is too inconsiderable to be taken into account in any calculations for which the lines and angles are given from the log and the compass. In this and the following sections, therefore, the earth will be considered as a sphere

† From λοξος and δρομος, an oblique course.

ging her latitude; and her departure for each instant ought to be considered as measured on the parallel which she is then crossing. The measure will not be correct, if it be taken wholly on the parallel which the ship has left, or on that upon which she has arrived. Suppose she proceed from A to C. (Fig. 18.) Let the whole distance be divided into indefinitely small portions Am, mn, nC. Draw the meridians PM, PM', PM'', PM'''; and the parallels AD, om, yn, BC. The departure for the first portion is om, for the second sn, for the third tC. And the whole departure is $om+sn+tC$; which, on account of the obliquity of the meridians, is *less* than $Bv+vt+tC=BC$ the meridian distance measured on the parallel upon which the ship has arrived, but *greater* than AD the meridian distance on the parallel which she has left.

41. The distance, departure, and difference of latitude, are measured in *geographical miles* or *minutes*; one of which is equal to the 60th part of a degree at the equator. As the circumference of the earth is about 25 thousand English miles, a degree is nearly $69\frac{1}{2}$ miles. So that a geographical or nautical mile is nearly $\frac{1}{6}$ greater than the common English mile. A *league* is three miles.

42. The peculiar nature of the *Rhumb-line* gives this important advantage in calculation, that the distance, departure, and difference of latitude, though they are curve lines, may be exactly given in length by the sides of a *right-angled plane triangle*, in which one of the angles is equal to the course. Suppose a ship proceeds from A to C, (Fig. 18.) describing the rhumb-line $AmnC$, on which the angles MAm, $M'mn$, $M''nC$ are equal. Let the whole distance be divided into portions so small, that the triangles Amo, mns, nCt, shall not differ sensibly from plane triangles. The meridians and parallels being drawn, the several differences of latitude are Ao, ms, nt; and the departures om, sn, tC. (Art. 40.)

In the straight line $A'C'$ (Fig. 19) make $A'm'=Am$, (Fig. 18.) $m'n'=mn$, $n'C'=nC$, and the angle $C'A'B'=mAo$. Draw $m'v'$ and $n't'$ parallel to $A'B'$; and $m'o'$, $n'y'$, and $C'B'$ perpendicular to $A'B'$. Then the triangles Amo, $A'm'o'$, mns, $m'n's'$, Ctn and $C't'n'$ are all similar to $A'B'C'$. The difference of latitude is

$$AB=Ao+ms+nt=A'o'+m's'+n't'=A'B'.$$

And the departure is

$$om+sn+tC=o'm'+s'n'+t'C'=B'C',$$

43. In plane sailing, then, the process of calculation is as accurate,* and as simple, as if the surface of the ocean were a plane. Let NS (Fig 20.) be a meridian line. If a ship sails from A to C, and if BC is perpendicular to NS; then

The *Course* is the *angle* at A; and the complement of the course, the angle at C;

The *Distance* is the *hypothenuse* AC;

The *Departure* is the *base* BC, which is always opposite to the course; and

The *Difference of Latitude* is the *perpendicular* AB, which is opposite to the complement of the course.

Of these four quantities, any *two* being given, the others may be found by rectangular trigonometry. (Trig. 116.) The parts given may be

1. The course and distance; or
2. The course and departure; or
3. The course and difference of latitude; or
4. The distance and departure;
5. The distance and difference of latitude; or
6. The departure and difference of latitude.

The solutions may be made by arithmetical computation, by Gunter's scale or sliding rule, or by geometrical construction. (Trig Sec. III, V, VI.) The first method is by far the most accurate. As the student is supposed to be already familiar with trigonometry, these operations will not be repeated here. In the geometrical construction, it will be proper to consider the upper side of the paper as *north*, and the lower side *south*. The right hand will then be east, and the left hand west.

CASE I.

44. Given { The course, And distance; } to find { The departure and Difference of latitude.

Here we have the hypothenuse and angles given, to find the base and perpendicular. (Trig. 134.)

Making then the distance radius,

Rad : Dist : : { Sin Course : Departure
Cos Course : Diff. Lat.

Example. I.

A ship sails from A (Fig. 20.) SW by S, 38 miles to C. Required her departure and difference of latitude?

* See Note B.

The course is 3 points. or 33° 45′ (Art. 35.)

R : 38 : : { Sin 33° 45′ : 21.1 = Depart.
Cos. 33° 45′ : 31.6 = Diff. Lat.

Example 2.

A ship sails S 29° E, 34 leagues. Her departure and difference of latitude are required.

Ans. 16.5 and 29.7 leagues.

The proportions in this and the following cases may be varied, by making different sides radius, as in Trigonometry Sec. III.

CASE II.

45. Given { The course, And departure; } to find { The distance, and Difference of latitude.

Making the distance radius, Trig. 137.)

Sin Course : Depart : : { Rad : Distance
Cos. Course : Diff. Lat.

Example 1.

A ship leaving a port in latitude 42° N, has sailed S 37° W, till she finds her departure 62 miles. What distance has she run, and in what latitude has she arrived?

Sin 37° : 62 : : { Rad : 103 = Distance.
Cos. 37° : 82.3 = Diff. Lat

The difference of latitude is 82.3 miles, or 1° 12′.3. (Art. 41.) This is to be *subtracted* from the original latitude of the ship, because her course was *towards* the equator. The remainder is 40° 47′.7, the latitude on which she has arrived.

Example 2.

A ship leaves a port in latitude 63° S, and runs N 54° E, till she makes a harbour where her departure is found to be 74 miles; how great is the distance of the two places, and what is the latitude of the latter?

The distance is 91½ miles; and the latitude of the latter place is 62° 06′.2.

CASE III.

46. Given { The course, and Diff. of latitude; } to find { The distance, And departure.

Making the distance radius,

Cos Course : Diff. Lat : : { Rad : Distance.
Sin Course : Departure.

Example.

A ship sails S 50° E, from latitude 7° N, to latitude 4° S. Required her distance and departure.

As the two latitudes are on different sides of the equator, the distance of the parallels is evidently equal to the *sum* of the given latitudes. This is 11° or 660 miles. The distance is 1026.8 miles, and the departure 786½.

CASE IV.

47. Given {The distance, And departure;} to find {The course, and Diff. of latitude.}

Making the distance radius, (Trig. 135.)

Dist : Rad : : Depart : Sin Course,

Rad : Dist : : Cos Course : Diff. Lat.

Example.

A ship having left a port in Lat. 3° N, and sailing between S and E 400 miles, finds her departure 180 miles. What course has she steered, and what is her latitude?

Her latitude is 2° 57′½ S, and her course S 26° 44′½ E.

CASE V.

48. Given. {The distance, and Diff. of latitude;} to find {The course, And departure.}

Making the distance radius,

Dist : Rad : : Diff. Lat : Cos Course,

Rad : Dist : : Sin Course : Departure.

Example.

A vessel sails between N and E 66 miles, from Lat. 34° 50′ to Lat. 35° 40′. Required her course and departure.

The course is N 40° 45′ E, and the departure 43.08 miles.

CASE VI.

49. Given. {The departure, and Diff. of latitude;} to find {The course, And distance.}

Making the difference of latitude radius, (Trig, 139.)

Diff. Lat : Rad : : Depart : Tan Course,

Rad : Diff. Lat : : Sec Course : Distance.

Example.

A ship sails from the equator between S and W, till her

latitude is 5° 52′, and her departure 264 miles. Required her course and distance.

The course is S 36° 52′$\frac{1}{3}$ W, and the distance is 440 miles.

Examples for practice.

1. Given a ship's course S 46° E, and departure 59 miles; to find the distance and difference of latitude.

2. Given the distance 68 miles, and departure 47; to find the course and difference of latitude.

3. Given the course SSE, and the distance 57 leagues; to find the departure and difference of latitude.

4. Given the course NW by N, and the difference of latitude 2° 36′; to find the distance and departure.

5. Given the departure 92, and the difference of latitude 86: to find the course and distance.

6. Given the distance 123, and the difference of latitude 97; to find the course and departure.

THE TRAVERSE TABLE.

50. To save the labour of calculation, tables have been prepared, in which are given the departure and difference of latitude, for every degree of the quadrant, or for every quarter of a degree. These are called *Traverse tables*, or tables of *Departure and Latitude*. The distance is placed in the left hand column, the departure and difference of latitude directly opposite, and the degrees if above 45° or 4 points, at the top of the page, but if under 45°, at the bottom. The titles at the top of the columns correspond to the courses at the top; and the titles at the bottom, to the courses at the bottom; the difference of latitude for a course *greater* than 45°, being the same as the departure for one which is as much *less* than 45°. See Trig. 104.

If the given distance is greater than any contained in the table, it may be *divided into parts*, and the departure and difference of latitude found for each of the parts. The *sums* of the numbers thus found will be the numbers required.

The departure and difference of latitude for *decimal parts* may be found in the same manner as for whole numbers, by supposing the decimal point in each of the columns to be moved to the left, as the case requires.

With the aid of a traverse table, all the cases of plane sailing may be easily solved by inspection.

Ex. 1. Given the course 33° 45′; and the distance 38 miles; to find the departure and difference of latitude.

Under 33°¾, and opposite 38, will be found the difference of latitude 31.6, and the departure 21.11; the same as in page 21.

2. Given the course 57°, and the distance 163.

The departure and diff. of lat. for 100 are 83.87 and 54.46

for 63	52.84	34.31
for 163	136.71	88.77

3. Given the course 39°, and the distance 18.23.

The departure and diff. of lat. for 18. are 11.33 and 13.99

for .23	0.14	0.18
for 18.23	11.47	14.17

4. Given the course 41° 15′, and the departure 60.

Under 41°¼, and against the departure 60, will be found the difference of latitude 68.42 and the distance 91.

5. Given the distance 63, and the departure 56.

Opposite the distance 63, find the departure 56; in the adjoining column will be the latitude 28.85, and at the bottom, the course 62°¾.

6. Given the departure 72, and the difference of latitude 37.

Opposite these numbers in the columns of latitude and departure, will be found the distance 81, and at the foot of the columns, the course 62°¾.

51. The traverse table is useful, not only for taking out departure and difference of latitude; but for finding by inspection the sides and angles of *any right-angled triangle* whatever. In plane sailing, the distance is the hypothenuse, (see Fig. 20.) the difference of latitude is the perpendicular, the departure is the base, and the course is the acute angle

at the perpendicular. If then the hypothenuse of any right-angled triangle whatever, be found in the column of distances, in the traverse table; the perpendicular will be opposite in the latitude column, and the base in the departure column; the angle at the perpendicular, being at the top or bottom of the page.

Ex. 1. Given the hypothenuse 24, and the angle at the perpendicular $54°\frac{1}{2}$; to find the base and perpendicular by inspection.

Opposite 24 in the distance column, and over $54°\frac{1}{2}$ will be found the base 19.54 in the departure column, and the perpendicular 13.94 in the latitude column.

2. Given the angle at the perpendicular $37°\frac{1}{4}$, and the base 46; to find the hypothenuse and perpendicular.

Under $37°\frac{1}{4}$. look for 46 in the departure column; and opposite this will be found the perpendicular 60.5 in the latitude column, and the hypothenuse 76 in the distance column.

3. Given the perpendicular 36, and the base 30.21; to find the hypothenuse and angles.

Look in the columns of latitude and departure, till the numbers 36 and 30.21 are found opposite each other; these will give the hypothenuse 47, and the angle at the perpendicular 40°.

SECTION II.

PARALLEL AND MIDDLE LATITUDE SAILING.

52. By the methods of calculation in plane sailing, a ship's course, distance, departure. and difference of latitude are found. There is one other particular which it is very important to determine, the *difference of longitude*. The departure gives the distance between two meridians in *miles*. But the situations of places on the earth, are known from their latitudes and longitudes; and these are measured in *degrees*. The lines of longitude, as they are drawn on the globe, are farthest

from each other at the equator, and gradually converge towards the poles. A ship, in making a hundred miles of departure, may change her latitude in one case 2 degrees, in another 10, and in another 20. It is important, then, to be able to convert departure into difference of longitude; that is, to determine how many degrees of longitude answer to any given number of miles, on any parallel of latitude. This is easily done by the following

THEOREM.

53. AS THE COSINE OF LATITUDE,
TO RADIUS;
SO IS THE DEPARTURE,
TO THE DIFFERENCE OF LONGITUDE.

By this is to be understod, that the cosine of the latitude is to radius; as the distance between two meridians measured on the given parallel, to the distance between the same meridians measured on the equator.

Let P (Fig. 21.) be the pole of the earth, A a point at the equator, L a place whose latitude is given, and LO a line perpendicular to PC. Then CL or CA is a semi-diameter of the earth, which may be assumed as the radius of the tables; PL is the complement of the latitude, and OL the sine of PL, that is, the *cosine* of the latitude.

If the whole be now supposed to revolve about PC as an axis, the radius CA will describe the equator, and OL the given parallel of latitude. The circumferences of these circles are as their semi diameters OL and CA, (Sup. Euc. 8. 1.) And this is the ratio which any portion of one circumference has to a like portion of the other. Therefore OL is to CA, that is, the cosine of latitude is to radius, as the distance between two meridians measured on the given parallel, to the distance between the same meridians measured on the equator.

Cor. 1. Like portions of different parallels of latitude are to each other, as the cosines of the latitudes.

Cor. 2. A degree of longitude is commonly measured on the equator. But if it be considered as measured on a parallel of latitude, the *length* of the degree will be as the cosine of the latitude.

[illegible] owing [illegible] ongi- tude for each degree of latitude.

D.L.	Miles.	D.L.	Miles.	D.L.	Miles	D.L.	Miles	D.L.	Miles.	D.L.	Miles.
1	59.99	16	57.67	31	51.43	46	41.68	61	29.09	76	14.52
2	59.96	17	57.38	32	50.88	47	40.92	62	28.17	77	13.50
3	59.92	18	57.06	33	50.32	48	40.15	63	27.24	78	12.47
4	59.85	19	56 73	34	49.74	49	39.36	64	26 30	79	11.45
5	59.77	20	56.38	35	49.15	50	38.57	65	25.36	80	10.42
6	59.67	21	56 01	36	48.54	51	37.76	66	24.40	81	9.39
7	59 55	22	55.63	37	47 92	52	36.94	67	23.44	82	8.35
8	59 42	23	55.23	38	47.28	53	36.11	68	22.48	83	7.31
9	59.26	24	54.81	39	46.63	54	35.27	69	21.50	84	6.27
10	59.08	25	54 38	40	45.96	55	34.41	70	20.52	85	5.23
11	58.89	26	53.93	41	45.28	56	33.55	71	19.53	86	4.19
12	58.68	27	53.46	42	44.59	57	32.68	72	18 54	87	3.14
13	58.46	28	52.97	43	43.88	58	31.80	73	17.54	88	2.09
14	58.22	29	52 47	44	43.16	59	30.90	74	16.54	89	1.05
15	57.95	30	51.96	45	42.43	60	30.00	75	15.53	90	0.00

The length of a degree of longitude in different parallels is also shown by the *Line of Longitude*, placed over or under the line of chords, on the plane scale. See Trig. (165.)

The sailing of a ship on a parallel of latitude* is called *Parallel Sailing* In this case, the departure is equal to the distance. The difference of longitude may be found by the preceding theorem; or if the difference of longitude be given, the departure may be found by inverting the terms of the proportion. (Alg. 380. 3.)

55. The *Geometrical Construction* is very simple. Make CBD (Fig. 22.) a right angle, draw BC equal to the departure in miles, lay off the angle at C equal to the latitude in degrees, and draw the hypothenuse CD for the difference of longitude. The angles C, and the sides BC and CD, of this triangle, have the same relations to each other, as the latitude, departure, and difference of longitude.

For Cos C : BC::R : CD (Trig. 121.)
And Cos Lat : Depart::R : Diff. Lon. (Art. 53.

* See Note C.

56. The parts of the triangle may be found by *inspection* in the traverse table. (Art. 51.) The angle opposite the departure is D the complement of the latitude, and the difference of longitude is the hypothenuse CD. If then the departure be found in the departure column under or over the given number of degrees in the co-latitude, the difference of longitude will be opposite in the distance column.

Example I.

A ship leaving a port in Lat. 38° N. Lon. 16° E. sails west on a parallel of latitude 117 miles in 24 hours. What is her longitude at the end of this time?

Cos 38° : Rad : : 117 : $148'\frac{1}{2}$=2° $28'\frac{1}{2}$ the difference of longitude.

This subtracted from 16° leaves 13° $31'\frac{1}{2}$ the longitude required.

Example II.

What is the distance of two places in Lat. 46° N. if the longitude of the one is 2° 13′ W. and that of the other 1° 17′ E.?

As the two places are on opposite sides of the first meridian, the difference of longitude is 2° 13′+1° 17′=3° 30′, or 210 minutes. Then

Rad : Cos 46° : : 210 : 145.88 miles, the departure, or the distance between the two places.

Example III.

A ship having sailed on a parallel of latitude 138 miles, finds her difference of longitude 4° 3′ or 243 minutes. What is her latitude?

Diff. Lon. 243 : Dep. 138 : : Rad : Cos Lat. 55° $23'\frac{3}{4}$.

Example IV.

On what part of the earth are the degrees of longitude *half* as long as the equator?

Ans. In latitude 60.

Middle Latitude Sailing.

57. By the method just explained, is calculated the difference of longitude of a ship *sailing on a parallel of latitude.* But instances of this mode of sailing are comparatively few. It is necessary then to be able to calculate the longitude when the course is *oblique.* If a ship sail from A to C, (Fig. 18.) the departure is equal to $om+sn+tC$. But the sum of these small lines is *less* than BC, and *greater* than AD. (Art. 40.) The departure, then, is the meridian distance measured not on the parallel from which the ship sailed, nor on that upon which she has arrived, but upon one which is between the two. If the exact situation of this intermediate parallel could be determined, by a process sufficiently simple for common practice, the difference of longitude would be easily obtained. The parallel usually taken for this purpose, is an *arithmetical mean* between the two extreme latitudes. This is called the *Middle Latitude.* The meridian distance on this parallel is not exactly equal to the departure. But for small distances, the errour is not material, except in high latitudes.

The middle latitude is equal to *half the sum* of the two extreme latitudes, if they are both north or both south : but to *half their difference.*, if one is north and the other south.

58. In middle latitude sailing, all the calculations are made in the same way as in plane sailing, excepting the proportions in which the *difference of longitude* is one of the terms. The departure is derived from the difference of longitude, and the difference of longitude from the departure, in the same manner as in parallel sailing, (Arts. 53, 54.) only substituting in the theorem the term *middle latitude* for latitude.

Theorem I.

As the cosine of middle latitude,
To radius ;
So is the departure,
To the difference of longitude.

59. The learner will be very much assisted in stating the proportions, by keeping the geometrical construction steadily in his mind. In Fig. 20 we have the lines and angles in plane sailing, and in Fig. 22, those in parallel sailing. By

bringing these together, as in Fig. 23, we have all the parts in middle latitude sailing. The two right angled triangles, being united at the common side BC, which is the departure, form the oblique angled triangle ACD.

60. The angle at D is the complement of the middle latitude. (Art. 55.) Then in the triangle ACD, (Trig. 143.)

Sin D : AC : : Sin A : DC ; that is,

THEOREM II.

AS THE COSINE OF MIDDLE LATITUDE,
TO THE DISTANCE ;
SO IS THE SINE OF THE COURSE,
TO THE DIFFERENCE OF LONGITUDE.

61. The two preceding theorems, with the proportions in plane sailing, are sufficient for solving all the cases in middle latitude sailing. A third may be added, for the sake of reducing two proportions to one.

In the triangle BCD (Fig. 23.) Cos BCD : R : : BC : CD
And in the triangle ABC, AB : R : : BC : Tan A.

The means being the same in these two proportions, the extremes are reciprocally proportional. (Alg. 387.) We have then

Cos BCD : AB : : Tan A : CD ; that is,

THEOREM III.

As the cosine of middle latitude,
To the difference of latitude ;
So is the tangent of the course,
To the difference of longitude.

Among the other data in middle latitude sailing, one of the extreme latitudes must always be given.

Example I.

At what distance, and in what direction, is Montock Point from Martha's Vineyard ; the former being in Lat. 41° 04′N. Lon. 72° W., and the latter in Lat. 41° 17′ N. Lon. 70° 48′ W. ?

Here are given the two latitudes and longitudes, to find the course and distance.

The difference of longitude is 72′
The difference of latitude 13′
The middle latitude 41° 10′½

Beginning with the triangle in which there are two parts given, by theorem I,

R : Diff. Lon : : Cos Mid. Lat : Depart.=54.2

And by plane sailing, Case VI,

Diff. Lat : Rad : : Depart : Tan Course=76° 30′$\frac{2}{3}$

Or to find the course at a single statement, by theorem III,

Diff. Lat : Cos Mid. Lat : : Diff. Lon : Tan Course=76° 30′$\frac{2}{3}$

To find the distance by plane sailing, Case III,

Cos Course : Diff. Lat : : Rad : Dist.=55.73

Example II.

A ship leaving New-York light-house in Lat 40° 28′ N. and Lon. 74° 08′ W. sails S. E. 67 miles in 24 hours. Required her latitude and longitude at the end of that time.

By plane sailing,

Rad : Dist : : Cos Course : Diff. Lat.=47′.4

The latitude required, therefore, is 39° 40′.6, and the middle latitude 40° 04′.3.

Then by theorem II,

Cos Mid. Lat : Dist : : Sin Course : Diff. Lon.=61′.9

Or by theorem III,

Cos Mid. Lat : Diff. Lat : : Tan Course : Diff. Lon.=61′.9

The longitude required is 73° 06′.1.

Example III.

A ship leaving a port in Lat 49° 57′ N. Lon. 5° 14′ W. sails S. 39° W. till her latitude is 45° 31′. Required her longitude and distance.

Ans. 10° 34′.3 W. and 342.3 miles.

Example IV.

A ship sailing from Lat. 49° 57′ N. and Lon. 5° 14′ W. steers west of south, till her longitude is 23° 43′, and her departure 789 miles. Required her course, distance, and latitude.

Course 51° 5′ W.
Latitude 39° 20′ N.
Distance 1014 miles.

SECTION III.

MERCATOR'S SAILING.

ART. 62. THE calculations in middle latitude sailing are simple, and sufficiently accurate for short distances, particularly near the equator. But they become quite erroneous, when applied to great distance, and to high latitudes. The only method in common use, which is strictly accurate, is that called *Mercator's Sailing*, or Wright's Sailing. This is founded on the construction of a *chart*, published in 1556 by Gerard Mercator. About forty years after, Mr. Edward Wright gave demonstrations of the principles of this chart, and applied them to the solution of problems in navigation.

63. In the construction of Mercator's chart, the earth is supposed to be a sphere. Yet the meridians, instead of converging towards the poles. as they do on the globe, are drawn *parallel* to each other. The distance of the meridians, therefore, is every where too great, except at the equator. To compensate this, the degrees of *latitude* are proportionally enlarged. On the artificial globe, the parallels of latitude are drawn at equal distances. But on Mercator's chart, the distances of the parallels increase from the equator to the poles, so as every where to have the same ratio to the distances of the meridians, which they have on the globe. Thus in latitude 60°, where the distance of the meridians must be *doubled*, to make it the same as at the equator, a degree of latitude is also made twice as great as at the equator. The dimensions of places are extended in the projection, in proportion as they are nearer the poles. The diameter of an island in latitude 60° would be represented twice as great as if it were on the equator, and its area four times as great.

* Robertson's Navigation, London Phil. Trans. for 1666 and 1696, Hutton's Dictionary, Introduction to Hutton's Mathematical Tables, Bowditch's Practical Navigator, Emerson's and M'Laurin's Fluxions, M'Kay's Navigation, Emerson's Prin. Navig. Barrow's Navigation.

64. *Table of Meridional Parts.* If a meridian on a sphere be divided into degrees or minutes, the portions are all *equal.* But in Mercator's projection, they are extended more and more as they are farther from the equator. To facilitate the calculations in navigation, *tables* have been prepared, which contain the length of any number of degrees and minutes on this extended meridian, or the distance of any point of the projection from the equator. These are called tables of *Meridional Parts.* The common method of computing them is derived from the following proposition.

65. ANY MINUTE PORTION OF A PARALLEL OF LATITUDE,
IS TO A LIKE PORTION OF THE MERIDIAN;
AS RADIUS,
TO THE SECANT OF THE LATITUDE.

For, by the theorem in parallel sailing, (Art. 53.) the cosine of latitude is to radius, as the departure to the difference of longitude measured on the equator; that is, as a part of the parallel of latitude, to a like part of the equator. But on a sphere, the equator and meridian are equal.

Therefore Cos Lat : Rad::*a part of the parallel: a like part of the meridian.*

But Cos Lat : Rad::Rad : Sec Lat. (Trig. 93. 3.)

By equality of ratios then, (Alg. 384.)
A part of the parallel : *a like part of the merid*::Rad : Sec Lat.

By like parts of the parallel of latitude and the meridian are here meant minutes, seconds or other portions of a degree. The proposition is true when applied either to the circles on a sphere, or to the lines in Mercator's projection. For the parts of the latter have the same ratio to each other, as the parts of the former. (Art. 63.) The divisions of Mercator's meridian, however, should be made very small; for the measure of each part is supposed to be taken *at* the parallel of latitude, and not at a distance from it. In the common tables, the meridian is divided into *minutes.*

66. Suppose then that the length of each minute of a degree of Mercator's meridian is required. By the proposition in the last article,

1′ *of the parallel* : 1′ *of the meridian*::Rad : Sec. Lat.

But in this projection, the parallels of latitude are all *equal.*

(Art. 63.) Whatever be the latitude, then, the first term of the proportion is equal to a minute at the equator, or a geographical mile; and if this is assumed as the radius of the trigonometrical tables. (Trig. 100.) the first and third terms are equal, and therefore the second and fourth must be equal also. (Alg. 395.) That is, *the length of any one minute of Mercator's meridian is equal to the natural secant of the latitude of that part of the meridian.*

The *first* The *second* The *third*	minute of the meridian is equal to the secant of	*one* minute, *two* minutes, *three* minutes,

&c. &c.

The table of meridional parts is formed by adding together the several minutes thus found.* Beginning from the equator, an arc of the meridian

of *two* minutes=sec 1′+sec 2′,
of *three* minutes=sec 1 +sec 2′+sec 3′,
of *four* minutes=sec 1′+sec 2′+sec 3′+sec 4′,
&c. &c.

See the table at the end of this number.

To find from the table the length of any given number of degrees and minutes, look for the degrees at the top of the page, and the minutes on the side; then against the minutes, and under the degrees, will be the length of the arc in nautical miles.

67. *Meridional Difference of Latitude.* An arc of Mercator's meridian contained between *two* parallels of latitude, is called meridional difference of latitude. It is found by subtracting the meridional parts for the less latitude from the meridional parts for the greater, if both are north or south; or by adding them, if one latitude is north and the other south.

Thus the lat. of Boston is	42° 23′	Merid. parts	2813
Baltimore	39° 23	Merid. parts	2575
Proper difference of lat.	3°	Merid. diff. of lat.	238

68. If one latitude and the meridional difference of latitude be *given*, the *proper* difference of latitude is found by reversing this process.

* See Note D.

When the two latitudes are on the same side of the equator, subtracting the meridional difference of latitude from the meridional parts for the greater, will give the meridional parts for the less; or adding the meridional difference to the parts for the less latitude, will give the parts for the greater. But if the two latitudes are on opposite sides of the equator, subtracting the parts for the one latitude from the meridional difference, will give the parts for the other.

Thus the meridional difference of latitude between New-York and New-Orleans is			793
The lat. of N. Orleans is	29° 57′	Merid. parts	1885
The lat of New-York	40 42	Merid. parts	2678

69. *Solutions in Mercator's Sailing.* The Solutions in Mercator's sailing are founded on *the similarity of two right-angled triangles, in one of which the perpendicular sides are the proper difference of latitude and the departure; and in the other the meridional difference of latitude and the difference of longitude.*

According to the principle of Mercator's projection, the enlargement of each minute portion of the meridian is proportioned to the enlargement of the parallel of latitude which crosses it. (Art. 63.) Any part of the meridian before it is enlarged, is *proper difference of latitude;* and after it is enlarged, is *meridional difference of latitude.* A part of the parallel, before it is enlarged, is *departure;* and after it is enlarged, is equal to the corresponding *difference of longitude;* because in this projection, the distance of the meridians is the same on any parallel, as at the equator, where longitude is reckoned.

If then we take a small portion of the distance which a ship has sailed, as Am, (Fig. 18.)

Prop. Dif Lat. Ao : Depart. om : : Merid. Dif Lat : Dif. Lon.

In the triangle ABC, (Fig. 24.) let the angle at A = the course oAm, (Fig. 18.) AB=the proper difference of latitude. AC=Am+mn+nC the distance, and BC= om+sn+tC the departure. Then as the triangles Aom, msn, ntC are each similar to the triangle ABC, (Fig. 24.) the difference of latitude for any one of the small distances as Am, is to the corresponding departure; as the *whole difference of latitude* AB to the *whole departure* BC.

P. Dif. Lat. AB : Dep. BC :: Mer. Dif. Lat. { for Am : for mn : for nC : } : Dif. Lon. { for Am, for mn, for nC. }

But the whole meridional difference of latitude for the distance AC, is equal to the sum of the differences for Am, mn, and nC; and the whole difference of longitude is equal to the sum of the differences for Am, mn, and nC. Therefore, (Alg. 388. Cor. 1.)

Prop. Dif. Lat. AB : Dep. BC::Merid. Dif. Lat : Dif. Lon.

Extend AB, (Fig. 24.) making Al equal to the meridional difference of latitude corresponding to the proper difference of latitude AB; from L draw a line parallel to BC, and extend AC to intersect this in D. Then is DL the *difference of longitude*. For it has been shown that the difference of longitude is a *fourth proportional* to the proper difference of latitude, the departure, and the meridional difference of latitude; and by similar triangles,

AB : BC::AL : LD.

70. To solve all the cases, then, in Mercator's sailing, we have only to represent the several quantities by the parts of two similar right-angled triangles, as ABC and ALD, (Fig. 24.) and to find their sides and angles. In the smaller triangle ABC the parts are the same as in plane sailing, and the calculations are made in the same manner. The sides AL and DL are added for finding the difference of longitude; or when the difference of longitude is given, to derive from it one of the other quantities. The *course* is common to both the triangles, and the complement of the course is either ACB or ADL. The hypothenuse AD is not one of the quantities which are given or required in navigation.

71. In the similar triangles ABC, ALD, (Fig. 24.)

AB : AL::BC : LD; that is,

Theorem I.

As the proper difference of latitude,
To the meridional difference of latitude;
So is the departure,
To the difference of longitude.

72. In the triangle ALD, if AL be made radius,

Rad : Tan A::AL : DL; that is,

Theorem II.

As radius,
To the tangent of the course;
So is the meridional difference of latitude,
To the difference of longitude.

By this theorem, the difference of longitude may be calculated, without previously finding the departure.

73. In Mercator's, as well as in middle latitude sailing, *one latitude* must always be given. This is requisite in converting proper difference of latitude and meridional difference of latitude into each other. (Arts. 67, 68.)

74. When the difference of latitude is *very small*, the difference of longitude will be more correctly found by middle latitude sailing, than by Mercator's sailing; unless a table is used in which the meridional parts are given to *decimals*. Mercator's sailing is strictly correct in *theory*. But the common tables are not carried to a degree of exactness, sufficient to mark very minute differences. On the other hand, the errour of middle latitude sailing is *diminished*, as the difference of latitude is lessened.

Example I.

The latitudes of Montock and Martha's Vineyard are { 41° 4′N. / 41 17N. }

Their longitudes { 72° W. / 70 48′ W. }

Required the course and distance from one to the other.

Lat. of Martha's Vin,	41° 17′	Merid. parts	2724
of Montock	41 04	Merid. parts	2707
Proper Diff. of Lat.	13′	Mer. Dif. Lat.	17 (Art. 67.)

The difference of longitude is 1° 12′=72 miles.

To find the course by theorem II, (Fig. 24.)
Merid. Diff. Lat : Diff. Lon : : Rad : Tan Course=76° 43.′

To find the distance by plane sailing,
Cos Course : Prop. Diff. Lat. : : Rad : Dist.=56.58.

The results by middle latitude sailing, page 31, are a little different, as that method is not perfectly accurate.

Example II.

A ship sailing from the Lizard in Lat. 49° 57′ N. Lon. 5° 15′ W. proceeds S. 39° W till her latitude is found by observation to be 45° 31′ N. What is then her longitude, and what distance has she run?

Here are given the difference of latitude and the course, to find the distance and the difference of longitude.

The proper difference of latitude is 4° 26′=266′
The meridional difference of latitude 396

Then by plane sailing,

Cos Course : Prop. Diff. Lat.::Rad : Dist.=342.3.

And by theorem II,

Rad : Tan Course::M. Dif. Lat : Dif. Lon.=320′.7=5°20′.7

This added to the longitude of the Lizard 5° 14′ gives the longitude of the ship 10° 34′.7 W.

Example III.

A ship sailing from Lat. 49° 57′ N. and Lon. 5° 14′ W. steers west of south, till her latitude is 39° 20′ N. and her departure 789 miles. Required her course, distance. and longitude.

The proper difference of latitude is 10° 37′=637′
The meridional difference of latitude 899

Then by theorem I, (Fig. 24)

P. Dif. Lat. : M. Diff. Lat::Dep : Diff. Lon.=1113′.5=18°33′.5

The longitude of the ship is therefore 23° 47′½

And by plane sailing,

Prop Diff. Lat : Rad::Depart : Tan Course=51° 5′
Rad : Prop. Diff. Lat.::Sec Course : Distance=1014 miles.

Example IV.

A ship sailing from a port in Lat. 14° 45′ N. Lon. 17° 33′ W. steers S. 28° 7′½ W. till her longitude is found by observation to be 29° 26′ W. Required her distance and latitude.

The difference of longitude is 11° 53′=713′.

By theorem II,

Tan Course : Rad::Dif. Lon : M. Dif. Lat.=1334 S.

Lat. of the port	14° 45′ N.	Merid. parts	895 N.
of the ship	7 18 S.	Merid. parts	439 S. (Art. 68.)
Diff. of Lat.	22° 3′=1323′		

By plane sailing,

Cos Course : Diff. Lat::Rad : Distance=1500 miles.

Example V.

A ship sails 300 miles between north and west, from Lat. 37° N. to 41° N. What is her course and difference of longitude?

The course is N. 36° 52′ W., and the difference of longitude 3° 52′.

Example VI.

A ship sails S. 67° 30′ E. from Lat. 50° 10′ S till her departure is 957 miles. What is her distance, difference of latitude, and difference of longitude?

The distance is 1036 miles.
The difference of latitude 6° 36′.4
The difference of longitude 26° 53′

Example VII.

A ship sailing from Lat. 26° 13′ N. proceeds S. 27° W. 231 miles. What is her difference of latitude and difference of longitude?

Example III.

A ship sailing from Lat. 14° S. 260 miles, between south and west, makes her departure 173 miles. What is her course, difference of latitude, and difference of longitude?*

* See Note F.

SECTION IV.

TRAVERSE SAILING.

ART. 75. BY the methods in the preceding sections, are found the difference of latitude, departure, &c. for a *single course*. But it is not often the fact that a ship proceeds from one port to another in a direct line. Variable and contrary winds frequently render a change of direction necessary every few hours. The irregular path of the ship sailing in this manner, is called a *traverse*.

Resolving a traverse is reducing the compound course to a single one. This is commonly done at sea every noon. From the several courses and distances in the log-book, the departure, difference of latitude, &c. are determined for the whole 24 hours. In the same manner, the courses of several successive *days* are reduced to one, so as to ascertain, at any time, the situation of the ship. The following methods by construction and by calculation, are sufficiently accurate for short distances, at least near the equator.

76. *Geometrical construction of a traverse.* To construct a traverse, draw a meridian line and lay down the first course and distance; from the end of this, lay down the *second* course and distance; from the end of that, a *third* course, &c. Then draw a line connecting the extremities of the first and last of these, to show the whole distance, and the direction of the ship from the point of starting.

This will be easily understood by an example.

Example I.

A ship sails from a port in Lat. 32° N., and in 24 hours makes the following courses;

1. N. 25° E. 16 miles,
2. S. 54° E. 11,
3. N. 13° W 7,
4. N. 61° E. 5,
5. N. 38° W. 18,

It is required to find the departure, difference of latitude, distance, and course, for the whole traverse.

On A as a centre (Fig. 25.) describe a circle and draw the meridian NAS. Then considering the upper part as north, the right hand east, and the left hand west, draw the lines A1, A2, A3, A4, and A5, to correspond with the several courses; that is, make the angle NA1=25°, SA2=54°; NA3=13°, NA4=61°, and NA5=38°.

Make A1B=16, BC=11 and parallel to A2, CD=7 and parallel to A3, DF=5 and parallel to A4, FG=18 and parallel to A5; join AG, and draw GP perpendicular to NS.

Then if the surface of the ocean be considered as a plane, G is the place of the ship at the end of the 24 hours, AG the *distance* from port, PG the *departure*, AP the *difference of latitude*, and GAP the *course*. The angles may be measured by a line of chords, and the distances taken from a scale of equal parts.* (Trig. 148, 161, 2.)

The distance is	32.3 miles.
The departure	7.38
The difference of lat.	31 45
The course	13° 12′

77. *Resolving a traverse, by Calculation or Inspection.* When a ship sails on different courses for a short time, the difference of latitude, at the end of that time, is equal to the difference between the sum of the northings and the sum of the southings, and the departure is nearly equal to the difference between the sum of the eastings and the sum of the westings. See Arts. 78. 79, If then the difference of latitude and the departure for each course be found by calculation or inspection, and placed in separate columns in a table; the difference of latitude for the whole time may be obtained exactly, and the departure nearly, by addition and subtraction; and the corresponding distance and course may be determined by trigonometrical calculation or inspection, as in the last case of plane sailing. (Art. 49.)

The following table contains the courses, distances. departure, and difference of latitude in the preceding example. See Fig. 25.

* See Note F.

TRAVERSE TABLE.

Courses.	*Distances.*	*Diff. Lat.*		*Departure.*	
		N.	S.	E.	W.
1. N. 25° E.	AB 16	14.50		6.76	
2. S. 54° E.	BC 11		6.47	8.90	
3. N. 13° W.	CD 7	6.82			1.57
4. N. 61° E	DF 5	2.42		4.37	
5. N. 38° W.	FG 18	14.18			11.08
		37.92	6.47	20.03	12.65
		6.47		12.65	
N 13° 12⅓ E.	AG. 32.3	31.45		7.38	

The sum of the northings is 37.92. Subtracting from this the southing 6.47, we have the difference of latitude AP 31.45 N.

The sum of the eastings is 20.03. Subtracting from this the sum of the westings 12.65 we have the departure GP 7.38 E. Then (Art. 49.)

Diff. Lat : Rad::Depart : Tan Course NAG=13° 12′⅓
Rad : Diff. Lat::Sec. Course : Distance AG=32.3

The latitude of the port is	32° N.
The difference of latitude	0° 31′.45 N.
The latitude of the ship	32° 31′.45 N.
The meridional difference of lat.	37.5

Then by Mercator's sailing,

Rad : Tan Course::Merid. Diff. Lat : Diff. Lon.=8′.8

Example II.

A ship sailing from a port in Lat. 42° N. makes the following courses and distances.

1. S. 13° E. 21 miles,
2. S. 18° W. 16,
3. N. 84° E. 9,
4. S. 67° E. 12,
5. N. 78° E. 14,
6. S. 12° W. 35.

The difference of latitude, departure, &c. are required.

The departure is	26′.19 E.
The diff. of latitude,	1° 10′$\frac{3}{4}$ S.
The diff of longitude,	35′.07
The direct course, S.	20° 18′$\frac{2}{3}$ E.
The distance,	75$\frac{1}{2}$ miles.

Accurate method of resolving a traverse.

78. The preceding method of resolving a traverse is frequently used at sea, because it is simple, and in most cases is sufficiently accurate for a run of 24 hours. But it is founded on the assumption, that when a ship sails from one place to another by *several courses*, she makes the *same departure*, as if she had proceeded by a *single course* to the same place. This is not strictly true. Suppose a vessel, instead of sailing directly from A to C, (Fig. 18.) proceeds by one course from A to H, and then by a different course from H to C. In the compound course, the whole departure, is $bd+g\text{H}+t\text{C}$; (Art. 40.) which on account of the obliquity of the meridians, is *less* than $om+sn+t\text{C}$, the departure on the single course. If the compound course had been on the other side of the single one, nearer the equator, the departure would have been *greater*.

79. But the *difference of latitude* is the same, whether the ship proceeds from one place to the other, on a single course, or on several. The difference of latitude AB (Fig. 18.)= $\text{A}o+ms+nt=\text{A}b+dg+\text{H}t$. The *difference of longitude* is also the same, whether the course is single or compound. For the difference of longitude is the distance between the meridians of the two places measured on the equator.

If then the difference of latitude and difference of longitude be calculated for each part of the compound course; the whole difference of latitude and difference of longitude will be found by addition and subtraction; and from these may be determined the direct course and distance. The difference of longitude for each course may be obtained independently of the departure, by theorem II. of Mercator's sailing.

It will facilitate the calculation of the longitude, to place in the traverse table, the latitudes at the beginning and end of each of the courses, the corresponding meridional parts, and the meridional differences of latitude.

In the following example, the courses and distances are the same as in Art. 76. Ex. 1. The port from which the ship is supposed to sail, is in latitude 32° N.

TRAVERSE TABLE.

Courses.	Dist.	Diff Lat.		Latitudes.	Merid. Parts.	Merid. Dif. Lat.	Diff. Long.	
		N.	S.				E.	W.
				32°	2028	17.5		
1. N. 25° E	16	14.50	6.47	32 14'.50	2045 5	7.5	8.16	
2. S. 54° E.	11			32 8 03	2038	7.8	10.32	
3. N 13° W.	7	6.82		32 14.85	2045.8	2 5		1.80
4 N. 61° E	5	2.42		32 17.27	2048 3	17.2	4.51	
5. N. 38° W.	18	14 18		32 31 45	2065 5			13.44
		37.92	647				22.99	15.24
		6.47					15.24	
11° 40′ 37″	32.12	31.45					7.75	

The difference of longitude is here found to be 7.75, and in Art. 77, 8′.8 the errour there being 1.05.

To find the direct course and distance from the port to the place of the ship.

Merid. Dif. Lat : Dif. Lon : : Rad : Tan Course = 11° 40′ 37″
Rad : Prop. Dif. Lat : : Sec Course : Distance = 32.12.

By comparing the results here with those in Art. 77, it will be seen that a small errour was introduced there, both into the *course* and the *distance*, by making them dependent on the departure ; which being obtained from the several courses, is not the same as for a single course. (Art. 78.)

Ex. 2 A ship sailing from a port in latiude 78°. 15′ N. makes the following courses and distances.

1. N. 67° 30′ W. 154 miles.
2. S. 45 W. 96
3. N. 50 37½ W. 89
4. N. 11 15 E. 110
5. N. 36 33¾ W. 56
6. S. 19 41¼ E. 78

Required the difference of latitude, the difference of longitude, and the distance the ship must have sailed, to reach the same place on a single course.

The difference of latitude is	2° 7′
The difference of longitude	22° 29′
The direct course	N. 63° 1′ W.
The distance	279.9 miles.

SECTION V.

MISCELLANEOUS ARTICLES.

I. The Plane Chart.

80. THE Charts commonly used in navigation are either *Plane Charts*, or *Mercator's Charts*. The latter are generally to be preferred. But plane charts will answer for short distances, such as the extent of a harbour or small bay.

In the construction of the plane chart, that part of the surface of the globe which is represented on it, is supposed to be a *plane*. The meridians are drawn parallel; and the lines of latitude at equal distances. Islands, coasts, &c. are delineated upon it, by laying down the several parts according to their known latitudes and longitudes.

81. On a chart extending a small distance, each side of the *equator*, the meridians ought to be at the same distance from each other, as the parallels of latitude. A similar construction is frequently applied to different parts of the globe. But this renders the chart much more incorrect than is necessary. A circular island in latitude 60 would, by such a construction, be thrown into a figure whose length from east to west would be twice as great, as from north to south; the comparative distance of the meridians being made twice as great as it ought to be. (Art. 53. Trig. 96. cor.)

But when the chart extends only a few degrees, if the distance of the meridians is proportioned to the distance of the parallels of latitude, *as the cosine of the mean latitude to radius;* (Art. 53.) the representation will not be materially incorrect. The meridian distance in the *middle* of the chart will be exact. On one side, it will be a little too great; and on the other, a little too small.

82. *To construct a Plane Chart*, then, on one side of the paper draw a scale of equal parts, which are to be counted as degrees or minutes of latitude, according to the proposed extent of the chart. Through the several divisions, draw the

parallels of latitude, and at right angles to these, draw the meridians in such a manner, that their distance from each other shall be to the distance of the parallels of latitude, as the cosine of the latitude of the middle of the chart, to radins.

After the lines on all the sides are graduated, the positions of the several places which are to be laid down, may be determined, by applying the edge of a rule or strip of paper, to the divisions for the given degree of longitude on each side, and another to the divisions for the degree of latitude. In the intersection of these, will be the point required.

The distance which a ship must sail, in going from one place to another, on a single course, may be nearly found, by applying the measure of the interval between the two places, to the scale of miles, of latitude on the side of the chart.*

II. Construction of Mercator's Chart.

83. In Mercator's chart, the meridians are drawn at equal distances, and the parallels of latitude at unequal distances, proportioned to the meridional differences of latitude. (Arts. 63, 67.) To construct this chart, then, make a scale of equal parts on one side of the paper, for the lowest parallel of latitude which is to be laid down, and divide it into degrees and minutes. Perpendicular to this, and through the dividing points for degrees, draw the lines of longitude. For the second proposed parallel of latitude, find from the table, (Art. 67.) the meridional difference of latitude between that and the parallel first laid down, and take this number of minutes from the scale on the chart, for the interval between the two parallels. In the same manner, find the interval between the second and third parallels, between the third and fourth, &c. till the projection is carried to a sufficient extent.

Places whose latitudes and longitudes are known, may be laid down in the same manner as on the plane chart, by the intersection of the meridians and lines of latitude passing through them.

If the chart is upon a small scale, the least divisions on the graduated lines may be *degrees* instead of minutes; and the meridians and parallels may be drawn for every fifth or every tenth degree. But in this case, it will be necessary to di-

*See Note G.

vide the meridional differences of latitude by 60, to reduce them from minutes to degrees.

84. *The Line of Meridional Parts* on *Gunter's scale* is divided in the same manner as Mercator's Meridian, and corresponds with the *table* of meridional parts; except that the numbers in the latter are *minutes*, while the divisions on the other are *degrees*. Directly beneath the line of meridional parts, is placed a line of *equal* parts. The divisions of the latter being considered as degrees of longitude, the divisions of the former will be degrees of latitude adapted to the same scale. The meridional *difference* of latitude is found, by extending the compasses from one latitude to the other.

A chart may be constructed from the scale, by using the line of equal parts for the degrees of longitude, and the line of meridional parts for the intervals between the parallels of latitude.

85. It is an important property of Mercator's chart, that all the rhumb-lines projected on it are *straight* lines. This renders it, in several respects, more useful to navigators, than even the artificial globe. By Mercator's sailing, theorem II. (Art. 72.)

$$\text{Merid. Diff. Lat} : \text{Diff. Lon} :: \text{Rad} : \text{Tan Course}$$

So that, while the course remains the same, the ratio of the meridional difference of latitude to the difference of longitude is *constant*. If A, C, C′, and C″ (Fig. 26.) be several points in a rhumb-line, AB AB′, and AB″, the corresponding meridional differences of latitude, and BC, B′C′, B″C″, the differences of longitude; then

$$AB : BC :: AB' : B'C' :: AB'' : B''C''.$$

Therefore ABC, AB′C′, and AB″C″, are similar triangles, and ACC′C″ is a right line. (Euc. 32. 6.)

III. Oblique Sailing.

86. The application of oblique angled trigonometry to the solution of certain problems in navigation, is called oblique sailing. It is principally used in bays and harbours, to determine the bearings of objects on shore, with their distances from the ship and from each other. A few examples will be sufficient here, in addition to those already given under heights and distances.

One of the cases which most frequently occurs, is that in which the distance of a ship from land is to be determined, when leaving a harbour to proceed to sea. This is necessary, that her difference of latitude and departure may be reckoned from a fixed point, whose latitude and longitude are known.

The distance from land is found, by taking the bearing of an object from the ship, then running a certain distance, and taking the bearing again. The course being observed, there will then be given the angles and one side of a triangle, to find either of the remaining sides.

Example I.

The point of land C (Fig. 27.) is observed to bear N 67° 30′ W. from A. The ship then sails S. 67° 30′ W. 9 miles from A to B; and the direction of the point from B is found to be N. 11° 15′ E. At what distance from land was the ship at A?

Let NS and N′S′ be meridians passing through A and B. Then subtracting CAN and BAS each $67°\frac{1}{2}$ from 180, we have the angle CAB=45°. And subtracting CBN′ $11°\frac{1}{4}$ from BAS or its equal ABN′, we have ABC=$56°\frac{1}{4}$. The angle at C is therefore 78° 45′. And

Sin C : AB::Sin B : AC=7.63 miles.

Example II.

New-York light-house on Sandy-Point is in Lat. 40° 28′ N. Lon. 74° 8′ W. A ship observes this to bear N. 76° 16′ W., and after sailing S 35° 10′ W. 8 miles, finds the bearing to be N. 17° 13′ W. Required the latitude and longitude of the ship, at the first observation.

The latitude is 40° $26'\frac{1}{4}$.
The longitude 73 $58\frac{1}{2}$

In this example, as the difference of latitude is small, the difference of longitude is best calculated by middle latitude sailing. (Art. 74.)

Example III.

A merchant ship sails from a certain port S. 51° E. at the rate of 8 miles an hour. A privateer, leaving another port 7 miles N. E. of the first, sails at the rate of 10 miles an hour. What must be the course of the privateer, to meet the ship, without a change of direction in either?

Ans. S. 7° 43′ E.

Example IV.

Two light-houses are observed from a ship sailing S. 38° W. at the rate of 5 miles an hour. The first bears N. 21° W., the other N. 47° W. At the end of two hours, the first is found to bear N. 5° E., the other N. 13° W. What is the distance of the light-houses from each other?

Ans. 6 miles and 30 rods.

IV. Current Sailing.

87. When the measure given by the log-line is taken as the rate of the ship's progress, the *water* is supposed to be at *rest*. But if there is a tide or current, the log being thrown upon the water, and left at liberty, will move with it, in the same direction, and with the same velocity. The rate of sailing, as measured by the log, is the motion *through the water*.

If the ship is steered in the direction of the current, her whole motion is equal to the rate given by the log, *added* to the rate of the current. But if the ship is steered in opposition to the current, her absolute motion is equal to the *difference* between the current, and the rate given by the log. In all other cases, the current will not only affect the velocity of the ship, but will change its direction.

Suppose that a river runs directly south, and that a boat in crossing it is steered before the wind, from west to east. It will be carried down the stream as fast, as if it were merely floating on the water in a calm. And it will reach the opposite side as soon, as if the surface of the river were at rest. But it will arrive at a different point of the shore.

Let AB (Fig. 28.) be the direction in which the boat is steered, and AD the distance which the stream runs, while the boat is crossing. If DC be parallel to AB, and BC parallel to AD; then will C be the point at which the boat proceeding from A will strike the opposite shore, and AC will be the distance. For it is driven across by the wind, to the side BC, in the same time that it is carried down by the current, to the line DC.

In the same manner, if A*m* be any *part of* AB, and *mn* be the corresponding progress of the stream, the distance sailed will be A*n*. And if the velocity of the ship and of the stream continue uniform, A*m* is to *mn*, as AB to BC; so that A*n*C is a *straight line.* (Euc 32. 6.) The lines AB, BC, and AC, form the three sides of a triangle. Hence,

88. If the direction and rate of a ship's motion through the water, be represented by the position and length of one side of a triangle, and the direction and rate of the current, by a second side; the absolute direction and distance will be shown by the third side.

Example I.

If the breadth of a river running south (Fig. 28.) be 300 yards, and a boat steers S. 75 E. at the rate of 10 yards in a minute, while the progress of the stream is 24 yards in a minute; what is the actual course, and what distance must the boat go in crossing?

Cos BAP : AP::R : AB=310.6
And 10 : 24::AB : BC=745.44

Then in the triangle ABC,

$(BC+AB) : (BC-AB) :: \text{Tan } \frac{1}{2}(BAC+BCA) : \text{Tan } \frac{1}{2}(BAC-BCA)=17^\circ\ 33'\ 50''$

The angle BAC is $55^\circ\ 3'\ 50''$ Then
Sin BAC : BC::Sin ABC : AC=879 the distance.
And DAC=BCA=$19^\circ\ 56'\ 10''$ the course.

Example II.

A boat moving through the water at the rate of five miles an hour, is endeavouring to make a certain point lying S. $22\frac{1}{2}^\circ$ W. while the tide is running S. $78\frac{3}{4}^\circ$ E. three miles an hour. In what direction must the boat be steered, to reach the point by a single course?

Ans. S. $58^\circ\ 33'$ W.

89. But the most simple method of making the calculation for the effect of a current, in common cases, especially in resolving a traverse, is to consider the direction and rate of the current as *an additional separate course and distance;* and to find the corresponding departure and difference of latitude. A boat sailing from A (Fig. 28.) by the united action of the wind and current, will arrive at the same point, as if it were first carried by the wind alone from A to B, and then by the current alone from B to C.

Example I.

A ship sails S. 17° E. for 2 hours, at the rate of 8 miles an hour ; then S. 18° W. for 4 hours, at the rate of 7 miles an hour ; and during the whole time, a current sets N. 76° W. at the rate of 2 miles an hour. Required the direct course and distance.

		Dist.	N.	S.	E.	W.
First Course	S. 17° E.	16		15.3	4.68	
Second do.	S. 18° W.	28		26.6		8.65
Current	N. 76° W.	12	2.9			11.64
				41.9		20.29
				2.9		4.68
			Dif. Lat.	39.	*Dep.*	15.61

The course is 2[illegible]° 48′ 50″, and the distance 42 miles.

Example II.

A ship sails S. E. at the rate of 10 miles an hour by the log, in a current setting E. N. E. at the rate of 5 miles an hour. What is her true course? and what will be her distance at the end of two hours?

The course is 66° 13′, and the distance 25.56 miles.

V. Hadley's Quadrant.

90. In the preceding sections, has been particularly explained the process of determining the place of a ship from her course and distance, as given by the compass and the log. But this is subject to so many sources of errour, from variable winds, irregular currents, lee-way, uncertainty of the magnetic needle, &c. that it ought not to be depended on, except for short distances, and in circumstances which forbid the use of more unerring methods. The mariner who hopes to cross the ocean with safety, must place his chief reliance, for a knowledge of his true situation from time to time, on observations of the *heavenly bodies*. By these the latitude and longitude may be generally ascertained, with a sufficient degree of exactness. It belongs to astronomy to explain the methods of making the calculations. The subject will not be anticipated in this place, any farther than to give a description of the quadrant of reflexion, commonly called *Hadley's Quadrant*,* by which the altitudes of the heavenly bo-

*See Note H.

dies, and their distances from each other, are usually measured at sea. The superiority of this, over most other astronomical instruments, for the purposes of navigation, is owing to the fact, that the observations which are made with it, *are not materially affected by the motion of the vessel.*

91. In explaining the construction and use of this quadrant, it will be necessary to take for granted the following simple principles of Optics.

1. The progress of light when it is not obstructed, or turned from its natural course by the influence of some contiguous body, is in *right lines.* Hence a minute portion of light called a ray, may be properly represented by a line.

2. Any object appears in the direction in which the light from that object *strikes the eye.* If the light is not made to deviate from a right line, the object appears in the direction in which it really is. But if the light is reflected, as by a common mirror, the object appears not in its true situation, but in the direction of the glass, from which the light comes to the eye.

3. *The angle of reflection is equal to the angle of incidence;* that is, the angles which the reflected and the incident rays make with the surface of the mirror, are equal; as are also the angles which they make with a perpendicular to the mirror.

92. From these principles is derived the following propositiou; *When light is reflected by two mirrors successively, the angle which the last reflected ray makes with the incident ray, is* DOUBLE *the angle between the mirrors.*

If C and D (Fig. 29.) be the two mirrors, a ray of light coming from A to C, will be reflected so as to make the angle DCM=ACB; and will be again reflected at D, making HDM=CDE. Continue BC and ED to H, draw DG parallel to BH, and continue AC to P. Then is CPM the angle which the last reflected ray DP makes with the incident ray AC; and DHM is the angle between the mirrors.

By the preceding article, with Euc. 29. 1 and 15. 1,

$$GDC=DCM=ACB=PCM$$

And $HDM=EDC=EDG+GDC=DHM+PCM$

But by Euc. 32. 1 and 15. 1,

$$CPM+PCM=DHM+HDM=2DHM+PCM$$

Therefore $CPM=2DHM$

Cor. 1. If the two mirrors make an angle of a certain number of degrees, the *apparent direction* of the object will be changed twice as many degrees. The object at A, seen by the eye at P, without any mirror, would appear in the direction PA. But after reflection from the two mirrors, the light comes to the eye in the direction DP, and the apparent place of the object is changed from A to R.

Cor. 2. If the two mirrors be *parallel*, they will make no alteration in the apparent place of the object.

93. The principal parts of Hadley's Quadrant are the following;

1. A *graduated arc* AB (Fig. 17.) connected with the radii AC and BC.

2. An *index* CD, one end of which is fixed at the centre C, while the other end moves over the graduated arc.

3. A plane mirror called the *index glass*, attached to the index at C. Its plane passes through the centre of motion C, and is perpendicular to the plane of the instrument; that is, to the plane which passes through the graduated arc, and its centre C.

4. Two other plane mirrors at E and M, called *horizon glasses*. Each of these is also perpendicular to the plane of the instrument. The one at E, called the *fore horizon glass*, is placed parallel to the index glass when the index is at 0. The other, called the *back horizon glass*, is perpendicular to the first and to the index at 0. This is only used occasionally, when circumstances render it difficult to take a good observation with the other.

A *part* of each of these glasses is covered with quicksilver, so as to act as a mirror; while another part is left transparent, through which objects may be seen in their true situation.

5. Two *sight vanes* at G and L, standing perpendicular to the plane of the instrument. At one of these, the eye is placed to view the object, by looking on the opposite horizon glass. In the fore sight vane at G, there are two perforations, one directly opposite the transparent part of the fore horizon glass, the other opposite the silvered part. The back sight vane at L has only one perforation, which is opposite the centre of the transparent part of the back horizon glass.

6. *Coloured glasses* to prevent the eye from being injured by the dazzling light of the sun. These are placed at H, be-

tween the index mirror and the fore horizon glass. They may be taken out when necessary, and placed at N between the index mirror and the back horizon glass.

94. This instrument, which is in form an *octant*, is called a *quadrant*, because the graduation extends to 90 degrees, although the arc on which these degrees are marked is only the eighth part of a circle. The light coming from the object is first reflected by the index glass C, (Fig. 17.) and thrown upon the horizon glass E, by which it is reflected to the eye at G. If the index be brought to 0, so as to make the index glass and the horizon glass *parallel;* the object will appear in its true situation. (Art. 92. Cor. 2.) But if the index glass be turned, so as to make with the horizon glass an angle of a certain number of degrees; the apparent direction of the object will be changed *twice as many degrees.*

Now the graduation is adapted to the apparent change in the situation of the object, and not to the motion of the index. If the index move over 45 degrees, it will alter the apparent place of the object 90 degrees. The arc is commonly graduated a short distance on the other side of 0 towards P. This part is called the *arc of excess.*

95. The quadrant is used at sea, to measure the angular distances of the heavenly bodies from each other, and their elevations above the horizon. One of the objects is seen in its true situation, by looking through the transparent part of the horizon glass. The other is seen by reflection, by looking on the silvered part of the same glass. By turning the index, the apparent place of the latter may be changed, till it is brought in contact with the other. The motion of the index which is necessary to produce this change, determines the distance of the two objects.*

96. *To find the distance of the moon from a star.* Hold the quadrant so that its plane shall pass through the two objects. Look at the star through the transparent part of the horizon glass, and then turn the index till the nearest edge of the image of the moon is brought in contact with the star. This will measure the distance between the star and one *edge* of the moon. By adding the semi-diameter of the moon, we shall have the distance of its *centre* from the star.

* For the *adjustments* of the quadrant, see Vince's Practical Astronomy, Mackay's Navigation, or Bowditch's Practical Navigator.

The distance of the sun from the moon, or the distance of two stars from each other, may be measured in a similar manner.

97. *To measure the altitude of the sun above the horizon.* Hold the instrument so that its plane shall pass through the sun, and be perpendicular to the horizon Then move the index till the lower edge of the image of the sun is brought in contact with the horizon, as seen through the transparent part of the glass.

The altitude of any other heavenly body may be taken in the same manner.

98. To measure altitudes by the *back observation.* When the index stands at *o*. the index glass is at right angles with the back horizon glass. (Art. 93.) The apparent place of the object, as seen by reflection from this glass, must therefore be changed 180 degrees ; (Art. 92 Cor. 1.) that is, it must appear in the opposite point of the heavens. In taking altitudes by the back observation, if the object is in the east, the observer faces the west ; or if it be in the south, he faces the north ; and moves the index, till the image formed by reflexion is brought down to the horizon.

This method is resorted to, when the view of the horizon in the direction of the object is obstructed by fog, hills, &c.

99. *Dip or Depression of the Horizon.* In taking the altitude of a heavenly body at sea, with Hadley's Quadrant, the reflected image of the object is made to coincide with the most distant visible part of the *surface of the ocean.* A plane passing through the eye of the observer, and thus touching the ocean, is called the *marine horizon* of the place of observation If BAB′ (Fig. 13.) be the surface of the ocean, and the observation be made at T, the marine horizon is TA. But this is different from the *true horizon* at T, because the eye is elevated above the surface. Considering the earth as a sphere, of which C is the centre, the true horizon is TH perpendicular to TC. The marine horizon TA falls *below* this. The angle ATH is called the *dip* or *depression* of the horizon. This varies with the height of the eye above the surface. Allowance must be made for it, in observations for determining the altitude of a heavenly body above the true horizon.

In the right-angled triangle ATC, the angle ACT is equal to the angle of depression ATH ; for each is the complement of ATC. The side AC is the semi-diameter of the earth, and the hypothenuse CT is equal to the same semi-diameter added to BT the height of the eye. Then

AC : R : : TC : Sec ACT=ATH the depression.*

100. *Artificial Horizon.* Hadley's Quadrant is particularly adapted to measuring altitudes *at sea.* But it may be made to answer the same purpose on land, by means of what is called an artificial horizon. This is the level surface of some fluid which can be kept perfectly smooth. Water will answer, if it can be protected from the action of the wind, by a covering of thin glass or talc which will not sensibly change the direction of the rays of light. But quicksilver, Barbadoes tar, or clear molasses, will not be so liable to be disturbed by the wind. A small vessel containing one of these substances, is placed in such a situation that the object whose altitude is to be taken may be reflected from the surface. As this surface is in the plane of the horizon, and as the angles of incidence and reflection are equal, (Art. 91.) the image seen in the fluid must appear as far *below* the horizon, as the object is *above*. The distance of the two will, therefore, be *double* the altitude of the latter. This distance may be measured with the quadrant, by turning the index so as to bring the image formed by the instrument to coincide with that formed by the artificial horizon.

101. *The Sextant* is a more perfect instrument than the quadrant, though constructed upon the same principle. Its arc is the sixth part of a circle, and is graduated to 120 degrees. In the place of the sight vane, there is a small telescope for viewing the image. There is also a magnifying glass, for reading off the degrees and minutes. It is commonly made with more exactness than the quadrant, and is better fitted for nice observations, particularly for determining longitude, by the angular distances of the heavenly bodies.

A still more accurate instrument for the purpose is the *Circle of Reflexion.* For a description of this, see Borda on the Circle of Reflexion, Rees' Cyclopedia, and Bowditch's Practical Navigator.

* See Note I, and Table II.

SURVEYING.

SECTION I.

SURVEYING A FIELD BY MEASURING ROUND IT.

ART. 105. THE most common method of surveying a field is to measure the length of each of the sides, and the angles which they make with the meridian. The lines are usually measured with a chain, and the angles with a compass.

106. *The Compass.* The essential parts of a Surveyor's Compass are a graduated circle, a magnetic needle, and sight-holes for taking the direction of any object. There are frequently added a spirit level, a small telescope, and other appendages. The instrument is called a Theodolite, Circumferentor, &c. according to the particular construction, and the uses to which it is applied.

For measuring the angles which the sides of a field make with each other, a graduated circle with sights would be sufficient. But a needle is commonly used for determining the position of the several lines with respect to the meridian. This is important in running boundaries, drawing deeds, &c. It is true, the needle does not often point directly north or south. But allowance may be made for the variation, when this has been determined by observation. See Sec. V.

107. *The Chain.* The Surveyor's or Gunter's chain is four rods long, and is divided into 100 *links*. Sometimes a half chain is used, containing 50 links. A rod, pole, or perch, is $16\frac{1}{2}$ feet. Hence

1 Link $=7.92$ inches $=\frac{2}{3}$ of a foot nearly.
1 Rod $=$ 25 links $=16\frac{1}{2}$ feet.
1 Chain $=100$ links $=66$ feet.

108. The measuring unit for the *area* of a field is the *acre*, which contains 160 square rods. If then the contents in square rods be divided by 160, the quotient will be the number of acres. But it is commonly most convenient to make the computation for the area in square *chains or links*, which are decimals of an acre. For a square chain $=4\times4=16$ square rods, which is the tenth part of an acre. And a square link $=\frac{1}{100}\times\frac{1}{100}=\frac{1}{10000}$ of a square chain $=\frac{1}{100000}$ of an acre. Or thus,

625 links, or	$272\frac{1}{4}$ feet	=	1 square rod,
10000	4356	=	1 chain or 16 rods,
25000	10890	=	1 rood or 40 rods,
100000	43560	=	1 acre or 160 rods.

109. The contents, then, being calculated in chains and links; if *four* places of decimals be cut off, the remaining figures will be square *chains*; or if *five* places be cut off, the remaining figures will be *acres*. Thus the square of 16.32 chains, or 1632 links, is 2663424 square links, or 266.3424 square chains, or 26.63424 acres. If the contents be considered as square chains and *decimals*, removing the decimal point one place to the left will give the acres.

110. In surveying a piece of land, and calculating its contents, it is necessary, in all common cases, to suppose it to be reduced to a *horizontal level*. If a hill, or any uneven piece of ground, is bought and sold; the quantity is computed, not from the irregular surface, but from the *level base* on which the whole may be considered as resting. In running the lines, therefore, it is necessary to reduce them to a level. Unless this is done, a correct plan of the survey can never be exhibited on paper.

If a line be measured upon an ascent which is a regular *plane*, though oblique to the horizon; the length of the corresponding level base may be found, by taking the angle of elevation.

Let AB (Fig. 30.) be parallel to the horizon, BC perpendicular to AB, and AC a line measured on the side of a hill. Then, the angle of elevation at A being taken with a quadrant, (Art. 4.)

$$R : \text{Cos } A :: AC : AB, \text{ that is,}$$

As radius, to the cosine of the angle of elevation;
So is the oblique line measured, to the corresponding horizontal base.

If the chain, instead of being carried parallel to the surface of the ground, be kept constantly parallel to the *horizon*; the line thus measured will be the base line required. The line AB (Fig. 30.) is evidently equal to the sum of the parallel lines *ab*, *cd*, and *e*C.

Plotting a Survey.

111. When the sides of a field are measured, and their bearings taken, it is easy to lay down a plan of it on paper. A north and south line is drawn, and with a line of chords, a protractor, or a sector, an angle is laid off, equal to the angle which the first side of the field makes with the meridian, and the length of the side is taken from a scale of equal parts. (Trig. 156—161.) Through the extremity of this, a second meridian is drawn parallel to the first, and another side is laid down; from the end of this, a third side, &c. till the plan is completed. Or the plot may be constructed in the same manner as a *traverse* in navigation. (Art. 76.) If the field is correctly surveyed and plotted, it is evident the extremity of the last side must coincide with the beginning of the first.

Example I.

Draw a plan of a field, from the following courses and distances, as noted in the field-book;

				Ch	Links
1.	N.	78°	E.	2	46
2.	S.	16°	W.	3	54
3.	N.	83°	W.	2	72
4.	N.	12°	E.	2	13
5.	N.	60½°	E.	0	95

Let A (Fig. 31.) be the first corner of the field.

Thro' A, draw the merid NS, make BAN=78°, & AB=2.46
Thro' B, draw N'S' par. to NS, make S'BC=16°, & BC=3.54
Thro' C, draw N''S'' par. to NS, make DCN=83°, & CD=2.72
&c. &c.

112. To avoid the inconvenience of drawing parallel lines, the sides of a field may be laid down *from the angles which they make with each other*, instead of the angles which they make with the meridian. The position of the line BC (Fig. 31.) is determined by the angle ABC, as well as by the angle S'BC. When the several courses are given, the angles

which any two contiguous sides make with each other, may be known by the following rules.

1. If one course is North and the other South, one East and the other West; *subtract the less from the greater.*

2. If one is North and the other South, but both East or West; *add them together.*

3. If both are North or South, but one East and the other West; *subtract their sum from 180 degrees.*

4. If both are North or South, and both East or West; *add together 90 degrees, the less course, and the complement of the greater.*

The reason of these rules will be evident by applying them to the preceding example. (Fig. 31.)

The first course is BAN, which is equal to ABS′. (Euc. 29.1.) If from this the second course CBS′ be subtracted, there will remain the angle ABC.

If the second course CBS′, or its equal BCN″, be added to the third course DCN; the sum will be the angle BCD.

The sum of the angles CDS, NDE, and CDE, is 180 degrees. (Euc. 13.1.) If then the two first be subtracted from 180 degrees, the remainder will be the angle CDE.

Lastly, let EP be perpendicular to NS. Then the sum of the angles DES, PES, and AEP the complement of AEN, is equal to the angle DEA.

We have then the angle $ABC=62°$, $DEA=131\frac{1}{2}°$,
$BCD=99°$, $EAB=162\frac{1}{2}°$.
$CDE=85°$,

With these angles, the field may be plotted without drawing parallels, as in Trig. 173.

Finding the Contents of a Field.

113. There are in common use two methods of finding the contents of a piece of land, one by dividing the plot into *triangles*, the other by calculating the *departure and difference of latitude* for each of the sides.

When a survey is plotted, the whole figure may be divided into triangles, by drawing diagonals from the different angles. The lengths of the diagonals, and of the perpendiculars on the bases of the triangles, may be measured on the same scale of equal parts from which the sides of the field were laid down. The area of each of the triangles is equal to half the product of its base and perpendicular; and their sum is the area of the whole figure. (Mens. 13.)

Example I.

Let the plan Fig. 32 be the same as Fig. 31, the sides of which, with their bearings, are given in art. 111.

Then the triangle ABC=BC×½AP =3.84 sq. chains.
ACE=AC×½EP′=1.53
DCE=EC×½DP″=2.89

The contents of the whole=8.26

114. This method cannot be relied on, where great accuracy is required, if the lines are measured by a scale and compasses only. But the parts of the several triangles may be found by *trigonometrical calculation*, independently of the projection; and then the area of each may be computed, either from two sides and the included angle, or from the three sides. (Mens. 9, 10.)

The sides of the field and their bearings being given by the survey, the angles of the original figure may all be known. (Art. 112.) Then in the triangle ABC (Fig. 32.) we have the sides AB and BC, with the angle ABC, to find the other parts. (Trig. 153.) And in the triangle CDE, we have the sides DC and DE, with the angle CDE. Subtracting the angle BAC from BAE, we shall have CAE; and subtracting DEC from DEA, we shall have CEA. There will then be given, in the triangle ACE, the side EA and the angles. (Trig. 150.)

The sides and bearings, as given in art. 111, are

1. AB	N.	78°	E.	2.46 chains.
2. BC	S.	16	W.	3.54
3. CD	N.	83	W.	2.72
4. DE	N.	12	E.	2.13
5. EA	N.	60½	E.	0.95

Then by Mensuration, art. 9,

R : Sin ABC: :AB×BC : 2 *area* ABC=7.69 sq. chains
R : Sin AEC: :AE×EC : 2 *area* AEC=3.06
R : Sin CDE: :CD×DE : 2 *area* CDE=5.77

2)16.52

Contents of the whole field, 8.26

Or the areas of the several triangles may be found by the rule in Mensuration, art. 10, viz. If a, b, and c, be the sides of any triangle, and h=half their sum;

$$\textit{The area} = \sqrt{h \times (h-a) \times (h-b) \times (h-c)}$$

Example II.

	Courses.	Ch.	Links.
1.	E.	26	34
2.	S. 10° 30′ E.	32	26
3.	N. 42 W.	13	35
4.	S. 58 W.	23	52
5.	N.	30	55

Contents of the field, 69.735 acres.

The method which has been explained, of ascertaining the contents of a piece of land by dividing it into triangles, is of use in cases which do not require a greater degree of accuracy, than can be obtained by the scale and compasses. But if the areas of the triangles are to be found by trigonometrical calculation, the process becomes too laborious for common practice. The following method is often to be preferred.

FINDING THE AREA OF A FIELD BY DEPARTURE AND DIFFERENCE OF LATITUDE.

115. Let ABCDE (Fig. 33.) be the boundary of a field. At a given distance from A, draw the meridian line NS. Parallel to this draw L′R′, AG, BH, and DK. These may be considered as portions of meridians passing through the points A, B, D, and E. For all the meridians which cross a field of moderate dimensions, may be supposed to be *parallel*, without sensible errour. At right angles to NS draw the parallels AL, BM, CO, EP, and DR. These will divide the figure LABCDR into the three trapezoids ABML, BCOM, and CDRO; and the figure LAEDR into the two trapezoids DEPR and EALP. The area of the field is evidently equal to the difference between these two figures.

The sum of the parallel sides of a trapezoid, multiplied into their distance, is equal to twice the area. (Mens. 12.) Thus

$$(AL+BM) \times AG = 2\,\textit{area}\ ABML.$$

Now AL is a given distance, and BM=AL+BG. But BG is the *departure*, and AG the *difference of latitude*, corresponding to AB one of the sides of the field. (Arts. 39, 40.) And by art 44,

$$\text{Rad : Dist. AB} :: \begin{cases} \text{Sin BAG : Depart. BG} \\ \text{Cos BAG : Diff. Lat. AG} \end{cases}$$

Or the departure and difference of latitude may be taken from the *Traverse Table*, as in Navigation. (Art. 50.)

In the same manner, from the sides BC, CD, DE, and EA, may be found the departure CH, CK, DR′, AL′, and the differences of latitude BH, DK, ER′, and EL′. We shall then have the parallel sides of each of the trapezoids, or the distances of the several corners of the field from the meridian NS. For

$$BM=AL+BG, \qquad DR=CO-CK,$$
$$CO=BM+CH, \qquad EP=DR-DR'.$$

If the field be measured in the direction ABCDE, the differences of latitude AG, BH, and DK, will be *Southings*, while R′E and E′L will be *Northings*. The former are the breadths of the three trapezoids which form the figure LABCDR; and the latter are the breadths of the two trapezoids which form the figure LAEDR. The difference, then between the sum of the products of the northings into the corresponding meridian distances, and the sum of the products of the southings into the corresponding meridian distances, is twice the area of the field.

It will very much facilitate the calculation, to place in a *table* the several courses, distances, northings, southings, &c We have, then, the following

RULE.

116. *Find the northing or southing, and the easting or westing, for each side of the field, and place them in distinct columns in a table. To these add a column of Meridian Distances, for the distance of one end of each side of the field from a given meridian; a column of Multipliers to contain the pairs of meridian distances for the two ends of each of the sides; and columns for the north and south Areas.* See Fig. 23, and the table for example I.

Suppose a meridian line to be drawn without the field, at any given distance from the first station; and place the assumed distance at the head of the column of Meridian Distances. To this add the first departure, if both be east or both west; but subtract, if one be east and the other west; and place the sum or difference in the column of Meridian Distances, against the first course. To or from the last number, add or subtract the second departure, &c. &c.

For the column of Multipliers, add together the first and second numbers in the column of Meridian Distances; the second and third, the third and fourth, &c. placing the sums opposite the several courses.

Multiply each number in the column of Multipliers into its corresponding northing or southing, and place the product in the column of north or south areas. The difference between the sum of the north areas, and the sum of the south areas, will be twice the area of the field.

This method of finding the contents of a field, as it depends on departure and difference of latitude, which are calculated by right-angled trigonometry, is sometimes called *Rectangular Surveying.*

117. If the assumed meridian pass through the eastern or western extremity of the field, as L′ER′ (Fig. 33.) the distance EP will be reduced to nothing, and the figures AEL′ and EDR′ will be *triangles* instead of trapezoids. If the survey be made to begin at the point E, *cipher* is to be placed at the head of the column of meridian distances, and the first number in the column of multipliers will be the same, as the first in the column of meridian distances. See example II.

118. When there is a *re-entering angle* in a field, situated with respect to the meridian as CDE; (Fig. 34.) the area EDM, being included in the figure BCRA, will be *repeated* in the column of south areas. But, as it is also included in the figure DCRM, it will be contained in the column of north areas. Therefore the *difference* between the north areas and the south areas, will be twice the area of the field, in this case, as well as in others.

119. If any side is directly *east or west*, there will be no difference of latitude, and consequently no number to be placed against this course, in the columns of north and south areas. See example II. Course 1. AB (Fig. 34.)

The number in the columns of areas will be wanting also, when any side of the field coincides with the assumed meridian. See example II. Course 5. EA (Fig. 34.)

120. In finding the departure and difference of latitude from the traverse table, the numbers for the *links* may be looked out separately; care being taken to remove the decimal point two places to the left, because a link is the 100th part of a chain.

Thus if the course be 29°, and the distance 23.46 chains;

The dif. of lat. & depart.	for 23 chains are	20.12 and	11.15
	for 46 links	.40	.22
	for 23.46	20.52	11.37

Example I. See Fig. 33.

Courses.	*Dist.*	*Diff. Lat* N.	S.	*Departure.* E.	W.	*M. D.* A L, O. E.	*Mult*	*N. Areas*	*S. Areas*
1. BAG S. 64° E.	AB 30 ch.		AG 13 15	GB 26.96		BM 46.96	AL+BM 66.96		ABML 880.5240
2. CBH S. 14° E.	BC 10		BH 9.70	HC 2.42		CO 49.38	BM+CO 96.34		2BMOC 934 4980
3. CDK S. 35° W.	CD 30		KD 24.57		CK 17 21	DR 82.17	CO+DR 81 55		2CDRO 2003.6835
4. KDE N. 65° W.	DE 20	R'E 8.45			DR' 18.13	EP 14.04	DR+EP 48.21	2 DEPR 390.4745	
5. L'EA N. 8° 42' E	EA 39.42	EL' 38.97		L'A 5.96		AL 20.	EP+AL 34.04	2EPLA 1326.5388	
		47.42	47 42	35.34	35.31			1717.1133	3818.7055

Twice the figure ABCDRL is 3818.7055 square chains;
Twice the figure AEDRL 1717.1133

The difference 2101.5922
The contents of the field 1050.7961 sq. ch. or 105.0796 acres. (Art. 109.)

Example II. See Fig. 34.

Courses.	Dist.	Diff. Lat. N.	Diff. Lat. S.	Departure. E.	Departure. W.	M. Dist. 00	Mult.	N. Areas.	S. Areas.
1. E.	AB 26.34	00	00	26.34		AB 26.34	00	00	00
2. S. 10½°E.	BC 32.26		31.72	5.88		CR 32.22	AB+CR 58.56		2ABCR 1857.5232
3. N. 42°W.	CD 18.35	13.64			12.27	DM 19.95	CR+DM 52.17	2CDMR 711.5988	
4. S. 58°W.	DE 23.52		12.47		19.95	00	DM 19.95		2DME 248.7765
5 N.	EA 30.55	30.55		00	00	00	00		
		44.19	44.19					711.5988	2106.2997

The contents of the field $=\frac{1}{2}(2106.3-711.6)=697.35$ sq. ch.
Or 69.735 acres.

In this example, the meridian distance of the first station A being nothing, *cipher* is placed at the head of the column of meridian distances. (Art. 117.) The first side AB being directly east and west, has no difference of latitude, and therefore the number in the column of areas against this course is wanting, as it is against the fifth course, which is directly north. (Art. 119.) The number against the fourth course, in the column of multipliers, is only the length of the line DM; the figure DME being a *triangle*, instead of a trapezoid.

Example III.

Find the contents of a field bounded by the following lines;

1. N. 35° 30′ E. 15 ch. 50 links.
2. N. 72 45 E. 18 70
3. S. 70 45 E. 18 70
4. S. 53 W. 12 45
5. S. 83 15 E. 24 10
6. S. 31 15 W. 15 20
7. S. 62 45 W. 22 60
8. N. 73 30 W. 27 30
9. N. 17 25 W. 14 56

The area is 145½ acres.

121. When a field is correctly surveyed, and the departures and differences of latitude accurately calculated; it is evident the sum of the northings must be equal to the sum of the southings, and the sum of the eastings equal to the sum of the westings. If upon adding up the numbers in the departure and latitude columns, the northings are not found to agree nearly with the southings, and the eastings with the westings, there must be an errour, either in the survey or in the calculation, which requires that one or both should be revised. But if the difference be small, and if there be no particular reason for supposing it to be occasioned by one part of the survey rather than another: it may be apportioned among the several departures or differences of latitude, according to the different lengths of the sides of the field, by the following rule ;

As the whole perimeter of the field,
To the whole errour in departure or latitude ;
So is the length of one of the sides,
To the correction in the corresponding departure or latitude,

This correction, if applied to the column in which the sum of the numbers is too small, is to be *added;* but if to the other column, it is to be *subtracted.** See the example on the next page.

* See the fourth Number of the Analyst published at Philadelphia.

Example IV.

Courses.	Dist.	Diff. Lat.		Departure.		Cor. Lat.	Cor. Dep.	Cor. Diff. Lat.		Cor. Dep.		Mer. Dist.	Mult.	N. Areas.	S. Areas.
	Chains.	N.	S.	E.	W.			N.	S.	E.	W.	00			
1. N. 55¼° E.	18	10.26		14.79		+.06	+.08	10.32		14.87		14.87	14.87	153.46	
2. S. 62½° E.	14½		6.70	12.87		−.05	+.07		6.65	12.94		27.81	42.68		283.82
3. S. 40° W.	11		8.43		7.07	−.04	−.04		8.39		7.03	20.78	48.59		407.67
4. S 4¼° E.	14		13.96	1.04		−.05	+.06		13.91	1.10		21.88	42.66		593.40
5. N. 73¾° W.	12½	3.50			12.00	+.04	−.05	3.54			11.95	9.93	31.81	112.61	
6. S. 52° W.	9½		5.85		7.49	−.03	−.04		5.82		7.45	2.48	12.41		72.23
7. N. 7° W.	21	20.84			2.56	+.07	−.08	20.91			2.48	00	2.48	51.85	
Perimeter	100½	34.60	34.94	28.70	29.12			34.77	34.77	28.91	28.91			317.92	1357.12
			34.60		28.70										317.92
		Errour	.34	Errour	.42									Double area	1039.2

In this example, the whole perimeter of the field is $100\frac{1}{2}$ chains, the whole errour in latitude .34, the whole errour in departure .42. and the length of the first side 18. To find the corresponding errours,

$$100\tfrac{1}{2} : 18 :: \begin{cases} .34 : .06 \text{ the errour in latitude,} \\ .42 : .08 \text{ the errour in departure.} \end{cases}$$

The errour in latitude is to be added to 10.26 making it 10.32, as in the column of corrected northings; and the errour in departure is to be added to 14.79 making it 14.87, as in the column of corrected eastings. After the corrections are made for each of the courses, the remaining part of the calculation is the same as in the preceding examples.

122. If the length and direction of each of the sides of a field *except one* be given, the remaining side may be easily found by calculation. For the difference between the sum of the northings and the sum of the southings of the given sides, is evidently equal to the northing or southing of the remaining side; and the difference between the sum of the eastings and the sum of the westings of the given sides, is equal to the easting or westing of the remaining side. Having then the difference of latitude and departure for the side required, its length and direction may be found, in the same manner as in the sixth case of plane sailing. (Art. 49.)

Example V.

What is the area of a field of six sides, of which five are given, viz.

1. S. 56° E. 4.18 chains
2. N. 21 E. 4.80
3. N. 56 W. 3.06
4. S. 21 W. 0.13
5. N. 66°½ W. 1.44
6. ——— ———

The area is two acres.

Example VI.

1. N 38° W. 17.21 chains
2. N. 13 E. 21.16
3. N 72 E. 24.11
4. S. 41 E. 19 26
5. S. 11 W. 24.35.
6. ——— ———

123. *Plotting by departure and difference of Latitude.* A survey may be easily plotted from the northings and southings, eastings and westings. For this purpose, the column of *Meridian Distances* is used. It will be convenient to add also another column, containing the distance of each station from a given *parallel of latitude*, and formed by adding the northings and subtracting the southings, or adding the southings and subtracting the northings.

Let AT (Fig. 33.) be a parallel of latitude passing through the first station of the field. Then the southing TB or LM is the distance of B, the second station, from the given parallel. To this adding the southing BH, we have LO the distance of CO from LT. Proceeding in this manner for each of the sides of the field, and copying the 7th column in the table, p. 65, we have the following differences of latitude and meridian distances.

Diff. Lat.	*Merid. Dist.*
	AL 20
1. LM 13.15	BM 46.96
2. LO 22.85	CO 49 38
3. LR 47.42	DR 32.17
4. LP 38.97	EP 14.04

To plot the field, draw the meridian NS, and perpendicular to this, the parallel of latitude LT. From L set off the differences of latitude LM, LO, LR, and LP. Through L, M, O, R, and P, draw lines parallel to LT; and set off the meridian distances AL, BM, CO, DR, and EP. The points A, B, C, D, and E, will then be given.

124. When a field is a *regular figure*, as a parallelogram, triangle, circle, &c. the contents may be found by the rules in Mensuration, Sec. I. and II.

125. The area of a field which has been plotted, is sometimes found by *reducing the whole to a* TRIANGLE *of the same area.* This is done by changing the figure in such a manner as, at each step, to make the number of sides one less, till they are reduced to three.

Let the side AB (Fig. 35.) be extended indefinitely both ways. To reduce the two sides BC and CD to one, draw a line from D to B, and another parallel to this from C, to intersect AB continued. Draw also a line from D to the point of intersection G. Then the triangles DBC and DBG are equal. (Euc. 37. 1.) Taking from each the common part DBH, there remains BGH equal to DCH. If then the triangle DCH be thrown out of the plot, and BGH be added, we shall have the five-sided figure AGDEF equal to the six-sided figure ABCDEF.

In the same manner, the line EL may be substituted for the two sides AF and EF; and then DM, for EL and ED. This will reduce the whole to the triangle MGD, which is equal to the original figure. The area of the triangle may then be found by multiplying its base into half its height; and this will be the contents of the field.

In practice it will not be necessary actually to draw the parallel lines BD, GC, &c. It will be sufficient to lay the edge of a rule on C, so as to be parallel to a line supposed to pass through B and D, and to mark the point of intersection G.

126. If after a field has been surveyed, and the area computed, the chain is found to be *too long* or *too short;* the true contents may be found, upon the principle that similar figures are to each other as the squares of their homologous sides. (Euc. 20.6.) The proportion may be stated thus;

As the square of the true chain, to the square of that by which the survey was made;
So is the computed area of the field, to the true area.

Ex If the area of a field measured by a chain 66.4 feet long, be computed to be 32.6036 acres; what is the area as measured by the true chain 66 feet long?

Ans. 33 acres.

127. A plot of a field may be changed to a *different scale,* that is, it may be enlarged or diminished in any given ratio, by drawing lines parallel to each of the sides of the original plan.

To enlarge the perimeter of the figure ABCDE (Fig. 36.) in the ratio of *a*G to AG; draw lines from G through each of the angular points. Then beginning at *a*, draw *ab* parallel to AB, *bc* parallel to BC, &c.

It is evident that the *angles* are the same in the enlarged figure, as in the original one. And by similar triangles,

$$AG : aG :: BG : bG :: CG : cG :: \&c.$$

And

$$AG : aG :: AB : ab :: BC : bc :: \&c.$$

Therefore ABCDE and *abcde* are similar figures. (Euc. Def. 1. 6.)

In the same manner, the smaller figure *a'b'c'd'e'* may be drawn, so as to have its perimeter proportioned to ABCDE as *a'*G to AG.

SECTION II.

METHODS OF SURVEYING IN PARTICULAR CASES.

ART. 128. MEASURING round a field, in the manner explained in the preceding section, is by far the most common method of surveying. The following problems are sometimes useful. They may serve to verify or correct the surveys which are made by the usual method.

PROBLEM I.

To survey a field from TWO STATIONS.

129. FIND THE DISTANCE OF THE TWO STATIONS, AND THEIR BEARINGS FROM EACH OTHER; THEN TAKE THE BEARINGS OF THE SEVERAL CORNERS OF THE FIELD FROM EACH OF THE STATIONS.

In the field ABCDE, (Fig. 37.) let the distance of the two stations S and T be given, and their bearings from each other. By taking the bearing of A from S and T, or the angles AST and ATS, we have the direction of the lines drawn from the two stations to one of the corners of the field. The point A is determined by the *intersection* of these lines. In the same manner, the point B is determined, by the intersection of SB and TB; the point C, by the intersection of SC and TC; &c. &c. The sides of the field are then laid down, by connecting the points ABCD, &c.

The *area* is obtained, by finding the areas of the several triangles into which the field is divided by lines drawn from one of the stations. Thus the area of ABCDE (Fig. 37.) is equal to

$$ABT+BCT+CDT+DET+EAT$$

or to

$$ABS+BCS+CDS+DES+EAS$$

Now we have the base line ST given and the angles, in the triangle AST, to find AS and AT; in the triangle BST, to find BS and BT, &c. After these are found, we have two sides and the included angle in the triangles ABT, BCT, &c. from which the areas may be calculated. (Mens. 9.)

Example.

Let the station T (Fig. 37.) be N. 80° E. from S, the distance ST 27 chains, and the bearings of the several corners of the field from S and T as follows;

TA	N. 30° W.	SA	N. 17° E.
TB	N. 15 E.	SB	N 55 E.
TC	S. 53 E.	SC	S. 73 E.
TD	S. 55 W.	SD	S. 24 W.
TE	N. 70 W.	SE	N. 26 W.

These will give the following angles;

ATS= 70°	AST= 63°	ATB= 45°
BTS=115	BST= 25	BTC=112
CTS=133	CST= 27	CTD=108
DTS= 25	DST=124	DTE= 55
ETS= 30	EST=106	ETA= 40

From which, with the base line ST, are calculated the following lines and areas.

AT=32.89 chains.	ABT=206.45 sq. chains
BT=17.75	BCT=294.95
CT=35.84	CDT=740.7
DT=43.46	DET=665 1
ET=37.36	EAT=395.

Contents of the field, =230.22 acres.

The course and length of each of the *sides* of the field may be found, if necessary. After the parts mentioned above are calculated, there will be given two sides and the included angle, in the triangle ATB, to find AB, in BTC to find BC &c.

If the base line between the two stations be *too short*, compared with the sides of the field and their distances, the survey will be liable to inaccuracy. It should not generally be less than one tenth of the longest straight line which can be drawn on the ground to be measured.

130. It is not necessary that the base line, from the extremities of which the bearings are taken, should be *within* the field. It may be one of the sides, or it may be entirely *without* the field.

Let S and T (Fig. 38.) be two stations from which all the corners of a field ABCDE may be seen. If the direction and length of the base line be measured, and the bearings of the points A, B, C, D, and E, be taken at each of the stations; the areas of the several triangles may be found. The figure ABCTDE is equal to

DET+EAT+ABT+BCT

From this subtracting DCT, we have the area of the field ABCDE.

In this manner, a piece of ground may be measured which, from natural or artificial obstructions, is *inaccessible*. Thus an island may be measured from the opposite bank, or an enemy's camp, from a neighbouring eminence.

131. The method of surveying by making observations from two stations, is particularly adapted to the measurement of a *bay* or *harbour*.

The survey may be made on the water, by anchoring two vessels at a distance from each other, and observing from each the bearings of the several remarkable objects near the shore. Or the observations may be made from such elevated situations on the land as are favourable for viewing the figure of the harbour. If all the parts of the shore cannot be seen from two stations, three or more may be taken. In this case, the direction and distance of each from one of the others should be measured.

Problem. II.

To survey a field by measuring from ONE STATION.

132. TAKE THE BEARINGS OF THE SEVERAL CORNERS OF THE FIELD, AND MEASURE THE DISTANCE OF EACH FROM THE GIVEN STATION.

If the length and direction of the several lines AT, BT, CT, DT, and ET, (Fig. 37.) be ascertained; there will be given two sides and the included angle of each of the triangles ABT, BCT, CDT, DET, and EAT; from which their areas may be calculated, (Mens. 9.) and the sum of these will be the contents of the whole figure.

The station may be taken in one of the sides or angles of the field, as at C. (Fig. 32.) The lines
CD, CE, CA, CB, and the angles
DCE, ECA, ACB, being given,
the areas of the triangles may be found.

PROBLEM III.

To survey a field by the CHAIN ALONE.

133. MEASURE THE SIDES OF THE FIELD, AND THE DIAGONALS BY WHICH IT IS DIVIDED INTO TRIANGLES.

By measuring the sides (Fig. 32.)
AB, BC, CD, DE, EA,
and the diagonals CA and CE, we have the three sides of each of the triangles into which the whole figure is divided. They may therefore be constructed, (Trig. 172.) and their areas calculated. (Mens. 10.)

134. *To measure an* ANGLE *with the chain*, set off equal distances on the two lines which include the angle, as AB, AC, (Fig. 39.) and measure the distance from B to C. There will then be given the three sides of the isosceles triangle ABC, to find the angle at A by construction or calculation.

The chain may be thus substituted for the compass, in surveying a field by going round it, according to the method explained in the preceding section; or by measuring from one or two stations, as in problems I. and II.

PROBLEM IV.

To survey an irregular boundary by means of OFFSETS.

135. RUN A STRAIGHT LINE IN ANY CONVENIENT DIRECTION, AND MEASURE THE PERPENDICULAR DISTANCE OF EACH ANGULAR POINT OF THE BOUNDARY FROM THIS LINE.

The irregular field (Fig. 40.) may be surveyed, by taking the bearing and length of each of the four lines AE, EF, FI, IA, and measuring the perpendicular distances BB′, CC′, DD′, GG′, HH′, KK′. These perpendiculars are called *offsets*. It is necessary to note in a field book the parts into which the line that is measured is divided by the offsets, as in the following example. (See Fig. 40.)

Offsets on the left		*Courses and Distances.*	*Offsetts on the right.*
	Chains.	AE N.85°E. 12.74 ch.	
BB′	2.18	AB′ 3.25	
CC′	2.18	B′C′ 2.13	
DD′	1.23	C′D′ 1.12	
		D′E 6.24	
		EF S. 24° E. 7.23	
		FI N. 87 W. 13.34	
GG′	2.86	FG′ 3.84	
HH′	1.48	G′H′ 2.22	
		H′I 7.28	
		IA N. 26 W. 5.32	
		IK′	KK′ 2.94
		K′A	

As the offsets are perpendicular to the lines surveyed, the little spaces ABB′, BB′CC′, CC′DD′ &c. are either right angled triangles, parallelograms, or trapezoids. To find the *contents* of the field, calculate in the first place the area between the lines surveyed, as the trapezium AEFIA, (Fig. 40.) and then add the spaces between the offsets, if they fall within the boundary line; or subtract them, if they fall without, as AIK.

When any part of a side of a field is *inaccessible*, equal offsets may be made at each end, and a line run *parallel* to the boundary.

Problem V.

To measure the distance between any two points on the surface of the earth, by means of a series of triangles extending from one to the other.

136. Measure a side of one of the triangles for a BASE LINE, take the BEARING of this or some other side, and measure the ANGLES in each of the triangles.

If it be required to find the distance between the two points A and I, (Fig. 41.) so situated that the measure cannot be taken in a direct line from one to the other; let a series of triangles be arranged in such a manner between them, that one side shall be common to the first and second, as BC, to the second and third as CD, to the third and fourth, &c. Then measure the length of BC for a base line, take the bearing of the side AB, and measure the angles of each of the triangles.

These data are sufficient to determine the length and bearing of each of the sides, and the distance and bearing of I from A. For in the two first triangles ABC and BCD, the angles are given and the side BC, to find the other sides. When CD is found, there are given, in the third triangle CDE, one side and the angles, to find the other side. In the same manner, the calculation may be carried from one triangle to another, till all the sides are found.

The *bearings* of the sides, that is, the angles which they make with the meridian, may be determined from the bearing of the first side, and the angles in the several triangles. Thus if NS be parallel to AM, the angle BAP, or its equal ABN subtracted from ABD leaves NBD; and this taken from 180 degrees leaves SBD.

From the bearing and length of AB may be found the *southing* AP, and the *easting* PB. In the same manner are found the several southings PP′, P′P″, P″P‴, P‴M. The sum of the southings is the line AM. And if the distance is so small, that the several meridians may be considered parallel, the difference between the sum of the eastings and the sum of the westings, is the perpendicular IM. We have then, in the right-angled triangle AMI, the sides AM and MI, to find the distance and bearing of I from A.

137. This problem is introduced here, for the purpose of giving the general outlines of those important operations which have been carried on of late years, with such admirable precision, under the name of *Trigonometrical Surveying.*

Any explanation of the subject, however, which can be made in this part of the course. must be very imperfect. In the demonstration of the problem, the several triangles are supposed to be in the same plane, and the distances of the meridians so small, that they may be considered parallel. But in practice, the ground upon which the measurement is to be made is very irregular. The stations selected for the angular points of the triangles, are such elevated parts of the country as are visible to a considerable distance. They should be so situated, that a signal staff, tower, or other conspicuous object in any one of the angles, may be seen from the other two angles in the same triangle. It will rarely be the case that any two of the triangles will be in the same plane, or any one of them parallel to the horizon. Reductions will therefore be necessary to bring them to a common level. But even this level is not a plane In the cases in which this kind of surveying is commonly practised, the measurement is carried over an extent of country of many miles. The several points, when reduced to the same distance from the centre of the earth, are to be considered as belonging to a *spherical* surface. To make the calculations then, if the line to be measured is of any considerable extent, and if nice exactness is required, a knowledge of Spherical Trigonometry is necessary.

138. The decided superiority of this method of surveying, in point of accuracy, over all others which have hitherto been tried, particularly where the extent of ground is great, is owing partly to the fact that almost all the quantities measured are *angles;* and partly to this, that for the single line which it is necessary to measure, *the ground may be chosen*, any where in the vicinity of the system of triangles. It would be next to impossible to determine the precise horizontal distance between two points, by carrying a chain over an irregular surface. But in the trigonometrical measurements which are made upon a great scale, there can generally be found, somewhere in the country surveyed, a level plane, a heath or a body of ice on a river or lake, of sufficient extent for a *base*. This is the only line which it is absolutely necessary to measure. It is usual however, to measure a second, which is called a line of *verification.* If the length of the base BC (Fig. 41.) and the angles be given, all the other lines in the figure may be found by trigonometrical calculation. But if GH be also measured, it will serve to detect any errour which may have been committed, either in taking the angles, or in computing the sides, of the series of triangles between BC and GH.

139. In measuring these lines, rods of copper or platina have been used in France, and glass tubes or steel chains in England. The results have in many instances been extremely exact. A base was measured, on Hounslow Heath, by General Roy, with glass rods. Several years after, it was re-

measured by Colonel Mudge, with a steel chain of very nice construction. The difference in the two measurements was less than *three inches* in more than five miles. Two parties measured a base in Peru of 6272 toises, or more than seven miles; and the difference in their results did not exceed *two inches*.

Exact as these measurements are, the exquisite construction of the instruments which have been used for taking the *angles*, has given to that part of the process a still higher degree of perfection. The amount of the errours in the angles of each of the triangles, measured by Ramsden's Theodolite, did not exceed *three seconds*. In the great surveys in France, the angles were taken with nearly the same correctness.

140. One of the most important applications of trigonometrical surveying, is in measuring *arcs of the meridian*, or of *parallels of latitude*, particularly the former. This is necessary in determining the *figure of the earth*, a very essential problem in Geography and Astronomy. A degree of zeal has been displayed on this subject, proportioned to its practical importance. Arcs of the meridian have been measured at great expense, in England, France, Lapland, Peru, &c. Men of distinguished science have engaged in the undertaking.

A meridian line has been measured, under the direction of General Roy and Colonel Mudge, from the Isle of Wight, to Clifton in the north of England, a distance of about 200 miles. Several years were occupied in this survey. Another arc passing near Paris, has been carried quite through France, and even across a part of Spain to Barcelona. In measuring this, several distinguished mathematicians and astronomers were engaged for a number of years. These two arcs have been connected by a system of triangles running across the English Channel, the particular object of which was to determine the exact difference of longitude between the observatories of Greenwich and Paris. Besides the meridian arcs, other lines intersecting them in various directions have been measured, both in England and France. With these, the most remarkable objects over the face of the country have been so connected, that the geography of the various parts of the two kingdoms is settled, with a precision which could not be expected from any other method.

141. The exactness of the surveys will be seen from a comparison of the lines of *verification* as actually measured, with the lengths of the same lines as determined by calculation. These would be affected by the amount of all the errours in measuring the base lines, in taking the angles, in computing the sides of the triangles, and in making the necessary reductions for the irregularities of surface. A base of verification measured on Romney Marsh in England, was found to differ but about *two feet* from the length of the same line, as deduced from a series of triangles extending more than 60 miles. A base of verification connected with the meridian passing through France, was found not to differ *one foot* from the result of a calculation which depended on the measurement of a base 400 miles distant A line of verification of more than 7 miles, on Salisbury Plain, differed scarcely *an inch* from the length as computed from a system of triangles extending to a base on Hounslow Heath.*

* See Note K.

SECTION III.

LAYING OUT AND DIVIDING LANDS.

ART. 142. TO those who are familiar with the principles of geometry, it will be unnecessary to give particular rules, for all the various methods of dividing and laying out lands. The following problems may serve as a specimen of the manner in which the business may be conducted in practice.

PROBLEM I.

To lay out a given number of acres in the form of a SQUARE.

143. REDUCE THE NUMBER OF ACRES TO SQUARE RODS OR CHAINS, AND EXTRACT THE SQUARE ROOT. This will give one side of the required field. (Mens. 7.)

Ex. 1. What is the side of a square piece of land containing $124\frac{1}{4}$ acres. Ans. 141 rods.

2. What is the side of a square field which contains $58\frac{3}{4}$ acres?

PROBLEM II.

To lay out a field in the form of a PARALLELOGRAM, *when one side and the contents are given.*

144. DIVIDE THE NUMBER OF SQUARE RODS OR CHAINS BY THE LENGTH OF THE GIVEN SIDE. The quotient will be a side perpendicular to the given side. (Mens. 7.)

Ex. What is the width of a piece of land which is 280 rods long, and which contains 77 acres?

Ans. 44 rods.

Cor. As a triangle is half a parallelogram of the same base and height, a field may be laid out in the form of a triangle whose area and base are given, by dividing twice the area by the base. The quotient will be the perpendicular from the opposite angle. (Mens. 8.)

PROBLEM III.

To lay out a piece of land in the form of a parallelogram, the length of which shall be to the breadth in a given RATIO.

145. As the *length* of the parallelogram to its breadth ;
So is the *area*, to the area of a *square* of the same breadth.

The side of the square may then be found by problem I, and the length of the parallelogram by problem II.

If BCNM (Fig. 42.) is a square in the right parallelogram ABCD, or in the oblique parallelogram ABC'D', it is evident that AB is to MB or its equal BC, as the area of the parallelogram to that of the square.

Ex. If the length of a parallelogram is to its breadth as 7 to 3, and the contents are $52\frac{1}{2}$ acres, what is the length and breadth.

PROBLEM IV.

The area of a parallelogram being given, to lay it out in such a form, that the length shall exceed the breadth by a given DIFFERENCE.

146. Let x=BC the breadth of the parallelogram ABCD (Fig. 42.) and the side of the square BCNM.
d=AM the difference between the length and breadth.
a=the area of the parallelogram.
Then $a=(x+d)\times x=x^2+dx$. (Mens. 4.)
Reducing this equation, we have

$$\sqrt{a+\frac{1}{4}d^2}-\frac{1}{2}d=x.$$

That is, to the area of the parallelogram, add one fourth of the square of the difference between the length and the breadth, and from the square root of the sum, subtract half the difference of the sides ; the remainder will be the breadth of the parallelogram.

Ex. If four acres of land be laid out in the form of a parallelogram, the difference of whose sides is 12 rods, what is the breadth ?

PROBLEM V.

To lay out a TRIANGLE *whose area and angles are given.*

147. CALCULATE THE AREA OF ANY SUPPOSED TRIANGLE WHICH HAS THE SAME ANGLES. THEN

As THE AREA OF THE ASSUMED TRIANGLE,
To THE AREA OF THAT WHICH IS REQUIRED ;
So IS THE SQUARE OF ANY SIDE OF THE FORMER,
To THE SQUARE OF THE CORRESPONDING SIDE OF THE LATTER.

If the triangles B′CC′ and BCA (Fig. 43.) have equal angles, they are similar figures, and therefore their areas are as he squares of their like sides, for instance, as $\overline{AC}^2 : \overline{CC'}^2$. (Euc. 19. 6.) The *square* of CC′ being found, extracting the *square root* will give the line itself

To lay out a triangle of which *one side* and the area are given, divide twice the area by the given side; the quotient will be the length of a perpendicular on this side from the opposite angle. (Mens. 8.) Thus twice the area of ABC (Fig. 45.) divided by the side AB, gives the length of the perpendicular CP.

148. This problem furnishes the means of cutting off, or laying out, a given quantity of land in various forms.

Thus from the triangle ABC, (Fig. 43.) a smaller triangle of a given area may be cut off, by a line parallel to AB. The line CC′ being found by the problem, the point C′ will be given, from which the parallel line is to be drawn.

149. If the directions of the lines AE and BD, (Fig. 44.) and the length and direction of AB be given ; and if it be required to lay off a given area, by a line parallel to AB ; let the lines AE and BD be continued to C. The angles of the triangle ABC with the side AB being given, the area may be found. From this subtracting the area of the given trapezoid, the remainder will be the area of the triangle DCE ; from which may be found, as before, the point E through which the parallel is to be drawn.

If the trapezoid is to be laid off on the *other side* of AB, its area must be *added* to ABC, to give the triangle D′CE′.

150. If a piece of land is to be laid off from AB, (Fig. 45. by a line in a given direction as DE, *not parallel* to AB ;

let AC parallel to DE be drawn through one end of AB The required trapezium consists of two parts, the triangle ABC, and the trapezoid ACED. As the angles and one side of the former are given, its area may be found. Substracting this from the given area, we have the area of the trapezoid, from which the distance AD may be found by the preceding article.

151. If a given area is to be laid off from AB, (Fig. 46.) by a line proceeding from a *given point* D; first lay off the trapezoid ABCD. If this be too small, add the triangle DCE; but if the trapezoid be too large, subtract the triangle DCE'.

PROBLEM VI.

To divide the area of a triangle into parts having given ratios to each other, by lines drawn from one of the angles to the opposite base.

152. DIVIDE THE BASE IN THE SAME PROPORTION AS THE PARTS REQUIRED.

If the triangle ABC (Fig. 47.) be divided by the lines CH and CD; the small triangles, having the same height, are to each other as the bases BH, DD, and AD. (Euc. 1. 6.)

PROBLEM VII.

To divide an irregular piece of land into any two given parts.

153. *Run a line at a venture, near to the true division line required, and find the area of one of the parts. If this be too large or too small, add or subtract, by the preceding articles, a triangle, a trapezoid, or a trapezium, as the case may require.*

A field may sometimes be conveniently divided by reducing it to a triangle, as in Art. 125, (Fig. 35.) and then dividing the triangle by problem VI.

SECTION IV.

LEVELLING.

Art. 154. IT is frequently necessary to ascertain how much one spot of ground is higher than another. The practicability of supplying a town with water from a neighbouring fountain, will depend on the comparative elevation of the two places above a common level. The direction of the current in a canal will be determined by the height of the several parts with respect to each other.

The art of levelling has a primary reference to the level surface of water. The surface of the ocean, a lake, or a river, is said to be level when it is at rest. If the fluid parts of the earth were perfectly *spherical*, every point in a level surface would be at the same distance from the centre. The difference in the heights of two places above the ocean would be the same, as the difference in their distances from the centre of the earth. It is well known that the earth, though nearly spherical, is not perfectly so. It is not necessary, however, that the difference between its true figure and that of a sphere should be brought into account, in the comparatively small distances to which the art of levelling is commonly applied. But it is important to distinguish between the *true* and the *apparent* level.

155. *The* TRUE LEVEL *is a* CURVE *which either coincides with, or is parallel to, the surface of water at rest.*

The APPARENT LEVEL *is a* STRAIGHT LINE *which is a* TANGENT *to the true level, at the point where the observation is made.*

Thus if ED (Fig. 48.) be the surface of the ocean, and AB a concentric curve, B is on a true level with A. But if AT be a tangent to AB, at the point A, the *apparent* level, as observed at A, passes through T.

156. When levelling instruments are used, the level is determined either by a *fluid*, or a *plumb-line*. The surface of the former is *parallel* to the horizon. The latter is *perpen-*

dicular. One of the most convenient instruments for the purpose is the *spirit level.* A glass tube is nearly filled with spirit, a small space being left for a bubble of air. The tube is so formed, that when it is horizontal, the air bubble will be in the middle between the two ends. To the glass is attached an index with sight vanes; and sometimes a small telescope, for viewing a distant object distinctly. The surveyor should also be provided with a pair of *levelling rods,* which are to be set up perpendicularly, at convenient distances, for the purpose of measuring the height from the surface of the ground to the horizontal line which passes through the spirit level.

If strict accuracy is aimed at, the spirit level should be in the *middle* between the two rods. Considering D′ED″D as the spherical surface of the earth, and B′AB″B as a concentric curve; a horizontal line passing through A is a *tangent* to this curve. If therefore AT′ and AT″ are *equal,* the points T′ and T″ are equally distant from the level of the ocean. But if the two rods are at T and T′, while the spirit level is at A, the height TD is greater than T′D′. The difference however will be trifling, if the distance of the stations T and T′ be small.

157. With these simple instruments, the spirit level and the rods, the comparative heights of any two places can be ascertained by a series of observations, without measuring their distance, and however irregular may be the ground between them. But when one of the stations is visible from the other, and their distance is known; the difference of their heights may be found by a *single observation,* provided allowance be made for atmospheric refraction, and for the difference between the true and the apparent level.

Problem I.

To find the difference in the heights of two places by levelling rods.

158. *Set up the levelling rods perpendicular to the horizon, and at equal distances from the spirit level; observe the points where the line of level strikes the rods before and behind, and measure the heights of these points above the ground; level in the same manner from the second station to the third, from the*

third to the fourth, &c. The difference between the sum of the heights at the back stations, and at the forward stations, will be the difference between the height of the first station and the last.

If the descent from H to H″ (Fig. 49.) be required, let the spirit level be placed at A, equally distant from the stations H and H′; observe where the line of level BF cuts the rods which are at H and H′, and measure the heights BH and FH′. The difference is evidently the descent from the first station to the second. In the same manner, by placing the spirit level at A′, the descent from the second station to the third may be found. The back heights, as observed at A and A′, are BH and B′H′; the forward heights are FH′ and F′H″.

Now FH′ − BH = the descent from H to H′,
And F′H″ − B′H′ = the descent from H′ to H″;

Therefore, by addition,

(FH′ + F′H″) − (BH + B′H′) = the whole descent from H to H″.

159. It is to be observed, that this method gives the *true* level, and not the apparent level. The lines BF and B′F′ are not parallel to each other; but one is parallel to a tangent to the horizon at N, the other to a tangent at N′. So that the points B and F are equally distant from the horizon, as are also the points B′ and F′. The spirit level may be placed at unequal distances from the two station rods, if a correction is made for the difference between the true and the apparent level, by problem II.

160. If the stations are numerous, it will be expedient to place the back and the forward heights in separate columns in a table, as in the following example.

	Back heights.		Fore heights.	
	Feet.	*In.*	*Feet.*	*In.*
1st. Observation	3	7	2	8
2.	2	5	3	1
3.	6	3	5	7
4.	4	2	3	2
5.	5	9	4	10
	22	2	19	4
	19	4		
Difference	2	10		

If the sum of the forward heights is less than the sum of the back heights, it is evident that the last station must be higher than the first.

PROBLEM II.

161. *To find the difference between the* TRUE *and the* APPARENT *level, for any given distance.*

If C (Fig. 12.) be the centre of the earth considered as a sphere, AB a portion of its surface, and T a point on an apparent level with A; then BT is the difference between the true and the apparent level, for the distance AT.

Let $2BC=D$, the diameter of the earth,
$AT=d$, the distance of T, in a right line, from A,
$BT=h$, the height of T, or the difference between the true and the apparent level.

Then by Euc. 36. 3,

$$(2BC+BT)\times BT=\overline{AT}^2;\ \text{that is,}\ (D+h)\times h=d^2$$

and reducing the equation,

$$h=\sqrt{\tfrac{1}{4}D^2+d^2}-\tfrac{1}{2}D.$$

Therefore, to find h the difference between the true and the apparent level, add together one fourth of the square of the earth's diameter, and the square of the distance, extract the square root of the sum, and subtract the semi-diameter of the earth.

162. This rule is exact. But there is a more simple one, which is sufficiently near the truth for the common purposes of levelling. The height BT is so small, compared with the diameter of the earth, that D may be substituted for $D+h$, without any considerable errour. The original equation above will then become

$$D\times h=d^2.\quad \text{Therefore}\ h=\frac{d^2}{D}$$

That is, *the difference between the true and the apparent level, is nearly equal to the square of the distance divided by the diameter of the earth.*

Ex. 1. What is the difference between the true and the apparent level, for a distance of one English mile, supposing the earth to be 7940 miles in diameter?

Ans. 7.98 inches, or 8 inches nearly.

In the equation $h=\frac{d^2}{D}$, as D is a *constant* quantity, it is evident that h varies as d^2. According to the last rule then, the difference between the true and the apparent level varies as the *square of the distance*. The difference for 1 mile being nearly 8 inches,

		In	Feet.	In
For 2 miles, it is	$8\times2^2=$	32=	2	8 nearly.
For 3 miles,	$8\times3^2=$		6	
For 4 miles,	$8\times4^2=$		10	8

&c. &c. See Table IV.

Ex. 2. An observation is made to determine whether water can be brought into a town from a spring on a neighbouring hill. At a particular spot in the town, the spring, which is $2\frac{1}{2}$ miles distant, is observed to be apparently on a level. What is the descent from the spring to this spot?

The descent is nearly 4 feet 2 inches for the whole distance, or 20 inches in a mile; which is more than sufficient for the water to run freely.

Ex. 3. A tangent to a certain point on the ocean, strikes the top of a mountain 23 miles distant. What is the height of the mountain? Ans. 352 feet.

163. One place may be *below* the apparent level of another, and yet *above* the true level. The difference between the true and the apparent level for 3 miles is 6 feet. If one spot then, be only two feet below the apparent level of another 3 miles distant, it will really be 4 feet higher.

If two places are on the same true level, it is evident that each is below the apparent level of the other.

PROBLEM III.

To find the difference in the heights of two places whose distance is known.

164. *From the angle of elevation or depression, calculate how far one of the places is above or below the apparent level of the other; and then make allowance for the difference between the apparent and the true level.*

By taking, with a quadrant, the elevation of the object whose distance is given, we have one side and the angles of a right angled triangle, to find the perpendicular height above a *horizontal plane.* (Art. 6.) Adding this to the difference between the true and the apparent level, we have the height of the object above the true level of the place of observation. When an angle of *depression* is taken, it will be necessary to *subtract* instead of adding.

Ex. 1. The angle of elevation of a hill, as observed from the top of another $4\frac{1}{2}$ miles distant, is found to be 7 degrees. What is the difference in the heights of the two hills?

Height of oue above the level of the other	2917.3 feet.
Difference of the levels	13.5
Difference in the heights of the hills	2930.8

Ex. 2. From the top of a tower, the angle of depression of a fort 4 miles distant, is found to be $3\frac{1}{4}$ degrees. What is the height of the tower above the fort?

Ans. 1189 feet.

If the operation of levelling is meant to be very exact, especially when extended to considerable distances, allowance should be made for atmospheric *refraction.**

* See Note A.

SECTION V.

THE MAGNETIC NEEDLE.*

ART. 166. THE direction in which a ship is steered, and the bearings of the sides of a field, are commouly determined by observing the angles which they make with the magnetic needle. This is a bar of steel to which the magnetic power has been communicated from some other artificial or natural magnet. When it is balanced on a pin, so as to turn freely in any direction, it points towards the north and south.

The *poles* of the needle are its two extremities ; and the vertical plane which passes through these, is called the *magnetic meridian.* The astronomical meridian passes through the poles of the earth. These two meridians rarely coincide. The needle does not often point directly north and south.

167. *The* DECLINATION *of the needle is the angle which it makes with a north and south line ;* or the angle between the magnetic and the astronomical meridians. It is said to be east or west, according as the north pole of the needle points east or west of the north pole of the earth.

The *variation* of the needle is properly the *change of its declination.* The term, however, is frequently used to signify the declination itself.

The declination of the needle is very different in different parts of the earth. In some places, it is 20 or 30 degrees ; in others, little or nothing. In the variation charts given by writers on magnetism, the declination is marked, as it is found by observation on different parts of the globe. Lines are drawn connecting all the points which have the *same* declination. Thus a line is drawn through the several places in which the declination is 10 degrees ; another through those

* Cavallo on Magnetism, Rees' Cyclopedia, Transactions of the Royal Society of London, the Royal Irish Academy, the American Philosophical Society, and the American Academy of Arts and Sciences.

in which it is 5 degrees, &c. These lines are very winding; yet they never *cross* each other, though they extend all over the globe. One of the lines of *no declination* passes through the middle parts of the United States. The declination is towards this line, in places which are on either side of it. Thus in New-England the declination is west, while on the Ohio it is east. It increases with the distance from the line of no declination.

168. The declination is not only different in different places, but different in the *same* place at different times. At London, about 230 years since, it was 11¼ degrees east. It gradually decreased till 1657, when the needle pointed directly north From that time it deviated more and more to the west, till in 1800 the declination became about 24 degrees. At present, it appears to be nearly stationary, both at London and Paris.

In New-England, the declination has been generally decreasing, for many years. At Boston, it was about 9 degrees in 1708, 8 degrees in 1742, 7 degrees in 1782, and 6¼ degrees in 1810; the rate of variation being about 1½′ in a year, or a degree in 40 years.*

The variation in the declination is by no means uniform. If the needle moves two minutes from the meridian in one year, it may move a greater or less distance the next year. Its progress is different in different places. In some it is moving east, and in others west; in some it is coming nearer to the meridian, in others going farther from it.

169. There is also a *diurnal* variation, which appears to be owing to a change of temperature. During the fore part of the day, the north end of the needle frequently moves a few minutes of a degree to the west. In the evening, it returns nearly to the same point from which it started. This diurnal variation is found to be the greatest in the summer months, when the action of the sun is the most powerful.

In addition to these various changes, there are local perturbations of the needle, occasioned probably by the attraction of ferruginous substances beneath the surface of the ground.

* See the observations of Mr. Bowditch, in the Memoirs of the American Academy of Arts and Sciences, Vol. III. Part II. p. 337.

170. So many irregularities must render the magnetic compass an inaccurate instrument, unless the state of the declination is ascertained by frequent observations. This is particularly necessary at sea, where the declination may be changed by a few hours sail.

The astronomical meridian is determined by the positions of the heavenly bodies. The situation of the sun at rising or setting being known, its distance from the magnetic meridian may be observed with an *azimuth compass*, which is a mariner's compass with the addition of sight vanes for taking the direction of any object.

171. On land, a true meridian line may be drawn by observations on the *pole star*. If this were exactly in the pole, it would be always on the meridian. But the star revolves round the pole, at a short distance. in a little less than 24 hours. In about 6 hours from its passing the meridian above the pole, it is at its greatest distance west; in about 12 hours, it is on the meridian beneath the pole, and in about 18 hours, at its greatest distance east. If the direction of the star can be taken, at the instant when it is on the meridian, either above or beneath the pole; a true north and south line may be found. This method, however, requires that the exact time of passing the meridian be known, and that the observations be made expeditiously.

172. But as the star comes very gradually to its greatest distance east or west, it is easy to observe these limits; and as the revolution is made in a circle round the axis of the earth, it is evident that the pole must be in the middle between the two extreme distances. To draw a true meridian line, then, *take the direction of the pole star when it is farthest west, and also when it is farthest east; and bisect the angle made by these two directions.*

When a meridian is once drawn, it may be rendered permanent, by fixing proper marks and the declination of the needle may then be ascertained at any time, by the surveyor's compass, or more accurately by the *variation compass*, which has a long needle, and a graduated arc of so large a radius as to admit of very accurate divisions.*

* See Note L.

NOTES.

NOTE A. Page 10 and 91.

A RAY of light, in coming from a distant object to the eye, through the air, is turned from a straight line into a curve which is concave towards the earth. The effect is to *elevate* the apparent place of the object, as each point appears in the direction in which the light from that point enters the eye. This change in the apparent situation is called *astronomical refraction*, when the heavenly bodies are concerned ; and *terrestrial refraction*, when the objects are on the earth. The measure of the latter is the angle at the eye, between a straight line drawn to the object, and a tangent to the curvilinear ray, as TAT′, (Fig. 50.) T being the place of the object, and T′ its apparent place as seen from A.

The refraction is very much affected by the state of the atmosphere ; changing with the temperature, as well as with the density indicated by the barometer. In the delicate observations made in the trigonometrical surveys in England and France, the terrestrial refraction was found to vary from $\frac{1}{4}$ to $\frac{1}{24}$ of the angle at the centre of the earth subtended by the distance of the object. The mean is $\frac{1}{14}$: thus if an object at T (Fig. 50.) as seen from A in the mean state of the atmosphere, appears to be raised to T′ ; the angle TAT′ is about $\frac{1}{14}$ of the angle ACT subtended by the distance AT. This angle is easily found from the arc AB, which is nearly equal to AT ; the whole circumference of the earth being to the arc, as 360 degrees to the angle required. The mean terrestrial refraction, as thus calculated, is 3″. 7 for a *mile*, and increases as the distance nearly ; the elevation of the object being supposed to be small in comparison with its distance. In measuring altitudes, the terrestrial refraction is to be *subtracted* from the observed angle of elevation.

The alteration in the *height* of the object, by the mean refraction, is equal to $\frac{1}{7}$ of the curvature of the earth for the given distance, or of the difference between the true and apparent levels, as calculated by the rule in Art. 162. If the angle of refraction were equal to *half* the angle at the centre of the earth subtended by the distance, the change in the height of the object would be just equal to the correction for the curvature. If an object at B (Fig. 12.) were raised by refraction, so as to be seen from A in the direction of the tangent AT; the change in the altitude would be equal to BT, which is the difference between the true and the apparent level of A. In this case, the angle BAT would be half ACB. (Euc. 32. 3 and 20. 3.) But as the angle of refraction is in fact only $\frac{1}{14}$ of the angle at the centre, the change in the altitude is only $\frac{1}{7}$ of the correction for curvature. The latter is about $\frac{2}{3}$ of a foot for a mile, and varies as the square of the distance. If then d be the distance in miles; the correction for the curvature will be $\frac{2}{3}d^2$, and the correction for refraction $\frac{2}{21}d^2$. See Table IV.

The greatest distance at which an object can be seen on the surface of the earth, as calculated by the rule in Art. 23, depends on the *apparent* altitude. This being to the real altitude as 7 to 6, and the distance being nearly as the square root of the altitude; the distance at which an object can be seen by the mean refraction, is to the distance at which it could be seen without refraction as $\sqrt{7} : \sqrt{6}$, or as 14 : 13 nearly. See Playfair's Astronomy, Sec. II. Vince's Astronomy, Chap. VII. and the accounts of the Trigonometrical Surveys in England and France.

Note B. p. 20.

The method of calculation in plane sailing is sometimes spoken of as inaccurate, as only approximating to the truth, in proportion to the smallness of the difference between a plane and that part of the ocean to which the calculation is applied. This view of the subject appears not to be strictly correct. It is true, that plane sailing is *incomplete*, as it does not ascertain the *longitude*. This belongs to middle latitude or Mercator's sailing. It is also true, that if a ship sails on *several courses*, the sum of the departures is not equal to the departure for the same distance on a single course, as would be the case on a plane. (Art. 78.) It is farther to be observed, that the departure, as calculated by plane sailing, is neither the meridian distance measured on the parallel of latitude from which the ship sails, nor that measured upon the parallel upon which she arrives. But the departure for a single course, as defined in Art. 40, and the difference of latitude, are as accurately calculated by plane sailing, as if the surface of the ocean were a plane. Let the whole distance be divided into portions so small, that one of the arcs shall differ less from its tangent than by any given quantity. Each of these portions is to the corresponding departure, as radius to the sine of the course ; and to the difference of latitude, as radius to the cosine of the course. Therefore the whole distance is to the whole departure, as radius to the sine of the course ; and to the whole difference of latitude, as radius to the cosine of the course. These proportions are exact, even for a spheroid, a cylinder, or any solid of revolution.

If there were any incorrectness in plane sailing, it would extend to Mercator's sailing also ; for one is founded on the other. In Mercator's sailing, the proper difference of latitude is to the meridional difference of latitude, as the *departure* to the difference of longitude. Now the departure is calculated by plane sailing ; and any errour in this must produce an errour in the longitude. Or if the longitude be found by the theorem in Art. 72, without previously calculating the departure ; yet the *table* of meridional parts which must be used, is founded on the ratio between the departure and the difference of longitude. Art. 65.

Note C. p. 27.

It is here supposed that the direction of the ship is at right angles with every meridian which she crosses. A number of curious questions have been started respecting sailing on a sphere : such as whether a due east and west line coincides with a parallel of latitude, &c. Most of these points are easily settled by proper definitions. But this is not the place to consider them, as they belong to spherical geometry.

NOTE D. p. 34.

As the length of a minute of Mercator's meridian, is equal to the secant of the latitude, it will be a little more exact to take the latitude of the *middle* of the arc, rather than that of one extremity. On the extended meridian, the first minute will then be made equal to the secant of $\frac{1}{2}'$, the second to the secant of $1\frac{1}{2}'$, the third to the secant of $2\frac{1}{2}'$, &c.

The method of calculating by natural secants, though useful in forming a *table* of meridional parts, is subject to this inconvenience, that to obtain the meridional parts for any number of degrees of latitude, it is necessary to find separately the parts for each of the minutes contained in the given arc, and then to add them together. There is a different method, by which the meridional parts for an arc of any extent, may be calculated independently of any other arc. A portion of Mercator's meridian, extending from the equator to a given latitude, the semi-diameter of the earth being 1, is equal to *the hyperbolic logarithm of the cotangent of half the complement of the latitude.* See the London Philosophical Transactions, vol. xix. No. 219, Vince's Fluxions, Art. 190, and the Introduction to Hutton's Mathematical Tables.

NOTE E. p. 39.

The distance which a ship sails, in going from one place to another on a rhumb line, is not the *nearest* distance; for this would be an arc of a great circle. To sail on a great circle, except on a meridian or the equator, she must be continually altering her course. If it were practicable to steer a vessel in this manner, the departure, difference of latitude, &c. might be calculated by spherical trigonometry.

NOTE F. p. 41.

A traverse may also be constructed like the plot of a field in surveying, either by drawing parallel lines, as in Art. 111, or from the angles given by the rules in Art. 112, or more simply, as in Art. 123, by departure and difference of latitude, when these have been found by calculation or inspection.

Note G. p. 46.

Plane sailing is sometimes represented as a method of calculation founded on the principles of the plane chart. But in the construction of this chart, a principle is assumed which is known to be erroneous. That part of the surface of the earth which is represented on it, is supposed to be a plane. This renders the construction more or less inaccurate. But in plane sailing, the calculations are strictly correct. The principle assumed here is not that the surface of the earth is a plane; but that, from the peculiar nature of the rhumb line, the distance, departure, and difference of latitude, where the course is given, have the *same ratio* to each other which they would have upon a plane.

Note H. p. 51.

The Quadrant of reflexion has received the name of *Hadley's Quadrant*, as the description of it which was first made public, was given by John Hadley, Esq. But he has not an undisputed claim to the first invention. His description of the instrument was communicated to the Royal Society of London, in May, 1731. It appears that the principle on which it is constructed had been suggested by Dr. Hooke, several years before. But the form which he proposed was not calculated to answer the purpose, as it admitted of only one reflexion. Sir Isaac Newton, however, who died in 1727, left among his papers a description of a quadrant with two reflexions, which is substantially the same as Hadley's. This was published in the Philosophical Transactions for 1742.

It is also stated that a quadrant similar to Hadley's had been contrived by Mr. Thomas Godfrey, of Philadelphia, before Hadley's description was communicated to the Royal Society.

Hooke's Posthumous Works, Hutton's Dictionary, Transactions of the Royal Society of London for 1731, 1734, and 1742, American Magazine for Aug. and Sept. 1758, Millar's Retrospect, i. p. 468, and Analectic Magazine, ix. 281.

Note I. p. 56.

The proportion in Art. 99, on account of the smallness of the height BT compared with the semi-diameter of the earth, is not very suitable for calculating the depression with exactness. The following rule, which includes the effect of refraction, is better adapted to the purpose. The depression is found by multiplying 59″ into the square root of the height in feet. See Vince's Astronomy, Art. 197, Rees' Cyclopedia, and Table II.

In taking the altitude of a heavenly body with Hadley's Quadrant, when the view of the ocean is unobstructed, the reflected image is made to coincide with the most remote visible point of the water. But when there is land in the direction in which the observation is to be made, the image is brought to the water's edge; and the dip is *increased*, in proportion as the distance of the land is diminished. See table III.

Note K. p. 81.

This is not the place for a detailed account of the various trigonometrical operations which have been undertaken, for the purpose of determining the length of a degree of latitude, in different parts of the earth. The subject belongs rather to astronomy, than to common surveying. It may not be amiss, however, to give a concise statement of the measurements which have been made in the present and the preceding century.

About the year 1700, Picard measured a degree between Paris and Amiens; and the arc was extended by Cassini to Perpignan, about 6 degrees south of Paris, and afterwards to the northward as far as Dunkirk. These measurements were made in the middle latitudes. To compare a degree here, with the length of one near the pole, and another on the equator, two expeditions were fitted out from France about the same time, one for Lapland, and the other for South-America. The latter sailed in May, 1735, for Peru, and after a series of the most formidable embarrassments, they succeeded, at the end of 8 years, in accomplishing their object. They measured an arc of the meridian, crossing the equator from 3° 7′ north latitude to about $3\frac{1}{2}°$ south. The other party, in 1736, under the direction of Maupertius, proceeded to the head of the gulf of Bothnia, and measured an arc of the meridian extending along the river Tornei, and crossing the polar circle. The difficulties which they experienced, in this frozen and desolate region, were scarcely inferiour to those with which the other adventurers were at the same time contending in South-America.

The determination of these three arcs, one in France, one in Peru, and one in Lapland, were sufficient to satisfy astronomers, that a degree of latitude near the poles is greater than one on the equator, and consequently that the equatorial diameter of the earth is longer than the polar. But it does not necessarily follow from this, that there is a regular increase in the length of a degree, from the lower to the higher latitudes. On the contrary, according to the survey which had been made by Picard and others, a degree was found to be *greater* in the south of France, than in the north. The zeal of astronomers was therefore excited to take farther measures, to determine what is the exact length of a degree, in various parts of the earth, and to ascertain whether in the influence of gravitation, there are local inequalities, which affect the astronomical observations.

La Caille, about this time, measured an arc of the meridian at the Cape of Good Hope. He was not, however, provided with such instruments as would insure a great degree of precision. Boscovich, a distinguished philosopher, measured an arc of two degrees in Italy, from Rimini to Rome. Between the years 1764 and 1768, Messrs. Mason and Dixon, under the direction of the Royal Society of London, measured an arc of the meridian of about one degree and an half, crossing the line between Pennsylvania and Maryland. As the country here is very level, the whole distance was measured, not by a combination of triangles, but in the first place with a chain, and afterwards with rods of fir. A degree was also measured in Piedmont, another in Austria, and a third in Hungary; the first by Beccaria, and the two latter by Liesganig.

But the most perfect of all the trigonometrical surveys upon a great scale, are those which have been made within a few years in England and France. The instruments used for taking the angles, particularly the theodolite of Ramsden, and the repeating circle of Borda, have been brought to a surprising degree of exactness. The re-measurement of the line from Dunkirk to Barcelona, a part of which had been several times surveyed before, was commenced in 1792, under the direction of the Academy of Sciences in Paris; for the purpose of obtaining a standard of measure of lengths, weights, capacity, &c. derived from a portion of the meridian. The northern part of the arc was measured by Delambre, and the southern part by Mechain, who lost his life in 1805, in attempting to extend the line beyond Barcelona, to the Balearic Islands in the Mediterranean. This line was afterwards continued by Biot to Formentera, the southernmost of these islands, which is in Lat. 38° 38′ 56″. The latitude of Dunkirk is 51° 2′ 9″. The whole arc is, therefore, more than 12 degrees, and is nearly bisected by the 45th parallel of latitude. This is connected with the line carried through England to Clifton in Lat. 53° 27′ 31″: making the whole extent nearly 15 degrees.

The arc which had been surveyed by Maupertius, on the polar circle, was re-measured by Swanberg and others, in 1802. There is a difference of 230 toises in the length of a degree, as calculated from these two measurements. In India, an arc of the meridian was measured, on the coast of Coromandel, in 1803, by Major Lambton.

According to these various measurements, we have the following lengths of a degree of latitude in different parts of the earth.

	Latitude.	*Toises.*	*Fathoms.*
1. In Peru, by Bouguer,	0.	56,750	60,480
2. In India, by Lambton,	11° 30′	56,756	60,487
3. At the Cape of Good Hope, by Le Caiele,	33 18	57,037	60,780
4. In Pennsylvania, by Mason and Dixon,	39 12	56,890	60,630
5. In Italy, by Boscovich,	43	56,980	60,725
6. In Piedmont, by Beccaria,	44 44	57,070	60,820
7. In France, by Delambre and Mechain,	45	57,011	60,760
8. In Austria, by Liesganig,	48 43	57,086	60,835
9. In England, by Roy and Mudge,	52 2	57,074	60,827
10. In Lapland, by Maupertius,	66 20	57,422	61,184
Do. by Swanberg,		57,192	60,952

On a comparison of all the measurements which have been made, it is found that a degree of latitude is greater near the poles, than in the middle latitudes; and greater in the middle latitudes, than near the equator. The earth is therefore compressed at the poles, and extended at the equator. But it does not appear that it is an exact spheroid, or a solid of revolution of any kind. If arcs of the meridian which are near to each other and of moderate length be compared, they will not be found to increase regularly from a lower to a higher latitude. On the southern part of the line which was measured in France, the degrees increase very slowly; towards the middle, very rapidly; and near the northern extremity, very slowly again. Similar irregularities are found in that part of the meridian which passes through England. These irregularities are too great to be ascribed to errours in the surveys. It is concluded, therefore, that the direction of the plumb line, which is used in determining the latitude, is affected by local inequalities in the action of gravitation, owing probably to the different densities of the substances of which the earth is composed. These inequalities must also have an influence upon the figure of the fluid parts of the globe, so that the surface ought not to be considered as exactly spheroidical.

See Col. Mudge's account of the Trigonometrical Survey in England. Gregory's Dissertations, &c. on the Trigonometrical Survey. Rees' Cyclopedia, Art. Degree. Playfair's Astronomy. Philosophical Transactions of London for 1768, 1785, 1787, 1790, 1791, 1795, 1797, 1800. Asiatic Researches, vol. viii. Puissant. "Traité de Géodesic." Maupertius. "Degré du Meridian entre Paris et Amiens." Do. "La-Figure de la Terre." Cassini. "Exposé des Operations, &c." Delambre. "Bases du système métrique." Swanberg. "Exposition des Operations faites en Lapponie." Laplace. "Traité de Mécanique Céleste."

Note L. p. 94.

One of the most simple methods of determining when the pole star is on the meridian, is from the situations of two other stars, Alioth and γ Cassiopeiæ, both which come on to the meridian a few minutes before the pole star, the one above and the other below the pole. Alioth, which is the star marked ϵ in the Great Bear, is on the same side of the pole with the pole star, and about 30 degrees distant. The star γ in the constellation Cassiopeiæ, is nearly as far on the opposite side of the pole. The right ascension of the latter in 1810 was 0h. 45m. 24s., increasing about 3½ seconds annually. The right ascension of Alioth was 12h. 45m. 36s., increasing about 2½ seconds annually. These two stars, therefore, come on to the meridian nearly at the same time. This time may be known by observing when the same vertical line passes through them both. The right ascension of the pole star in January 1810, was 0h. 54m. 36s., and increases 13 or 14 seconds in a year. So that this star comes to the meridian about 9 or 10 minutes after γ Cassiopeiæ. In very nice observations, it will be necessary to make allowance for nutation, aberration, and the annual variation in right ascension.

About 10 minutes after a line drawn from Alioth to γ Cassiopeiæ is parallel to the horizon, the pole star is at its greatest distance from the meridian. As this is the case only once in 12 hours, the two limits on the east side, and on the west side, cannot both be observed the same night, except at certain seasons of the year. But on any clear night, one observation may be made; and this is sufficient for finding a meridian line, if the distance of the star from the pole, and the latitude of the place be given. The angle between the meridian and a vertical plane passing through a star, or an arc of the horizon contained between these two planes, is called the *azimuth* of the star. And by spherical trigonometry, when the star is at its greatest elongation east or west,

As the cosine of the latitude,
To radius;
So is the sine of the polar distance,
To the sine of the azimuth.

The distance of the pole star from the pole in 1810, was 1° 42′ 19″.6, and decreases 19½ seconds annually.

To observe the direction of the pole star when its azimuth is the greatest, suspend a plumb line 15 or 20 feet long from a fixed point, with the weight swinging in a vessel of water, to protect it from the action of the wind. At the distance of 12 or 15 feet south, fix a board horizontal on the top of a firm post. On the board, place a sight vane in such a manner that it can slide a short distance to the east or west. A little before the time when the star is at its greatest elongation, let an assistant hold a lighted candle so as to illuminate the plumb line. Then move the sight vane, till the star seen through it is in the direction of the line. Continue to follow the motion of the star, till it appears to be stationary at its greatest elongation. Then fasten the sight vane, and fix a candle or some other object in the direction of the plumb line, at some distance beyond it.

As the declination of the needle is continually varying, the courses given by the compass in old surveys, are not found to agree with the bearings of the same lines at the present time. To prevent the disputes which arise from this source, the declination should always be ascertained, and the courses stated according to the angles which the lines make with the astronomical meridian.

It must be admitted, after all, that the magnetic compass is but an imperfect instrument. It is not used in the accurate surveys in England. In the wild lands in the United States, the lines can be run with more expedition by the compass, than in any other way. And in most of the common surveys, it answers the purpose tolerably well. But in proportion as the value of land is increased, it becomes important that the boundaries should be settled with precision, and that all the lines should be referred to a permanent meridian. The angles of a field may be accurately taken with a graduated circle furnished with two indexes. The bearings of the sides will then be given, if a true meridian line be drawn through any point of the perimeter.

EXPLANATION OF THE TABLES.

Table I contains the parts of Mercator's meridian, to every other minute. The parts for any odd minute may be found with sufficient exactness, by taking the arithmetical mean between the next greater and the next less. For the uses of this table, see Navigation, Sec. III.

Table II gives the depression or dip of the horizon at sea for different heights. Thus if the eye of the observer is 20 feet above the level of the ocean, the angle of depression is 4′ 24″. See Art. 99. This table is calculated according to the rule in note I, which gives the depression 59″ for one foot in altitude; allowance being made for the mean terrestrial refraction.

In Table III, is contained the depression for different heights and *different distances*, when the view of the ocean is more or less obstructed by land. Thus if the height of the eye is 30 feet, and the distance of the land $2\frac{1}{2}$ miles, the depression is 8′. See Note I.

Table IV contains the curvature of the earth, or the difference between the true and the apparent level. for different distances, according to the rule in art. 161. Thus for a distance of 17 English miles, the curvature is 192 feet.

Table V contains the distances at which objects of different heights may be seen from the surface of the ocean, in the mean state of the atmosphere. This is calculated by first finding the distance at which a given object might be seen, if there were no refraction, and then increasing this distance in the ratio of $\sqrt{7} : \sqrt{6}$. See Note A.

Table VI contains the polar distance, and the right ascension in time, of the pole star, from 1800 to 1820. From this it will be seen, that the right ascension is increasing at the rate of about 14 seconds a year, and that the north polar distance is decreasing at the rate of 19½ seconds a year. From the latitude of the place, and the polar distance of the star, its azimuth may be calculated, when it is at its greatest distance from the meridian. The time when it passes the meridian may be ascertained by finding the difference between the right ascension of the star and that of the sun. See Note L.

TABLE I.

MERIDIONAL PARTS.

M.	0°	1°	2°	3°	4°	5°	6°	7°	8°	9°	10°	11°	12°	M.
0	0	60	120	180	240	300	361	421	482	542	603	664	725	0
2	2	62	122	182	242	302	363	423	484	544	605	666	727	2
4	4	64	124	184	244	304	365	425	486	546	607	668	729	4
6	6	66	126	186	246	306	367	427	488	548	609	670	731	6
8	8	68	128	188	248	308	369	429	490	550	611	672	734	8
10	10	70	130	190	250	310	371	431	492	552	613	674	736	10
12	12	72	132	192	252	312	373	433	494	554	615	676	738	12
14	14	74	134	194	254	314	375	435	496	556	617	678	740	14
16	16	76	136	196	256	316	377	437	498	558	619	680	742	16
18	18	78	138	198	258	318	379	439	500	560	621	682	744	18
20	20	80	140	200	260	320	381	441	502	562	623	684	746	20
22	22	82	142	202	262	322	383	443	504	565	625	687	748	22
24	24	84	144	204	264	324	385	445	506	567	627	689	750	24
26	26	86	146	206	266	326	387	447	508	569	629	691	752	26
28	28	88	148	208	268	328	389	449	510	571	632	693	754	28
30	30	90	150	210	270	331	391	451	512	573	634	695	756	30
32	32	92	152	212	272	333	393	453	514	575	636	697	758	32
34	34	94	154	214	274	335	395	455	516	577	638	699	760	34
36	36	96	156	216	276	337	397	457	518	579	640	701	762	36
38	38	98	158	218	278	339	399	459	520	581	642	703	764	38
40	40	100	160	220	280	341	401	461	522	583	644	705	766	40
42	42	102	162	222	282	343	403	463	524	585	646	707	768	42
44	44	104	164	224	284	345	405	465	526	587	648	709	770	44
46	46	106	166	226	286	347	407	467	528	589	650	711	772	46
48	48	108	168	228	288	349	409	469	530	591	652	713	774	48
50	50	110	170	230	290	351	411	471	532	593	654	715	777	50
52	52	112	172	232	292	353	413	473	534	595	656	717	779	52
54	54	114	174	234	294	355	415	476	536	597	658	719	781	54
56	56	116	176	236	296	357	417	478	538	599	660	721	783	56
58	58	118	178	238	298	359	419	480	540	601	662	723	785	58
M.	0°	1°	2°	3°	4°	5°	6°	7°	8°	9°	10°	11°	12°	M.

TABLE I.

MERIDIONAL PARTS.

M.	13°	14°	15°	16°	17°	18°	19°	20°	21°	22°	M.
0	787	848	910	973	1035	1098	1161	1225	1289	1354	0
2	789	851	913	975	1037	1100	1164	1227	1291	1356	2
4	791	853	915	977	1039	1102	1166	1229	1293	1358	4
6	793	855	917	979	1042	1105	1168	1232	1296	1360	6
8	795	857	919	981	1044	1107	1170	1234	1298	1362	8
10	797	859	921	983	1046	1109	1172	1236	1300	1364	10
12	799	861	923	985	1048	1111	1174	1238	1302	1367	12
14	801	863	92[illegible]	987	1050	1113	1176	1240	1304	1369	14
16	803	865	927	989	1052	1115	178	1242	1306	1371	16
18	805	867	929	991	1054	1117	1181	1244	1308	1373	18
20	807	869	931	994	1056	1119	1183	1246	1311	1375	20
22	809	871	933	996	1058	1121	1185	1249	1313	1377	22
24	811	873	935	998	1060	1123	1187	1251	1315	1380	24
26	813	875	937	1000	1063	1126	1189	1253	1317	1382	26
28	816	877	939	1002	1065	1128	1191	1255	1319	1384	28
30	818	879	942	1004	1067	1130	1193	1257	1321	1386	30
32	820	882	944	1006	1069	1132	1195	1259	1324	1388	32
34	822	884	946	1008	1071	1134	1198	1261	1326	1390	34
36	824	886	948	1010	1073	1136	1200	1264	1328	1393	36
38	826	888	950	1012	1075	1138	1202	1266	1330	1395	38
40	828	890	952	1014	1077	1140	1204	1268	1332	1397	40
42	830	892	954	1016	1079	1142	1206	1270	1334	1399	42
44	832	894	956	1019	1081	1145	1208	1272	1336	1401	44
46	834	896	958	1021	1084	1147	1210	1274	1339	1403	46
48	836	898	960	1023	1086	1149	1212	1276	1341	1406	48
50	838	900	962	1025	1088	1151	1215	1278	1343	1408	50
52	840	902	964	1027	1090	1153	1217	1281	1345	1410	52
54	842	904	966	1029	1092	1155	1219	1283	1347	1412	54
56	844	906	969	1031	1094	1157	1221	1285	1349	1414	56
58	846	908	971	1033	1096	1159	1223	1287	1352	1416	58
M.	13°	14°	15°	16°	17°	18°	19°	20°	21°	22°	M.

TABLE I.

MERIDIONAL PARTS.

M	23°	24°	25°	26°	27°	28°	29°	30°	31°	32°	M.
0	1419	1484	1550	1616	1684	1751	1819	1888	1958	2028	0
2	1421	1486	1552	1619	1686	1753	1822	1891	1960	2031	2
4	1423	1488	1554	1621	1688	1756	1824	1893	1963	2033	4
6	1425	1491	1557	1623	1690	1758	1826	1895	1965	2035	6
8	1427	1493	1559	1625	1693	1760	1829	1898	1967	2038	8
10	1430	1495	1561	1628	1695	1762	1831	1900	1970	2040	10
12	1432	1497	1563	1630	1697	1765	1833	1902	1972	2043	12
14	1434	1499	1565	1632	1699	1767	1835	1905	1974	2045	14
16	1436	1502	1568	1634	1701	1769	1838	1907	1977	2047	16
18	1438	1504	1570	1637	1704	1772	1840	1909	1979	2050	18
20	1440	1506	1572	1639	1706	1774	1842	1912	1981	2052	20
22	1443	1508	1574	1641	1708	1776	1845	1914	1984	2054	22
24	1445	1510	1577	1643	1711	1778	1847	1916	1986	2057	24
26	1447	1513	1579	1645	1713	1781	1849	1918	1988	2059	26
28	1449	1515	1581	1648	1715	1783	1852	1921	1991	2061	28
30	1451	1517	1583	1650	1717	1785	1854	1923	1993	2064	30
32	1453	1519	1585	1652	1720	1787	1856	1925	1995	2066	32
34	1456	1521	1588	1654	1722	1790	1858	1928	1998	2069	34
36	1458	1524	1590	1657	1724	1792	1861	1930	2000	2071	36
38	1460	1526	1592	1659	1726	1794	1863	1932	2002	2073	38
40	1462	1528	1594	1661	1729	1797	1865	1935	2005	2076	40
42	1464	1530	1596	1663	1731	1799	1868	1937	2007	2078	42
44	1467	1532	1599	1666	1733	1801	1870	1939	2010	2080	44
46	1469	1535	1601	1668	1735	1803	1872	1942	2012	2083	46
48	1471	1537	1603	1670	1738	1806	1875	1944	2014	2085	48
50	1473	1539	1605	1672	1740	1808	1877	1946	2017	2088	50
52	1475	1541	1608	1675	1742	1810	1879	1949	2019	2090	52
54	1477	1543	1610	1677	1744	1813	1881	1951	2021	2092	54
56	1480	1546	1612	1679	1747	1815	1884	1953	2024	2095	56
58	1482	1548	1614	1681	1749	1817	1886	1956	2026	2097	58
M.	23°	24°	25°	26°	27°	28°	29°	30°	31°	32°	M.

TABLE I.

MERIDIONAL PARTS.

M.	33°	34°	35°	36°	37°	38°	39°	40°	41°	42°	M.
0	2100	2171	2244	2318	2393	2468	2545	2623	2702	2782	0
2	2102	2174	2247	2320	2395	2471	2548	2625	2704	2784	2
4	2104	2176	2249	2323	2398	2473	2550	2628	2707	2787	4
6	2107	2179	2252	2325	2400	2476	2553	2631	2710	2790	6
8	2109	2181	2254	2328	2403	2478	2555	2633	2712	2792	8
10	2111	2184	2257	2330	2405	2481	2558	2636	2715	2795	10
12	2114	2186	2259	2333	2408	2484	2560	2638	2718	2798	12
14	2116	2188	2261	2335	2410	2486	2563	2641	2720	2801	14
16	2119	2191	2264	2338	2413	2489	2566	2644	2723	2803	16
18	2121	2193	2266	2340	2415	2491	2568	2646	2726	2806	18
20	2123	2196	2269	2343	2418	2494	2571	2649	2728	2809	20
22	2126	2198	2271	2345	2420	2496	2573	2651	2731	2811	22
24	2128	2200	2274	2348	2423	2499	2576	2654	2733	2814	24
26	2131	2203	2276	2350	2425	2501	2578	2657	2736	2817	26
28	2133	2205	2279	2353	2428	2504	2581	2659	2739	2820	28
30	2135	2208	2281	2355	2430	2506	2584	2662	2742	2822	30
32	2138	2210	2283	2358	2433	2509	2586	2665	2744	2825	32
34	2140	2213	2286	2360	2435	2512	2589	2667	2747	2828	34
36	2143	2215	2288	2363	2438	2514	2591	2670	2750	2830	36
38	2145	2217	2291	2365	2440	2517	2594	2673	2752	2833	38
40	2147	2220	2293	2368	2443	2519	2597	2675	2755	2836	40
42	2150	2222	2296	2370	2445	2522	2599	2678	2758	2839	42
44	2152	2225	2298	2373	2448	2524	2602	2680	2760	2841	44
46	2155	2227	2301	2375	2451	2527	2604	2683	2763	2844	46
48	2157	2230	2303	2378	2453	2530	2607	2686	2766	2847	48
50	2159	2232	2306	2380	2456	2532	2610	2688	2768	2849	50
52	2162	2235	2308	2383	2458	2535	2612	2691	2771	2852	52
54	2164	2237	2311	2385	2461	2537	2615	2694	2774	2855	54
56	2167	2239	2313	2388	2463	2540	2617	2696	2776	2858	56
58	2169	2242	2316	2390	2466	2542	2620	2699	2779	2860	58
M.	33°	34°	35°	36°	37°	38°	39°	40°	41°	42°	M.

TABLE I.

MERIDIONAL PARTS.

M.	43°	44°	45°	46°	47°	48°	49°	50°	51°	52°	M.
0	2863	2946	3030	3116	3203	3292	3382	3474	3569	3665	0
2	2866	2949	3033	3118	3206	3295	3385	3478	3572	3668	2
4	2869	2951	3036	3121	3209	3298	3388	3481	3575	3672	4
6	2871	2954	3038	3124	3212	3301	3391	3484	3578	3675	6
8	2874	2957	3041	3127	3214	3303	3394	3487	3582	3678	8
10	2877	2960	3044	3130	3217	3306	3397	3490	3585	3681	10
12	2880	2963	3047	3133	3220	3309	3400	3493	3588	3685	12
14	2882	2965	3050	3136	3223	3312	3403	3496	3591	3688	14
16	2885	2968	3053	3139	3226	3316	3407	3499	3594	3691	16
18	2888	2971	3055	3142	3229	3319	3410	3503	3598	3695	18
20	2891	2974	3058	3144	3232	3322	3413	3506	3601	3698	20
22	2893	2976	3061	3147	3235	3325	3416	3509	3604	3701	22
24	2896	2979	3064	3150	3238	3328	3419	3512	3607	3704	24
26	2899	2982	3067	3153	3241	3331	3422	3515	3610	3708	26
28	2902	2985	3070	3156	3244	3334	3425	3518	3614	3711	28
30	2904	2988	3073	3159	3247	3337	3428	3521	3617	3714	30
32	2907	2991	3075	3162	3250	3340	3431	3525	3620	3717	32
34	2910	2993	3078	3165	3253	3343	3434	3528	3623	3721	34
36	2913	2996	3081	3168	3256	3346	3437	3531	3626	3724	36
38	2915	2999	3084	3171	3259	3349	3440	3534	3630	3727	38
40	2918	3002	3087	3173	3262	3352	3443	3537	3633	3731	40
42	2921	3005	3090	3176	3265	3355	3447	3540	3636	3734	42
44	2924	3007	3093	3179	3268	3358	3450	3543	3639	3737	44
46	2926	3010	3095	3182	3271	3361	3453	3547	3643	3741	46
48	2929	3013	3098	3185	3274	3364	3456	3550	3646	3744	48
50	2932	3016	3101	3188	3277	3367	3459	3553	3649	3747	50
52	2935	3019	3104	3191	3280	3370	3462	3556	3652	3750	52
54	2937	3021	3107	3194	3283	3373	3465	3559	3655	3754	54
56	2940	3024	3110	3197	3286	3376	3468	3562	3659	3757	56
58	2943	3027	3113	3200	3289	3379	3471	3566	3662	3760	58
M.	43°	44°	45°	46°	47°	48°	49°	50°	51°	52°	M.

TABLE I.

MERIDIONAL PARTS.

M.	53°	54°	55°	56°	57°	58°	59°	60°	61°	62°	M.
0	3764	3865	3968	4074	4183	4294	4409	4527	4649	4775	0
2	3767	3868	3971	4077	4186	4298	4413	4531	4653	4779	2
4	3770	3871	3975	4081	4190	4302	4417	4535	4657	4784	4
6	3774	3875	3978	4085	4194	4306	4421	4539	4662	4788	6
8	3777	3878	3982	4088	4197	4309	4425	4543	4666	4792	8
10	3780	3882	3985	4092	4201	4313	4429	4547	4670	4796	10
12	3784	3885	3989	4095	4205	4317	4433	4551	4674	4801	12
14	3787	3889	3992	4099	4208	4321	4436	4555	4678	4805	14
16	3790	3892	3996	4103	4212	4325	4440	4559	4682	4809	16
18	3794	3895	3999	4106	4216	4328	4444	4564	4687	4814	18
20	3797	3899	4003	4110	4220	4332	4448	4568	4691	4818	20
22	3800	3902	4006	4113	4223	4336	4452	4572	4695	4822	22
24	3804	3906	4010	4117	4227	4340	4456	4576	4699	4826	24
26	3807	3909	4014	4121	4231	4344	4460	4580	4703	4831	26
28	3811	3913	4017	4124	4234	4347	4464	4584	4707	4835	28
30	3814	3916	4021	4128	4238	4351	4468	4588	4712	4839	30
32	3817	3919	4024	4132	4242	4355	4472	4592	4716	4844	32
34	3821	3923	4028	4135	4246	4359	4476	4596	4720	4848	34
36	3824	3926	4031	4139	4249	4363	4480	4600	4724	4852	36
38	3827	3930	4035	4142	4253	4367	4484	4604	4728	4857	38
40	3831	3933	4038	4146	4257	4370	4488	4608	4733	4861	40
42	3834	3937	4042	4150	4260	4374	4492	4612	4737	4865	42
44	3838	3940	4045	4153	4264	4378	4495	4616	4741	4870	44
46	3841	3944	4049	4157	4268	4382	4499	4620	4745	4874	46
48	3844	3947	4052	4161	4272	4386	4503	4625	4750	4879	48
50	3848	3951	4056	4164	4275	4390	4507	4629	4754	4883	50
52	3851	3954	4060	4168	4279	4394	4511	4633	4758	4887	52
54	3854	3958	4063	4172	4283	4398	4515	4637	4762	4892	54
56	3858	3961	4067	4175	4287	4401	4519	4641	4766	4896	56
58	3861	3964	4070	4179	4291	4405	4523	4645	4771	4901	58
M.	53°	54°	55°	56°	57°	58°	59°	60°	61°	62°	M.

TABLE I.

MERIDIONAL PARTS.

M.	63°	64°	65°	66°	67°	68°	69°	70°	71°	72°	M.
0	4905	5039	5179	5324	5474	5631	5795	5966	6146	6335	0
2	4909	5044	5184	5328	5479	5636	5800	5972	6152	6341	2
4	4914	5049	5188	5333	5484	5642	5806	5978	6158	6348	4
6	4918	5053	5193	5338	5489	5647	5811	5984	6164	6354	6
8	4923	5058	5198	5343	5495	5652	5817	5989	6170	6361	8
10	4927	5062	5203	5348	5500	5658	5823	5995	6177	6367	10
12	4931	5067	5207	5353	5505	5663	5828	6001	6183	6374	12
14	4936	5071	5212	5358	5510	5668	5834	6007	6189	6380	14
16	4940	5076	5217	5363	5515	5674	5839	6013	6195	6387	16
18	4945	5081	5222	5368	5520	5679	5845	6019	6201	6394	18
20	4949	5085	5226	5373	5526	5685	5851	6025	6208	6400	20
22	4954	5090	5231	5378	5531	5690	5856	6031	6214	6407	22
24	4958	5095	5236	5383	5536	5695	5862	6037	6220	6413	24
26	4963	5099	5241	5388	5541	5701	5868	6043	6226	6420	26
28	4967	5104	5246	5393	5546	5706	5874	6049	6233	6427	28
30	4972	5108	5250	5398	5552	5712	5879	6055	6239	6433	30
32	4976	5113	5255	5403	5557	5717	5885	6061	6245	6440	32
34	4981	5118	5260	5408	5562	5723	5891	6067	6252	6447	34
36	4985	5122	5265	5413	5567	5728	5896	6073	6258	6453	36
38	4990	5127	5270	5418	5573	5734	5902	6079	6264	6460	38
40	4994	5132	5275	5423	5578	5739	5908	6085	6271	6467	40
42	4999	5136	5280	5428	5583	5745	5914	6091	6277	6473	42
44	5003	5141	5284	5433	5588	5750	5919	6097	6283	6480	44
46	5008	5146	5289	5438	5594	5756	5925	6103	6290	6487	46
48	5012	5151	5294	5443	5599	5761	5931	6109	6296	6494	48
50	5017	5155	5299	5448	5604	5767	5937	6115	6303	6500	50
52	5021	5160	5304	5454	5610	5772	5943	6121	6309	6507	52
54	5026	5165	5309	5459	5615	5778	5948	6127	6315	9514	54
56	5030	5169	5314	5464	5620	5783	5954	6133	6322	6521	56
58	5035	5174	5319	5469	5625	5789	5960	6140	6328	6528	58
M.	63°	64°	65°	66°	67°	68°	69°	70°	71°	72°	M.

TABLE I.

MERIDIONAL PARTS.

M.	73°	74°	75°	76°	77°	78°	79°	80°	81°	82°	M.
0	6534	6746	6970	7210	7467	7745	8046	8375	8739	9145	0
2	6541	6753	6978	7218	7476	7754	8056	8387	8752	9160	2
4	6548	6760	6986	7227	7485	7764	8067	8398	8765	9174	4
6	6555	6768	6994	7235	7494	7774	8077	8410	8778	9189	6
8	6562	6775	7001	7243	7503	7783	8088	8422	8791	9203	8
10	6569	6782	7009	7252	7512	7793	8099	8433	8804	9218	10
12	6576	6790	7017	7260	7521	7803	8109	8445	8817	9233	12
14	6583	6797	7025	7268	7530	7813	8120	8457	8830	9248	14
16	6590	6804	7033	7277	7539	7822	8131	8469	8843	9262	16
18	6597	6812	7041	7285	7548	7832	8141	8480	8856	9277	18
20	6603	6819	7048	7294	7557	7842	8152	8492	8869	9292	20
22	6610	6826	7056	7302	7566	7852	8163	8504	8883	9307	22
24	6617	6834	7064	7311	7576	7862	8174	8516	8896	9322	24
26	6624	6841	7072	7319	7585	7872	8185	8528	8909	9337	26
28	6631	6849	7080	7328	7594	7882	8196	8540	8923	9353	28
30	6639	6856	7088	7336	7603	7892	8207	8552	8936	9368	30
32	6646	6864	7096	7345	7612	7902	8218	8565	8950	9383	32
34	6653	6871	7104	7353	7622	7912	8229	8577	8963	9399	34
36	6660	6879	7112	7362	7631	7922	8240	8589	8977	9414	36
38	6667	6886	7120	7371	7640	7932	8251	8601	8991	9430	38
40	6674	6894	7128	7379	7650	7942	8262	8614	9005	9445	40
42	6681	6901	7136	7388	7659	7953	8273	8626	9018	9461	42
44	6688	6909	7145	7397	7668	7963	8284	8638	9032	9477	44
46	6695	6917	7153	7406	7678	7973	8295	8651	9046	9493	46
48	6702	6924	7161	7414	7687	7983	8307	8663	9060	9509	48
50	6710	6932	7169	7423	7697	7994	8318	8676	9074	9525	50
52	6717	6939	7177	7432	7706	8004	8329	8688	9088	9541	52
54	6724	6947	7185	7441	7716	8014	9341	8701	9103	9557	54
56	6731	6955	7194	7449	7725	8025	8352	8714	9117	9573	56
58	6738	6963	7202	7458	7735	8035	8364	8726	9131	9589	58
M.	73°	74°	75°	76°	77°	78°	79°	80°	81°	82°	M.

Depression of the Horizon of the Sea.

Height of the Eye in feet	Depression.
1	0′ 59″
2	1 24
3	1 42
4	1 58
5	2 12
6	2 25
7	2 36
8	2 47
9	2 57
10	3 7
11	3 16
12	3 25
13	3 33
14	3 41
15	3 48
16	3 56
17	4 3
18	4 10
19	4 17
20	4 24
22	4 37
24	4 49
26	5 1
28	5 13
30	5 23
35	5 49
40	6 14
45	6 36
50	6 57
60	7 37
70	8 14
80	8 48
90	9 20
100	9 50
120	10 47
140	11 39
160	12 27
180	13 12
200	13 55

Dip of the Sea, at different Distances from the Observer.

Dist. of land in sea miles.	Height of the Eye above the Sea, in feet.							
	5	10	15	20	25	30	35	40
0 ¼	11′	22′	34′	45′	56′	68′	79′	90′
0 ½	6	11	17	22	28	34	39	45
0 ¾	4	8	12	15	19	23	27	30
1 0	4	6	9	12	15	17	20	23
1 ¼	3	5	7	9	12	14	16	19
1 ½	3	4	6	8	10	11	14	15
2 0	2	3	5	6	8	10	11	12
2 ½	2	3	5	6	7	8	9	10
3 0	2	3	4	5	6	7	8	8
3 ½	2	3	4	5	6	6	7	7
4 0	2	3	4	4	5	6	7	7
5 0	2	3	4	4	5	5	6	6
6 0	2	3	4	4	5	5	6	6

TABLE IV.

Curvature of the Earth.

Dist. in miles.	Height.	Dist. in miles.	Height.
	Inches.		Feet.
¼	½	15	149
½	2	16	170
1	8	17	192
	Feet.	18	215
2	2.6	19	240
3	6.	20	266
4	10.6	25	415
5	16.6	30	599
6	23.9	35	814
7	32.5	40	1064
8	42.5	45	1346
9	53.8	50	1662
10	66.4	60	2394
11	80.2	70	3258
12	95.4	80	4255
13	112.	90	5386
14	130	100	6649

TABLE V.

Distances at which Objects can be seen at Sea.

Height in feet.	Distance in Eng. miles.	Height in feet.	Distance in Eng. miles,
1	1.3	60	10.2
2	1.9	70	11.1
3	2.3	80	11.8
4	2.6	90	12.5
5	2.9	100	13.2
6	3 2	200	18.7
7	3.5	300	22.9
8	3.7	400	26.5
9	4.	500	29.6
10	4.2	600	32.4
12	4.6	700	35.
14	4.9	800	37.4
16	5.3	900	39.7
18	5.6	1000	41.8
20	5 9	2000	59.2
25	6.6	3000	72.5
30	7.3	4000	83.7
35	7.8	5000	93.5
40	8.4	10000	133.
45	8.9	15000	163.
50	9.4	20000	188.

TABLE VI.

The Polar Distance and Right Ascension of the Pole Star.

	Polar Distance.	Ann. Var.	Right Ascension. h.	m.	s.	Ann. Var. s.
1800	1° 45′ 35″	—19″.5	0	52	24	+12.9
1801	1 45 15			52	37	
1802	1 44 56			52	50	
1803	1 44 36			53	3	
1804	1 44 17			53	16	
1805	1 43 57			53	29	
1806	1 43 38			53	42	
1807	1 43 18			53	55	
1808	1 42 58			54	9	
1809	1 42 39			54	22	s.
1810	1 42 19			54	36	+13 6
1811	1 42			54	50	
1812	1 41 40			55	04	
1813	1 41 21			55	18	
1814	1 41 1			55	33	
1815	1 40 42			55	47	
1816	1 40 23			56	02	
1817	1 40 04			56	17	
1818	1 39 45			56	32	
1819	1 39 25			56	46	s.
1820	1 39 05	—19″.4		57	01	+14.3

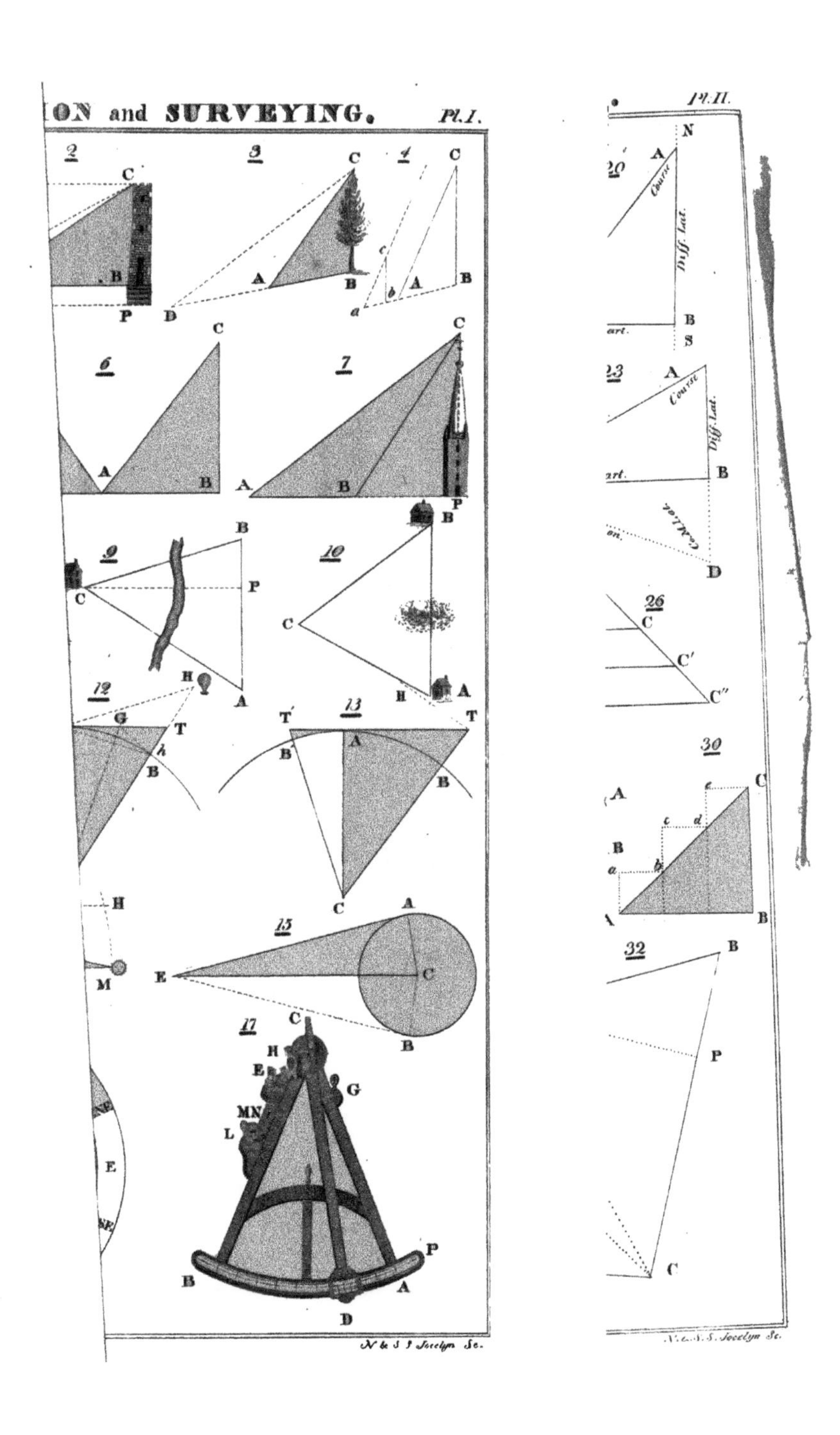
ION and SURVEYING.
Pl. I.
Pl. II.
N & S S Jocelyn Sc.
N & S S Jocelyn Sc.

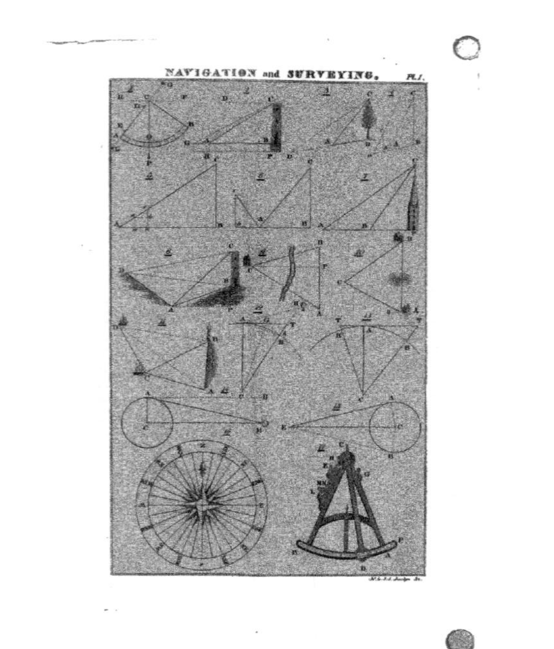
NAVIGATION and SURVEYING,
Pl. I.

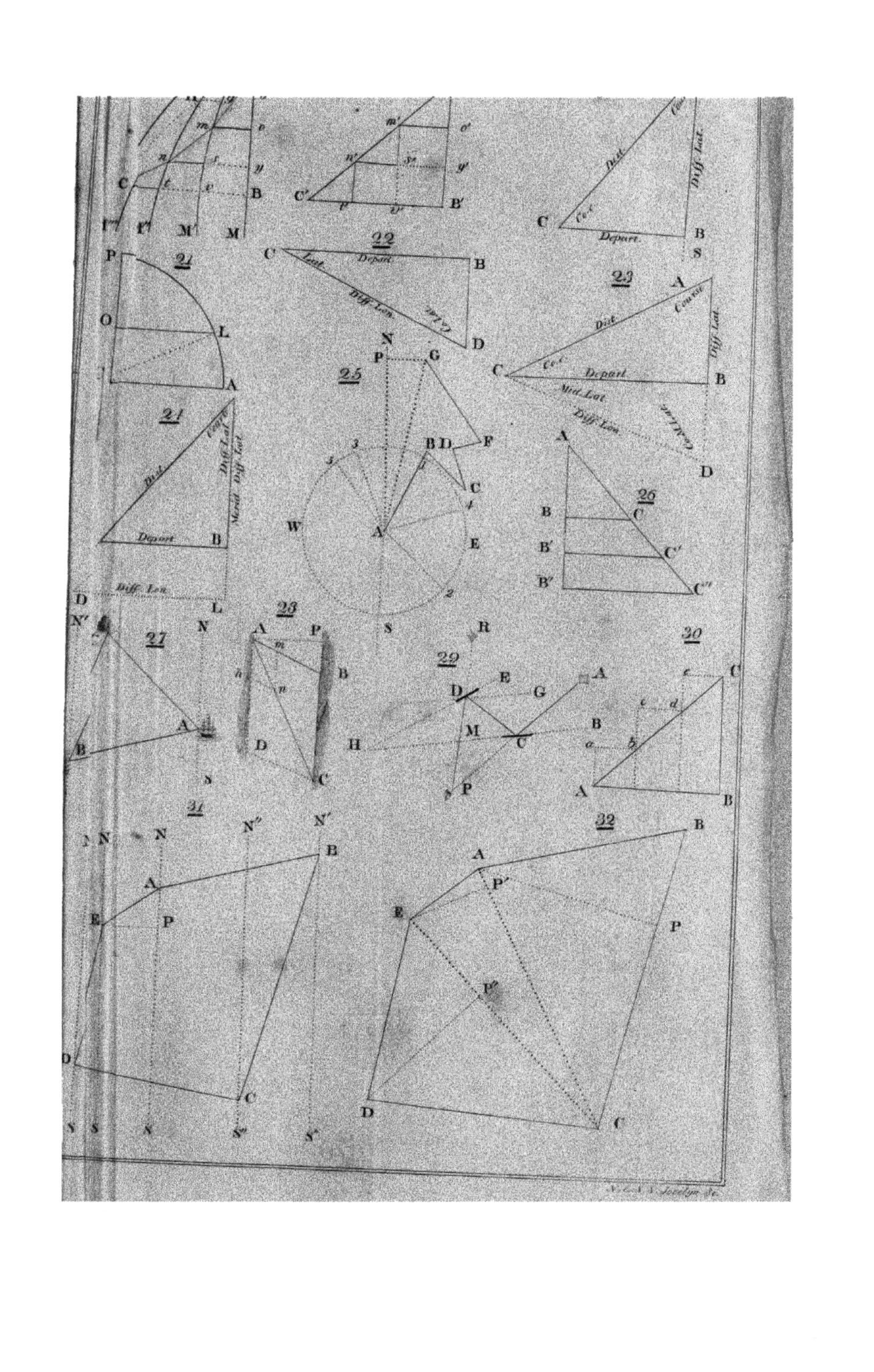

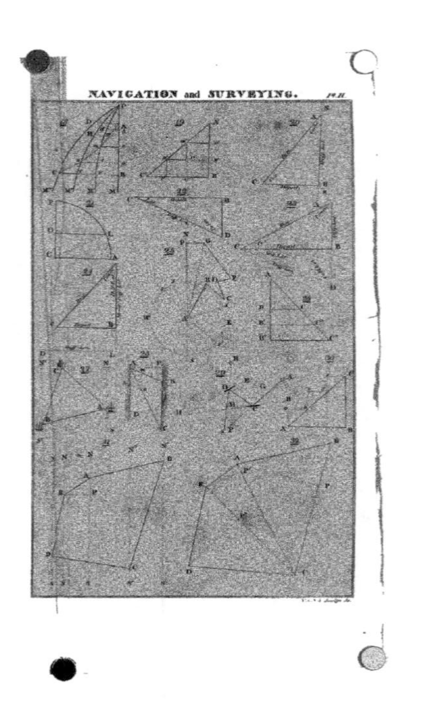
NAVIGATION and SURVEYING.

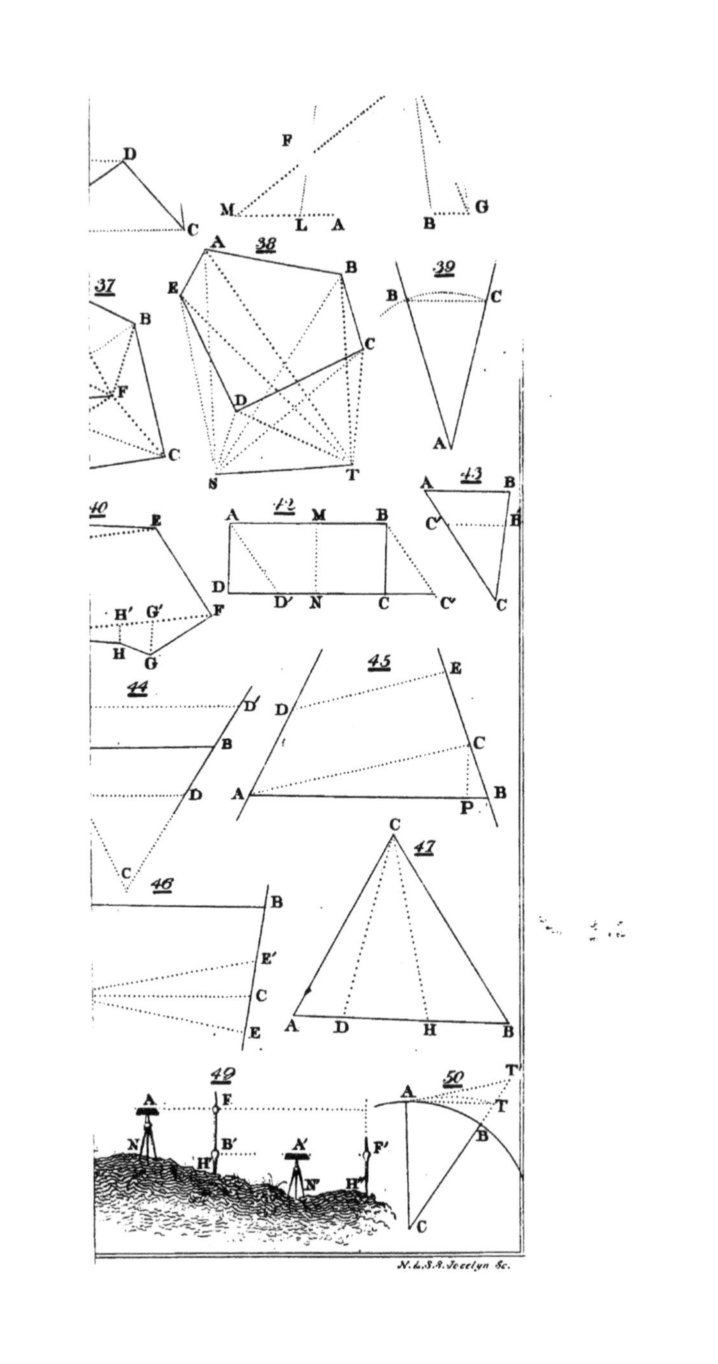
37
38
39
40
42
43
44
45
46
47
49
50
N. L. S. S. Jocelyn Sc.

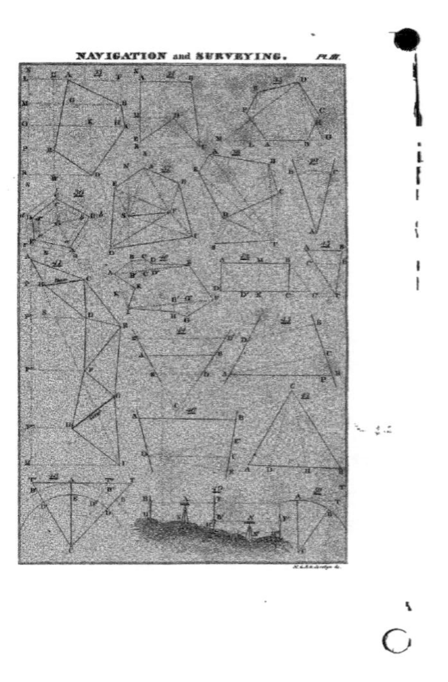
NAVIGATION and SURVEYING.

AN

ELEMENTARY

TREATISE

ON

ONIC SECTIONS, SPHERICAL GEOMETRY,

AND

SPHERICAL TRIGONOMETRY.

BY MATTHEW R. DUTTON,

PROFESSOR OF MATHEMATICS AND NATURAL PHILOSOPHY
IN YALE COLLEGE.

BEING THE FIFTH AND SIXTH PARTS

OF A COURSE OF MATHEMATICS ADAPTED TO THE METHOD OF
INSTRUCTION IN AMERICAN COLLEGES

NEW-HAVEN.

PUBLISHED BY HOWE & SPALDING.

S. Converse, printer.

1824.

ADVERTISEMENT.

IN preparing the following Treatise on Conic Sections, it was determined to adopt the general plan and arrangement of propositions in HUTTON's CONIC SECTIONS ; the deviations from that plan however have become so numerous and considerable as to leave but a slight resemblance to it. The authors whose works have been consulted, and in some cases freely used, are APPOLLONIUS, ARCHIMEDES, HAMILTON, SIMPSON, WEST, VINCE, EMERSON and BRIDGE ; especially the two authors last mentioned. This general acknowledgment is made, as it is inconvenient or impossible to specify more particularly what is due to each.

The authors whose works have supplied the principal materials for the Spherical Geometry and Trigonometry, are THEODOSIUS, SIMPSON, PLAYFAIR, and CAGNOLI ; the Projections and Examples are taken with slight alterations from the Mathematics of WEBBER, for, in the following work, there has been no affectation of originality, when there appeared no room for improvement.

M. R. D.

CONTENTS.

CONIC SECTIONS.

CONIC SECTIONS.

DEFINITIONS.

1. CONIC SECTIONS are the figures made by the mutual intersection of a cone and plane.

2. According to the different positions of the plane, five different figures or sections, are produced; viz, a *Triangle*, a *Circle*, a *Parabola*, an *Ellipse*, and a *Hyperbola*.

3. If the plane pass through the vertex of the cone, and any part of the base, the section will be a triangle.

Case 1. Let it pass through the vertex, perpendicular to the base, and coinciding with the axis of the cone; it will then coincide with the triangle by the revolution of which the cone is generated. The section therefore will be an isoceles triangle VCD, whose angle at the vertex will be double that of the generating triangle. Sup.3.11. Fig. 1

Case 2. Let the plane now cut the base obliquely. This section also will be a triangle, for every straight line drawn from the vertex of a cone to any point of the circumference of its base, is in the surface of the cone; since it coincides with the Hypothenuse of the revolving triangle by which the cone is generated.

4. If the plane cut the cone parallel to its base, the section will be a circle, as ABD—for this section will coincide with the revolution of a line perpendicular to the axis of the generating triangle, and will therefore be a circle, whose radius is equal to that perpendicular. Fig. 2.

5. If the cone be cut by a plane parallel to one of its sides, the section is called a *Parabola;* as ACD. Fig. 3.

As the plane in this case cuts one side of the cone parallel to its opposite side, it is evident that it can never meet the latter; the section therefore may be continued indefinitely, if the cone is supposed to be indefinitely extended. 1. 10

6. If the plane cut the cone obliquely, so as to make an angle with the base less than that made by the side of the cone, the section is called an *Ellipse*: as ABC. Fig. 4.

As the cutting plane in this case, is not parallel with the side of the cone V K, it will meet that side—that is it will intersect the surface of the cone on every side. This section therefore is complete and definite for every position of the plane.

7. If the cutting plane makes with the base of the cone an angle greater than that made by the side of the cone, the section is called an *Hyperbola*; as ACD. Fig. 5.

The cutting plane in this position will never meet the opposite side of the cone in the direction of V K, but it will meet it produced, on the other side of V, and if all the sides of the cone be supposed to be produced through the vertex V, forming an opposite and similar cone, and if the plane also be produced cutting this cone in the section B*cd*, the two sections ACD and B*cd* are called *Opposite Hyperbolas*.

Fig. 6. 1. 13. Cor.

8. If there be four cones, the angles at whose vertices are together equal to four right angles; and if their axes be all placed in one plane and their vertices all meet in a given point V, the sides of the cones will touch each other in the right lines LVF, BVH. If a plane perpendicular to that in which the axes of the cone are placed, and parallel to O P, the axis of the two cones BVL FVH, cut these cones in the opposite hyperbolas AD B*d*; if another plane, also perpendicular to that in which the axes of the cone are placed, pass through A or B parallel to KVM, the axes of the remaining cones, cutting opposite hyperbolas in these cones, these two pairs of opposite hyperbolas are mutually *conjugate* to each other. The two sections are commonly supposed to be placed in the same plane, the lines AB and *ab* bisecting each other, as in fig. 7.

If the angle at the vertex of each of the cones, (fig, 6) be a right angle, the cones will all be equal and similar, and the sections, being at the same distance from the vertex, will be equal and similar, and AB will be equal to *ab*; in this case they are called *Equilateral Hyperbolas*, and may be all cut by one plane, parallel to that in which the axes of the cones are placed; as represented in fig. 7.

9. As the properties of the triangle and circle are demonstrated in the Elements of Euclid, and are investigated without any reference to the sections of a cone, they are not usually reckoned among the conic sections ;—this term being appropriately applied to the three remaining sections,—viz. the Parabola, Ellipse and Hyperbola.

10. The *Vertices* of a conic section, are the points where the cutting plane meets the opposite sides of the cone ; or the sides of a vertical triangular section through the axis of the cone, and perpendicular to the plane of the given section.

In the Parabola, the plane does not meet the opposite side of the cone VH, (5) * This section therefore has but one vertex, as A . Fig. 3.

The Ellipse has two vertices, and opposite Hyperbolas have two ; as AB. Fig. 4. & 5.

11. The *Transverse axis*, is the straight line which connects the vertices ; as AB.

This definition is applicable to the Ellipse and opposite Hyperbolas, but not to the Parabola which has but one vertex. The *axis* of the Parabola, may be defined, the intersection of its plane with that of the vertical triangular section to which it is perpendicular. This definition is applicable to all the conic sections, since it evidently coincides with that before given for the Ellipse and Hyperbola. The line of their mutual intersection passes through the vertex of the Parabola and is of indeterminate length ; since the section itself may be indefinitely continued. Sup. 2. 3.

12. The *Centre* of a conic section is the middle of the transverse axis.

Hence the Parabola has no centre ; that of the Ellipse is within the section, and that of the Hyperbola without it, between the opposite Hyperbolas.

13. The *Conjugate axis* of a conic section, is a line drawn through the centre, at right angles to the transverse axis.

* This figure and others similar, refer to the articles numbered in this series of *definitions*.

Hence the Parabola has no conjugate axis. In the Ellipse it is within the section and terminated by the curve. The conjugate axis of the Hyperbola is without the section and is bounded by the curves of the conjugate Hyperbolas. Or more correctly, the conjugate axis is the transverse axis of the conjugate Hyperbolas. (8) Hence the two axes are mutually conjugates to each other.

The preceding definition, of the conjugate axis of a Hyperbola, is correct, in all cases in which it is applicable. It is limited however in its application, to those cases in which the Hyperbola, is cut by a plane parallel to the Axis of the cone.

The following limitation of the conjugate axis of a Hyperbola is general. It is a mean proportional between the diameters of the circular sections, which pass through the vertices of the transverse axis of the Hyperbola. Thus the conjugate axis of A B, is equal to a mean proportional between A Q and B R.

Fig. 5.

14. A *Diameter* of any conic section, is a straight line, drawn through the centre, and terminated in both directions by the curve.

This definition of a diameter evidently includes the axis.

This definition is inapplicable to the Parabola which has no centre. (12). A diameter of a Parabola, may be defined, *any line drawn parallel to its axis*, and terminated, like the axis, in one direction by the curve, while the other may be indefinitely extended, as C*d*.

Fig. 9.

15. The extremities of a diameter, or its intersections with the curve are called its *Vertices*.

16. The *Conjugate* to any diameter, is a line drawn through the centre of the section, and parallel to the tangent * at the vertex of the diameter. Thus D E is the conjugate of the diameter C H.

Fig. 8.

This definition evidently includes the conjugate *axis* (13.) If it is admitted, or assumed, that the tangent at the vertex of the transverse axis, is perpendicular to it.

* Euclid's definition of a *line touching a circle*, may be applied to a tangent of a Conic Section. It is a line which meets the curve, but does not cut it.

17. An *Ordinate* to any diameter is a right line, parallel to the tangent at its vertex, and terminated in one direction by the curve, and in the other, by the diameter as K G. Figures 8 and 9.

18. The *Absciss* to any ordinate is the part of the diameter, contained between its vertex and that ordinate, as C G and G H. Fig. 8 & 9.

Hence in the Ellipse and Hyperbola, every ordinate has two determinate abscisses. In the Parabola, there is but one, the other being supposed indefinitely extended. (11.) as G*d*. Fig. 9.

19. The *Parameter* of any diameter, sometimes called the *Latus Rectum*, is a third proportional to that diameter and its conjugate.

This definition is not applicable to the Parabola, (13.) but its Parameter is a third proportional to any absciss and its ordinate.

20. The *Focus* of a conic section, is the point in the axis where the ordinate is equal to half the parameter. As F where C F is equal to the semi-parameter of the section. Fig. 7. 8. & 9.

The Ellipse and Hyperbola have each two Foci as F, *f*, the Parbola has but one.

21. If tangents be drawn to the four vertices of the curves, in Conjugate Hyperbolas, forming an inscribed rectangle, the diagonals of this rectangle, are called the *Assymptotes* of the Hyperbolas. As L F and K H. Fig. 7.

22. *Similar* conic sections, are those of the same kind, whose transverse and conjugate axes have to each other the same ratio.

As this definition is not applicable to the Parabola, *Similar Parabolas*, are those whose *abscisses* and *ordinates* are to each other in the same ratio.

23. When a magnitude, which is supposed to vary according to a certain law, approaches continually towards equality with another given magnitude, so that ultimately the former differs from the latter less than by any assign-

able magnitude, then the latter is said *to limit*, or to be *the limit* of the former, and their *ultimate* or *limiting ratio*. is said to be a ratio of equality.

Thus Euclid, (Book XII Prop. 2,) demonstrates that a regular polygon inscribed in a circle, will, as the number of its sides is supposed to be increased, approach continually to equality with the circle, so as ultimately to differ from the circle less than by any assignable magnitude. By this truth, he demonstrates, that the areas of circles are as the squares of their diameters.

In this example, the circle is called *the limit* of the polygons. He also demonstrates that in this sense, the circle is also the *limit* of polygons circumscribed about it.

"The consideration of *limits* more or less disguised, must unavoidably enter into every investigation which has for its object, the mensuration of the circle." *Leslie.*

"This principle is general, and is the only one by which we can possibly compare curvilineal with rectilineal spaces, or the length of curve lines with the length of straight lines, whether we follow the methods of the ancient or of the modern geometers." *Playfair.*

General Scholium.

The cutting plane being parallel to the base, forming a circular section, if it be supposed to revolve about A, at the commencement of its motion, when it deviates in the least possible degree from parallelism with the base of the cone, the section will be an Ellipse, which in the language of mathematicians is *infinitely* near to a coincidence with the circular section. That is, it may differ from the circle less than by any assignable difference. Fig. 10.

If now the plane be supposed to revolve from B′ towards B″—in passing through B, where it is parallel to the base, the *circle* will be the section which forms the *transition* from the Ellipse immediately on one side of it, to that immediately on the other. It is the *limit* towards which each Ellipse approaches indefinitely, and may therefore be considered an Ellipse, whose transverse and conjugate axes are equal, and whose Foci unite in the centre. Fig. 10.

Again—if the plane be supposed to revolve towards a parallelism with the opposite side of the cone, the transverse axis of the Ellipse becomes extended, and the section approaches towards that of a Parabola, from which, *immediately* before the plane becomes parallel with the side of the cone, it differs less than by any assignable difference. In other words, the Parabola is the *limit* to which the Ellipse, by the revolution of the plane, continually and indefinitely approaches.—The Parabola, therefore, has been called an Ellipse whose axis is infinite, and whose centre is at an infinite distance.

Again—immediately after the plane ceases to be parallel to the side of the cone, the section is an Hyperbola, which is indefinitely near, in resemblance and coincidence, to the Parabola, which is the *limit* to which the Hyperbola approaches; it may therefore be called a Hyperbola whose axis is infinite. The Parabola is therefore the *limit* between the Ellipse and the Hyperbola, and holds the place of *transition* from one to the other.

Lastly—If a Hyperbolic section, perpendicular to the base of the cone, be supposed to move parallel to itself, it will at one point coincide with the axis of the cone. The section will then become a triangle. (def. 2.) Immediately before and after this coincidence, the section will be a Hyperbola, which is indefinitely near to the triangular section. The latter is the *limit* to which the Hyperbolic sections approach indefinitely on each side—it is the *transition* from one to the other, and may therefore be considered as a Hyperbola whose axis is infinitely small, or evanescent, and whose Foci unite with their vertices in the vertex of the triangle. The sides of this triangular section (which is the *limit* of the Hyperbolas, on either side of it,) evidently coincides with the two *assymptotes* of the Hyperbolas.

From the connexion and intimate relations thus seen to exist between all the Conic Sections, we should expect striking resemblances and analogies in their properties. It will be one object in the following treatise to point these out as they occur.

The most simple figures produced by the intersection of a cone and plane, are the *triangle* and *circle*; but for the reasons before mentioned (9) they are here omitted, and the properties of the Parabola, the section next in simplicity to them, will be first examined.

OF THE PARABOLA.

PROPOSITION I.

The squares of any two ordinates, are to each other as their abscisses.

Fig. 11. Let NVM be a triangular section through the axis of the cone, AGIH, a Parabolic section made by a plane, perpendicular to the former, AH the line of their mutual intersection, will be the axis of the Parabola, (11,) also let there be two circular sections parallel to the base of the cone, and cutting AH in the points F and H; FG and HI, the intersections of these planes with that of the Parabolic section, will be perpendicular to the plane of the triangular section, and therefore perpendicular to HA, and to FL and HN; they are therefore ordinates (17) both in the Parabola and in the circles.

Sup. 2. 18.

Sup. 2. def. 1.

The proposition then is $FG^2 : HI^2 :: AF : AH$

6. 4.	By sim. tri.	$AF : AH :: FL : HN$;
	but, because	AH is parallel to VM and KL to MN,
1. 34.	∴	KH is a parallelogram and $KF = MH$,
6. 1.	therefore	$AF : AH :: KF . FL : MH . HN$;
6. 6. Cor. & 3. 31.	but in the circle	$FG^2 = KF . FL$ and $HI^2 = MH . HN$,
5. 7.	therefore	$FG^2 : HI^2 :: AF : AH$.

Q. E. D.

Cor. 1.—This is true of ordinates *on both sides of the axis*, as the demonstration is equally applicable to ordinates on either side; and the ordinates in the circle are evidently bisected; therefore $FS = FG$, or GS is double of FG, it is therefore called a *double ordinate;* also the *curve* on each side of the axis is similar, the whole Parabola is bisected by the axis, and the two parts, if laid one upon the other, will coincide in every respect—for if they did not coincide, in any part, the corresponding ordinates on opposite sides of the axis, would be *unequal.* For a similar reason, the two parts of the tangent at the vertex, will coincide;—they therefore make equal angles with the axis—in other words, the tangent at the vertex of a Parabola, is perpendicular to the axis.

Schol.—If any straight line FG, be supposed to move parallel to itself, upon another straight line, as AH, to which it is perpendicular, and if the line FG vary in its length continually, so that its squares have always the same ratio to each other as its distances from the point A, the line FG will describe the half of a parabolic section. Or, (as FG^2 equals the rectangle GF.FS) if any line, as GS, be bisected by AH, to which it is perpendicular, and moving parallel to itself, vary in such a manner that the rectangles of its parts are as its distances from A, it will describe a Parabola. And, because, when the absciss increases, the rectangle GF . FS increases in the same ratio, the curve will always recede from the axis; but as this rectangle, or the square of FG, increases more rapidly than FG itself, or in the duplicate ratio of FG, the curve will recede less for any given increment of the absciss, as it is more distant from the vertex, or, in other words, it is always concave to the axis.

Cor. 2. (19)	$AF : FG :: FG : P =$ Parameter,	
therefore	$FG^2 = AF . P,$	6. 17.
also (19)	$AH : HI :: HI : P',$	
and	$HI^2 = AH . P',$	Con-
therefore	$FG^2 : HI^2 :: AF . P : AH . P',$	verse of 5. 14.
but (1)	$FG^2 : HI^2 :: AF : AH,$	6. 1.
therefore	$AF : AH :: AF . P : AH . P'.$	5. 11.
$\therefore$	$P = P'$, or the Parameter of the axis	

is a constant quantity; also the rectangle of this constant quantity, and any absciss of the axis, is equal to the rectangle of the equal parts of the double ordinate,—that is, to the square of the ordinate. This, expressed in algebraical language, is called *the equation of the curve.**

* Day's Algebra—Sec. 22—Let the Parameter be represented by a, the absciss by x, and the ordinate by y; then $ax = y^2$: this is termed "the equation of the Curve," of a Parabola.

SCHOL.—From this equality between the square of any ordinate, and the rectangle of its absciss and parameter, Appollonius of Perga, called this section the *Parabola.* (*Appollonii Conica, per J. Barrow* 10 *page. Lon.* 1675, *also Hamilton's* Conic *Sections* 84.)

PROPOSITION II

If any double ordinate to the axis, intersect any number of diameters, the parts of the diameters cut off by the double ordinate, are to each other, as the rectangles of the segments into which they divide the double ordinate.

Fig. 12. That is, AH : GL::KH . HI : KL . LI.

For (I. Cor. 2.) $P \cdot AH = HI^2$,

3. ax & and $P \cdot AF = FG^2$,

1. 6. therefore $P \cdot (FH=)\ GL = HI^2 - FG^2$,

2. 5. Cor. or $P \cdot GL = KL \cdot LI$,

5. 14. therefore $P \cdot AH : P \cdot GL :: (HI^2 =) KH \cdot IH : KL \cdot LI$;

6. 1. or $AH : GL :: KH \cdot HI : KL \cdot LI$.

Q. E. D.

The same demonstration is evidently applicable to any other diameter, as ED.

SCHOL.—If then any line, as AH, be supposed to move parallel to itself, on KI, to which it is always perpendicular, and to vary in its length continually, in the same ratio as the rectangles into which it divides KI, the space described by it will be a parabola; whose highest point or vertex will be above the middle of the base, because the rectangle is the greatest when the line is equally divided. 6. 27.

COR. 1.—As this is equally applicable to all double ordinates to the axis, the proposition may be stated universally, *the parts of all diameters cut off by any double ordinate to the axis, are to the parts cut off by any other double ordinates, as the rectangles of the segments into which they divide the former double ordinate, are to the rectangles of the segments into which they divide the latter.*

That is $GL : Gl :: KL \cdot LI : kl \cdot li$.

and $GL : Ed :: KL \cdot LI : kd \cdot di$.

This proposition evidently includes Proposition I. Also in all cases, the rectangle of the segments of any double ordinate to the axis, is equal to the rectangle of the Parameter of the axis, and the part of the diameter which is cut off by the double ordinate.

That is P . GL=KL . LI; and P . ED=KD . DI. (See dem. of Prop. 2d.)

This also includes Cor. 2, Prop. I. 6. 16.

Cor. 2. As the Parameter of the axis,
To the sum of any two ordinates,
So is the difference of those ordinates,
To the difference of their abscisses.

For the last step but two of the demonstration is,
P . GL=KL . LI.
therefore P : KL (or HI +FG) : : LI (or HI−FG) : GL.

PROPOSITION III.

If any double ordinate be produced and intersect any number of diameters *without* the curve, the external segments of these diameters, are to each other a[s t]he rectangles of the corresponding segments of the line.

That is $nI : bG : gn . no : gb . bo$. Fig. 12.

For (1) $(Fm^2 =)$ $HI^2 = AH . P$,
and $(fb^2 =)$ $FG^2 = AF . P$
$\therefore$ $HI^2 -$ $FG^2 (= Gm . mN) = FH . P = mI . P$, 2. 5.
and $FG^2 -$ $Fg^2 (= gb . bo) =$ $Ff . P = bG . P$;
therefore $Gm . mN : gb . bo : : mI . P : bG . P : : mI : bG$;
so $gn . no : gb . bo : : nI : bG$
or $nI : bG : : gn . no : gb . bo$.

Q. E. D.

Cor. I. Hence $gb . bo : Gm . mN : : bG : mI$.

Cor. 2. When *fa* approaches A*a*, the rectangle *gn* . *no*, approaches continually and indefinitely towards equality with the square o[f] A*a*, the intercepted part of the tangent at the vertex; when therefore *f* coincides with A, the property becomes $Gm . mN : Aa^2 : : mI : aI$
and $gb . [b]o : Ah^2 : : bG : hG$
$\therefore$ $Aa^2 : Ah^2 : : aI : hG$

That is, the squares of the intercepted parts of the tangent at the vertex, are as the external parts of the diameters intercepted.

PROPOSITION IV.

If a tangent be drawn to any point of the Parabola meeting the axis produced ; and if an ordinate to the axis be drawn from the point of contact ; then the absciss of that ordinate will be equal to the external part of the axis.

That is, if T C touch the curve at the point C, and CM be an ordinate to the point of contact,

Fig. 26.

then $AT = AM$

For from the point T draw a line cutting the curve in the two points E, H ; to which draw the ordinates D E, GH ;

Then by (I.) $AD : AG :: DE^2 : GH^2$,

6. 4. & 22. and by sim. tri. $TD^2 : TG^2 :: DE^2 : GH^2$,

5. 11. therefore $AD : AG :: TD^2 : TG^2$,

5. A. 17. and $AD : DG :: TD^2 : TG^2 - TD^2$,

2. 5. Cor. but $TG^2 - TD^2 = DG \,.\, \overline{TD + TG}$,

5. 7. ∴ $AD : DG :: TD^2 : DG \,.\, \overline{TD + TG}$,

6. 1. and the rect's. $AD \,.\, TD : TD \,.\, DG :: TD^2 : DG \,.\, \overline{TD + TG}$,

5. 16. consequently $AD \,.\, TD : TD^2 :: TD \,.\, DG : DG \,.\, \overline{TD + TG}$,

6. 1 therefore the bases $AD : TD :: TD : TD + TG$,

5. A. & E. and $AD : AT :: TD : TG$,

5. 16. " $AD : TD :: AT : TG$,

5. A. & E. " $AD : AT :: AT : AG$.

or AT is a mean proportional between AD and AG.

Now if the line T H be supposed to revolve about the point T, then as it recedes from the axis and approaches the tangent T C, the points E and H approach the point of contact, and when T H coincides with the tangent, the points E and H will unite in C. But when the points E and H approach towards each other in C; D and G also approach towards each other, and towards M. Consequently A D and A G approach towards the ratio of equality; and before they unite in M, they differ from each other less than by any assignable quantity; their *limiting ratio*, therefore, in M is a ratio of equality. But AT is always a mean proportional between them, it is always therefore greater than AD and less than A G, and like A M is a *limit* between A D and A G; it is therefore equal to AM. That is, the external part of the axis cut off by the tangent is equal to the absciss of the ordinate to the point of contact.

Q. E. D.

Cor. 1.—AM or AT is a mean proportional between AD and AG.

Cor. 2. As this is true of a tangent drawn from either extremity of a double ordinate, the tangents, CT, DT intersect the axis in the same point, and make equal angles with it. And conversely, a line bisecting the angle made by the tangents, or a line drawn from that point and bisecting the double ordinate, is the axis. Fig. 16.

Cor. 3.—If two or more Parabolas have the same axis and vertex, and any ordinate in one section be produced until it cut the other, and tangents to the curve be drawn from the points of intersection, these tangents will all meet the axis in the same point.

Since in each AT = AM.

Schol.—From this proposition it is easy to draw a tangent to the curve from any point, either in the *curve* or in the axis produced. If it be in the curve, as C, draw an ordinate from the point, and take a point T in the axis produced, at a distance from the vertex equal to the ordinate; a straight line connecting this point with the given point, will be the tangent.

If the given point be in the axis produced, take an absciss AD, equal to the distance of the point from the vertex, the ordinate to this absciss will cut the curve in the point of contact; the line which connects this with the given point, will be the tangent.

Cor. 4.—The *subtangent*, TM, is equal to twice the absciss.

PROPOSITION V.

If there be any tangent and double ordinate to the axis, drawn from the point of contact, and also any other diameter limited by the tangent and double ordinate, then shall the curve divide that diameter in the same ratio as the diameter divides the double ordinate.

Fig. 13. That is IE : EK : : CK : KL.

6. 4. For by sim. tri. CK : KI : : CD : DT or 2DA,
but (def. 19.) P : CD : : CD : DA,
5. 4. Cor. and P : CL : : CD : 2DA,
6. 22. therefore P : CK : : CL : KI;
but (II Cor. 2.) P : CK : : KL : KE;
5. 11. therefore CL : KL : : KI : KE;
5. 17. and CK : KL : : IE : EK;
or IE : EK : : CK : KL.

Q. E. D.

5. 14. Cor. 1.—If CK=KL, then IE=EK.

PROPOSITION VI.

The same being supposed as in the last proposition, the external parts of the diameters, between the curve and tangent, are to each other as the squares of the intercepted parts of the tangent.

Fig. 13. That is, $IE : TA :: CI^2 : CT^2$,

For (V) IE : EK : : CK : KL,
6. 1 And $IE : EK :: CK^2 : CK . KL$;
In the same way $TA : AD :: CD^2 : CD . DL$;
But (II) EK : AD : : CK . KL : CD . DL,
5. 22. Therefore $IE : TA :: CK^2 : CD^2$;
6. 4. And $IE : TA :: CI^2 : CT^2$.

Q. E. D.

PROPOSITION VII.

The distance from the Focus to the vertex, is equal to one fourth of the Parameter of the axis.

or $AF=\frac{1}{4}P$ Fig. 14.

for (19) $AF : FL :: FL : P$

but (20) $FL=\frac{1}{2}P \therefore AF=\frac{1}{2}FL=\frac{1}{4}P$ 5. C

Q. E. D.

Cor. 1.—If a tangent be drawn to the curve, from the point L, it is called the *Focal-tangent*, and the distance of its intersection with the axis from the Focus, equals the focal-ordinate. That is $Ft=(2FA=)\ FL$.

Cor. 2.—Also the distance of any point in this tangent from the Axis, equals the sum of the focal distance, and the corresponding absciss, that is $ON=Nt=NA+At=NA+AF$.

Cor. 3.—The tangent at the vertex, AP, intercepted by the Focal-tangent, is equal to half FL, (the ordinate to the Focus,)$=\frac{1}{4}P=AF=At$, so that A is the centre of a circle passing through t, P and F. 6. 4

Cor. 4.—If through the point T, where the Focal-tangent intersects the axis, a perpendicular to the axis be drawn it is called the *Directrix*, and it is evident that the distance of any point of the curve as M from the Directrix is equal to ON, or the ordinate drawn through this point, and produced to meet the Focal-tangent. Fig. 15.

For $MB=OT=ON=$(Cor. 2.) $Nt=$(Fig. 14.)$NA+AF$.

PROPOSITION VIII.

If a tangent to the curve meet the axis produced, then the distance of the Focus from the point of contact, is equal to the distance of the Focus from the intersection of the tangent and axis.

Fig. 14. That is, $FC=FT$.

For draw the ordinate CD, to the point of contact,

then (IV.) $AT=AD$,

Ax. 2. therefore $FT=AF+AD$,

2. 4. and $FT^2=AF^2+AD^2+2AF.AD$;

1 47. again $FC^2=FD^2+CD^2$,

but $FD=AD-AF$. and $FD^2=AF^2+AD^2 - 2AF.AD$;

and (I Cor. 2.) $CD^2=P.AD=4AF.AD$ (VII)

$\therefore$ $FC^2=AF^2+AD^2+2AF.AD$.

Ax. 1. therefore $FT^2=FC^2$ and $FT=FC$.

Q. E. D.

Cor. 1.—The triangle FCT, is isosceles, and the angle formed by the tangent, and a line drawn to the Focus, is equal to that formed by the tangent and the the axis, or any diameter; that is $FCT=FTC=KCM$. (1. 5.; 1. 29.)

Cor. 2.—If CG be drawn perpendicular to the tangent at C, it is called the *normal*, and FG, the distance of its intersection with the axis from the Focus equals FC, which by the proposition equals FT.

(1. 11.; 6. 2.) For draw FH, perpendicular to TC, it will bisect TC, ; therefore HF also, bisects TG, consequently $FG=FT=FC$.

Cor. 3.—Therefore F is the centre of a circle passing through T, C, G, and the angle at the centre TFC is double TGC; censequently FCK is double GCK; or $FCG=GCK$. (3. 9.; 1. 29.)

Schol.—It is a law in optics, that the angle made by a ray of reflected light, with a perpendicular to the reflecting surface, *is equal* to the angle which the incident ray makes with the same perpendicular. Hence, if a ray should fall upon the concave surface in the direction of KC, it would be reflected in CF; and all parallel rays, since they would coincide with diameters, would be reflected towards the same point, which is therefore called the Focus.

Cor. 4.—The *subnormal* DG equals the semi-parameter, or the ordinate at the Focus.

For (IV) (TD=)2AD : DC::DC : DG, — 6. 8.
but (def 19) AD : DC::DC : P=parameter.
therefore 2AD : AD::P : DG; — 5. 23.
∴ $DG=\frac{1}{2}P$. — 5. C
also GX being drawn perpendicular to CF, $CX=DG=\frac{1}{2}P$; for the triangles CDG, GXC are similar and equal.

Cor. 5.—A perpendicular to the axis drawn from the vertex, meets the tangent in the same point, H, in which a perpendicular from the Focus upon the tangent meets it. (See Cor. 2.)

For as (IV) AT=AD∴TH=HC. — 6. 2.

Hence also AH is a mean proportional between AF and AT, that is between AF and AD, the distances of the Focus and the ordinate to the point of contact, from the vertex; — 6. 8.

Or AF : AH : AT, are continued proportionals.

Also, FH, is a mean proportional between FA and FT, the distances of the Focus, from the vertex and the point of intersection of the tangent and axis produced.

Or FA : FH : FT, are continued proportionals.

Cor. 6.—The two tangents CHT and AHR mutually bisect each other. For as $TA=\frac{1}{2}TD$∴$AH=\frac{1}{2}DC=\frac{1}{2}AR$.

Cor. 7.—Hence FI, a line drawn from the focus perpendicular to the tangent and produced to meet the diameter which passes through the point of contact, is bisected by the tangent;—and, conversely, a line drawn from C bisecting FI, at right angles, is a tangent to the curve at that point.

COR. 8.—Hence (2nd step. dem.) FT (=FC)=FA+AT=FA+AD.

SCHOL. From this property the curve is easily described by points.

Let A be the vertex of the axis, draw any number of lines as EE perpendicular to the axis, then with the distances, TD TF, TO as radii, and the Focus as a centre, describe curves crossing the parallel lines in EE, and then draw the curve through all the points E, E, E, &c.

PROPOSITION IX.

If tangents be drawn to the vertex of the axis, and of any other diameter, meeting the axis and diameter produced, the triangles thus formed are equal.

Fig. 20. That is, AHT=CHE.

For, since (IV)TA=AD=EC, and AH=HE;

Therefore the triangles are similar and equal.

COR. 1.—If AM be parallel to CT, the triangle CDT

1,36 & 41. =AEM=CD. DA=MDAC=MATC.

COR. 2.—If PIN be perpendicular to AR, and PRL parrallel to CT. Then the triangle PIR=IA . AE.

For the triangles CDT PIR are similar;

6. 19. Therefore—triangle CDT : PIR : : CD^2 : PI^2,

6. 1. (I) : : DA : IA,

: : DA AE : IA . AE;

1, & 41. But CDT=DA . AE. ∴ PIR=IA AE.

COR. 3.—Hence to both PIR and IA . AE, add LRIN, and LPN=LRAE=LRTC.

COR. 4.—In a similar manner it may be shown, that, p i R=iA . AE=TC n i.

From p i R, and TC n i, take RL n i, and p n L=LRTC=LPN.

COR. 5.—Therefore Pp, parallel to CT. is bisected in L; or every diameter bisects its double ordinate.

SCHOLIUM.

The preceding proposition and the fifth, enable us to pass from the *axis*, to the other diameters of the section, and to prove the same things, concerning each of them, which have been demonstrated concerning the axis. The preceeding propositions therefore, except that which refers to the *Focus*, will appear to be *particular* truths, which are all included in the more *general* propositions that follow.

Care has been taken, that these *general properties* of all the diameters of the Parabola, be arranged in the same order, and expressed in the same form, in which the particular properties of the axis have been ; not only to assist the memory of the student, but to present more clearly the universal and striking analogies which are observabale in the different parts of the Parabola.

PROPOSITION X.

The squares of the ordinates of *any* diameter, are as its abscisses.

That is $QE^2 : RA^2 :: CQ : CR$. Fig. 17.

This has already been proved concerning the *axis*, (Prop. I ;) to shew that it is al so true of any other diameter, as CS, draw the tangent CP, and the externals EI, AT, parallel to the axis, or to the diameter CS;

Then because the ordinates QE, RA are parallel to the tangent CP, by the definition of them, therefore the figures IQ, TR, are parallelograms whose opposite sides are equal, $IE=CQ$, and $TA=CR$; also $QE=CI$, and $RA=CT$.

Therefore (VII) $CQ : CR :: QE^2 : RA^2$

Q. E. D.

COR. 1.—This is true of ordinates on either side of the diameter, because every diameter bisects all lines in the section which are parrallel to the tangent at its vertex. (5 Cor. Prop. IX.) Hence EV and AX are *double ordinates*: and, conversely, as all double ordinates are parallel to each other, and are bisected; any straight line which bisects two parallel straight lines in a Parabola, is a diameter.

Fig. 24. Hence a segment of a Parabola being given as ONI, it is easy to find its diameter. For draw any other line as C L parallel to IO, bisect them both, in M and D, the line MD, produced until it meets the curve, is the diameter of the segment of which I O is a double ordinate.

COR. 2.—Hence, (as in Cor. 2. Prop. I.) the parameter of any diameter being a third proportional to its absciss and ordinate, is a constant quantity, and the rectangle of any absciss, and the parameter of that diameter, is equal to the square of the ordinate, or in other words it is equal to the rectangle of the parts into which the diameter divides the double ordinate. (See Cor. 1.)

Fig. 17. Thus $CQ.\ P' = QE^2 = EQ.\ QV$,
And $CR.\ P'' = RA^2 = AR.\ RX$.

PROPOSITION XI.

Any straight line, in a Parabola, terminated by the curve, cuts off from all the diameters, which it intersects, parts which have the same ratio to each other, as the rectangles of the segments into which they divide the line.

Fig. 18. Draw another line as KI, parallel to EG, the line HF which bisects them both, is a diameter, (Cor. 1,) and the halves of these lines, as HI and FG are ordinates to this diameter, of which VH and VF are the abscisses;

Then $VH : GL :: KH\ .\ HI : KL\ .\ LI$.

For (X Cor. 2,) $VH . P' = HI^2$,
and " $VF . P' = FG^2$,
therefore $FH . P' = HI^2 - FG^2$,
or $FH . P' = KL . LI$,
∴ $FH . P' : VH . P' :: KL . LI : HI^2 = KH . HI$. 2 5 Cor. 5 D Converse.
and $FH : VH :: KL . LI : KH . HI$.
or $VH : FH$ (or GL) $:: KH . HI : KL . LI$.

Q. E. D.

Cor. 1.—As the demonstration is equally applicable to all lines, which are parallel to KI, the proposition may be stated universally —The parts of all diameters cut off by lines parallel to each other in the Parabola, are to each other, as the rectangles of the segments into which the diameters divide the parallel lines.

That is $VF : GL :: EF . FG : KL . LI$.

The proposition just stated evidently includes Prop. X and XI (since the rectangle of the equal parts of a double ordinate, is the same as the square of the ordinate) and therefore also I and II, which are included in them.

Cor. 2.—As the proposition is applicable to any parallel lines, as HL, MN, and DE, FG ; therefore by equality of ratios, the rectangles of corresponding segments into which parallel lines, in a Parabola divide each other, have to each other, constantly the same ratio; Fig. 22.

Viz.—The ratio of the parts of the diameters intersected by them.

That is $HR . RL : HI . IL :: DR . RE : FI . IG$.
and $DX . XE : FO . OG :: MX . XN : MO . ON$.

Cor. 3.—The 1 and 2 Corrollaries, are applicable to the segments of parallel lines, intercepting a diameter, or intersecting each other produced, *without* the section.

That is $KR . RH : KE . EH :: PR . RN : EL^2$. Fig. 21.
and $PD . DN : FL^2 :: CD^2 : CF^2$. (See Prop. 3.)

Fig. 18. Cor 4.—It is also true universally, that the rectangle of the parameter of any diameter, and the part of any other diameter cut off by a double ordinate to the former, is equal to the rectangle of the segments into which that double ordinate is divided by the second diameter.

That is $P' . GL = KL . LI$.
See Prop. XI Cor. 2. Also I Cor 2 and II Cor 1.

Cor. 5. As the parameter of any diameter,
Is to the sum of any two ordinates,
So is the difference of those ordinates
To the difference of their abscisses.

That is $P' : KL :: LI : FH = GL$.
(See II Cor. 2.)

PROPOSITION XII.

If there be a tangent to the curve, and any line be drawn from the point of contact to meet the curve in some other place, and if any diameter intersected by this line, be produced to meet the tangent, then shall the curve divide the diameter in the same ratio in which the diameter divides the line.

Fig. 19 That is $IE : Ek :: Ck : kP$,

6.4 & 22. For (VI) $IE : TP :: CI^2 : CT^2 :: Ck^2 : CP^2$.
6. 4. & 1. Also, sim. tri. $Ik : TP :: Ck : CP :: Ck^2 : Ck . CP$,
5. 32. Therefore $IE : Ik :: Ck . CP : CP^2 :: Ck : CP$,
5. A. & 17. And $IE : Ek :: Ck : kP$.

Q. E. D.

Cor. 1.—When $Ck = kP$, then $IE = Ek$.

Fig. 24. Cor. 2.—If TM be the diameter of which CL is a double ordinate, then $CD = DL$ and $DN = NT$, and as the demonstration is equally applicable to a tangent to either end of the double ordinate, the two tangents CT, and LT, meet the diameter in the same point.

Cor. 3. And, conversely, if any two tangents as CT, LT intersect each other, the diameter which passes through the point of intersection, bisects the line which connects the two points of contact, or this line is a double ordinate to that diameter.

Also, if the line which is drawn through the intersection of the tangents, bisects the line in the circle, the former is the diameter of which the latter is a double ordinate.

SCHOL.—Hence, (Cor. 1,) a tangent may be drawn from any point without the curve, as T. Draw TM parallel to the axis, take ND=NT, draw an ordinate from D, which will cut the curve in C the point of contact. The ordinate DC is to be drawn parallel to NR which may be drawn by Schol. Prop. IV.

PROPOSITION XIII.

If from any point of the curve a tangent and ordinate be drawn to any diameter, and another ordinate be produced to meet the tangent, then

As the difference of the ordinates,
Is to the difference added to the external part,
So is double the first ordinate,
To the sum of the ordinates.

That is HO : HP : : H*n* : HI. — Fig. 24.

For (def. 19,) P′ : DC : : DC : DN,
and P′ : 2DC : : DC : 2DN=DT; — 5. 4.
But sim. tri. DC : DT : : HP : HC, — 6. 4.
therefore HP : HC : : P′ : 2DC, — 5. 11.
or HP : P′ : : HC : (2DC=)CL; — 5. 16.
again II Cor. HO : P′ : : HC : HI,
therefore HO : HP : : (CL=)H*n* : HI. — 5. 23.

Q. E. D.

COR.—Hence, (by division, composition, inversion alternation &c.) PH is a mean proportional between PO and PI, or PO : PH : : PH : PI.

SCHOL.–Hence a tangant can easilybe drawn to the curve, from any point without it, as P. Draw PHI intersecting the curve in O and I, and take PH a mean proportional between PO and PI; then draw HC parallel to the axis,

PROPOSITION XIV.

The distance from the Focus to the vertex of any diameter, is equal to one fourth of its Parameter.

Fig. 23. That is, $P'=4FC$; or $\frac{1}{4}P=FC$.

For draw the ordinate MA parallel to the tangent CT, as also CD, perpendicular to the axis AD, and FH perpendicular to the tangent CT;

The (IV) $AD=AT=CM$,
and $P.AD=CD^2$; also $P'.CM=MA^2$
5. 14. therefore $P.AD:P'.CM::CD^2:MA^2$,
5. 23. and since $CM=AD \therefore P:P'::CD^2:MA^2=CT^2$,
6. 4. But by sim. tri. $FH:FT::CD:CT$,
6. 22. and $FH^2:FT^2::CD^2:CT^2$,
5. 11. therefore $P:P'::FH^2:FT^2$.
but (VIII. Cor. 5.) $FH^2=FA.FT$,
5. 7. $\therefore$ $P:P'::FA.FT:FT^2$;
6. 1. and $P:P'::FA:FT=FC$;
But (III) $P=4FA$,
5. C. therefore $P'=4FT$, or $FC=\frac{1}{4}P$.

Q. E. D.

Cor. 1.—If from each of the vertices of two diameters an ordinate be drawn to the other diameter, the two abscisses will be equal to each other. For $AD=AT=CM$.

Cor. 2.—The distance from the vertex of any diameter to the ordinate passing through the Focus, is equal to the distance from the vertex to the Focus; and is therefore equal to $\frac{1}{4}$ its parameter. That is. $CN=FT=FC$.

Cor. 3—Of any diameter, that double ordinate, which passes through the Focus, is equal to the parameter of the diameter.

For (def. 19.) $CN:NK::NK:P'$,
6. 1. But (Cor. 2) $CN=\frac{1}{4}P' \therefore CN.P'=\frac{1}{4}P'.P'=\frac{1}{4}P'^2$,
6. 17. " (19) $CN.P'=NK^2 \therefore NK^2=\frac{1}{4}P'^2$;
2. 8. or $P'^2=4NK^2$, and $P'=2NK$.

Cor. 4.—Hence, and from (Cor. 3, and VII. Cor. 4.) it appears, that if the directrix HTB be drawn, and any lines HE, HE parallel to the axis; then every parallel HE will be equal to EF, or $\frac{1}{4}$ of the parameter of the diameter to the point E. It is also equal to CN the absciss of the Focal ordinate. Fig. 15. Fig. 23.

Cor. 5.—From the fourth step of the demonstration, it appears that the parameters of different diameters, are as the squares of their ordinates, which are equally distant from their vertices.

Cor. 6.—The parameter of any diameter equals the parameter of the axis, added to four times the absciss of an ordinate drawn from the vertex of that diameter, to the axis.

That is $P' = P + 4AD$.

For by the last step of the demonstration,

$$P' = 4FT = 4FA + 4AT_{,} = 4FA + 4AD.$$

and $P = 4FA \therefore P' = P + 4AD$.

Cor. 7.—Also any two straight lines which intersect each other *in the Focus* are to each other as the rectangles of the segments into which they are divided.

That is. $CG : KE :: CF \cdot FG : KF \cdot FE$.

For by Cor. 3. these lines are the parameters of those diameters to which they are double ordinates.

Hence by (II Cor. 1.) P' or $CG \cdot AF = CF \cdot FG$,

and " P' or $KE \cdot AF = KF \cdot FE$,

therefore $CG \cdot AF : KE \cdot AF :: CF \cdot FG : KF \cdot FE$. 5. 14.

and $CG : KE :: CF \cdot FG : KF \cdot FE$.

And generally; if two straight lines intersect each other, *in any point* within the section, the rectangles of their segments are to each other as the parameters of those diameters to which they are double ordinates. 6. 1.

For CG and KE (Cor. 3.) are those parameters,

and (VIII Cor. 2.) $CF \cdot FG : cf \cdot fg :: KF \cdot FE : Kf \cdot fE$,

$\therefore$ $CG : KE :: cf \cdot fg : Kf \cdot fE$.

Cor. 8.—The subtangent $PT=NK=2CN=\frac{1}{2}P'$ the parameter of the diameter CN.

6. 8. For. $TD : TC :: TC : TP$,
but $TD=2AD=2CM$ & $TC=AM$,
$\therefore$ $2CM : AM :: AM : TP \therefore 2CM . TP=AM^2$,
but $CM . P'=AM^2 \therefore 2CM : TP (=2TP . CM)=P'CM$,
$\therefore$ $2TP=P'$ or $TP=\frac{1}{2}P'=NK=2CN$.

PROPOSITION XV.

If there be two tangents to the curve, intersecting each other, and if from the points of contact, lines be drawn to the focus, these lines make an angle which is double to that made by the tangents.

Fig. 25. That is, the angle $DFC=2DTC$.
Draw the axis, NMH, produce DT; to cut the axis in N,
Then (VIII, Cor. 3.) angle, $DFH=2DNH$;
And " $CFH=2CMH$;
Therefore " $DFC=2DNH+2CMH=2TNM+2TMN$;
That is, " $DFC=2DTM=2DTC$.

Q. E. D.

Cor.—Hence Dtc, is a right angle,
For $DFH+cFH=2$ right angles,
$\therefore$ $Dtc=a$ right angle.

OF THE ELLIPSE.

PROPOSITION I.

THE squares of any two ordinates to the axis, are to each other, as the rectangles of their abscisses.

Let RVB be a triangular section through the axis of the cone (3), AGIB, another section perpendicular to the former, forming an Ellipse ; AB, the mutual intersection of the two, will be the transverse axis of the Ellipse (11) ; let MIN, KGL be circular sections parallel to the base (4) ; FG, HI, the mutual intersections of these planes with that of the Ellipse, will be ordinates, both in the Ellipse, and in the circles, being perpendicular, both to MN and KE, and to AB. Fig. 1.

Then $FG^2 : HI^2 :: AF \cdot FB : AH \cdot HB$.

For sim. tri.	$AF : AH :: FL : HN$,	
And	$FB : HB :: KF : MH$;	
The rect. ∴	$AF \cdot FB : AH \cdot HB :: KF \cdot FL : MH \cdot HN$,	6. C.
But	$KF \cdot FL = FG^2$, and $MH \cdot HN = HI^2$,	3.2.&6.8.
Therefore	$AF \cdot FB : AH \cdot HB :: FG^2 : HI^2$.	
	Q. E. D.	5. 7.

SCHOL.—If then a straight line, placed perpendicularly upon another finite straight line, were to move parallel to itself, and to vary in its length continually, so that its squares should always have the same ratio to each other, as the rectangles of the parts into which it divides the other line, the figure formed by the moving line, would be the portion of an Ellipse between the axis and curve.

Cor. 1.—As the proposition and demonstration, are equally applicable to ordinates on opposite sides of the axes, having the same abscisses, those ordinates are equal or GS, is double of FG. Hence it is called a double ordinate.

Cor. 2.—The squares of the ordinates are the same as the rectangles of the equal segments of the double ordinates, therefore, the proposition may be thus expressed. *The rectangles of the segments of the double ordinates, are to each other as the rectangles of the abscisses, or segments into which they divide the transverse.*

Cor. 3.—Ordinates at equal distances from either vertex, are equal; since the abscisses and consequently the rectangles of the abscisses are equal. And conversely, if the ordinates are equal, their distances from the vertices are equal.

Cor. 4.—Hence the Foci are equally distant from either vertex, for their ordinates are equal, being each equal to the semi-parameter, (20), their distances from the centre are therefore equal.

Cor. 5.—Hence also every diameter is bisected in the centre. For since at equal distances from the centre, the ordinates are equal, the third sides and the remaining angles of the right angled triangles, C*de*, C*hg* are equal, therefore *ecg*, is a right line bisected in the centre.

Fig. 14
1. 4.
1. 14.

And conversely, the ordinates from the ends of any diameter upon the transverse axis are equally distant from the centre.

Cor. 6.—Hence also, the whole Ellipse, is by the transverse axis, divided into two equal parts, which if placed one upon the other will coincide in every respect. For if they should not coincide in any part, the ordinate in one, at that point, would be unequal to the corresponding ordinate in the other.

Cor. 7.—For similar reasons, the two parts of the tangent at the vertex, would coincide; that is, the tangent at the vertex makes equal angles with the axis, on each side of it. It is therefore perpendicular to the axis.

Cor. 8. If a circular section as P C O. pass through the centre of the Ellipse, the semi-conjugate Ca, is a mean proportional between OC and CP, the parts of the diameter of the *circular section* ; for in this circle Ca is an ordinate to this diameter.

Fig. 1

Also the whole conjugate axis, is a mean proportional between BR and AQ, the diameters of the circular sections, passing through the two vertices of the Ellipse.

For as AB=2AC,∴BR=2CP, and QA=2OC, therefore BR : ab : : ab : AQ.

5. 15.

Schol.—It has been already remarked, that if the cutting plane be supposed to revolve about A, the axis AB will become more extended, the abscisses FB and HB, will approach to a ratio of equality, and the section itself will approach indefinitely near to a Parabola ; and when it becomes parallel to the side of the cone, it will be a Parabola (5). If then the indefinite abscisses are considered equal, the preceding proposition will become,

$$AF : AH : : FG^2 : : HI^2$$

6. 1.

the same as the first proposition of the Parabola.

Again, if the cutting plane revolve in the opposite direction, it will come into coincidence with the circular section A Q, parallel to the base of the cone. As the ratio of the ordinates and abscisses remains the same during the revolution, we conlude, that it will be found also in the circle. This property of the circle is demonstrated by Euclid ; compare 31 Prop. in the 3d Book with the 8th in the 6th Book, from which it will appear, that if ordinates, or perpendiculars, be drawn to the diameter of a circle, their squares are respectively equal to the rectangles of their abscisses. The squares therefore have to each other, the same ratio which the rectangles have, and its being the ratio of *equality* is what distinguishes the circle from the Ellipse.

Note.—When the *converse* of propositions are stated in this work, the demonstrations are not added, because they are easily made out by a *reductio ad absurdum*, by which it is proved that to deny the converse of the proposition is in effect to deny the proposition itself of which it is the converse. (See Eucl. XVIII and XIX, XXIV and XXV, &c.)

PROPOSITION II.

As the square of the Transverse Axis,
Is to the square of the conjugate Axis,
So is the rectangle of the abscisses of the Transverse,
To the square of their ordinates.

Fig. 2. That is $AB^2 : ab^2 :: AD . DB : DE^2$.

For, (I) $AC . CB : AD . DB :: Ca^2 : DE^2$,
But if C be the centre, $AC . CB = AC^2$, and Ca is the semi-conjugate,
Therefore $AC^2 : AD . DB :: Ca^2 : DE^2$,
and $AC^2 : Ca^2 :: AD . DB : DE^2$,
5. 15. or $AB^2 : ab^2 :: AD . DB : DE^2$.

Q. E. D.

2. 5. Cor. 1.—Since the rectangle $AD . DB = CA^2 - CD^2$,

Therefore $AC^2 : aC^2 :: CA^2 - CD^2 : DE^2$;
or $AB^2 : ab^2 :: CA^2 - CD^2 : DE^2$.

6. 20. Cor. 2.—(19) $AB : ab : ab :$ Parameter$=P$,
Cor. 2. also $AB : P :: AB^2 : ab^2$,
5. 11. ∴ $AB : P :: AD . DB : DE^2$,
or The Transverse axis,
Is to its Parameter,
As the rectangle of its abscisses,
To the square of their ordinate.

Cor. 3.—If different Ellipses have the *same Transverse* axis the corresponding ordinates in each, will be to each other as their Conjugate axes.

For as the first and third terms in the proposition, will then be the same, the second will vary as the fourth;

Fig. 6. Pl. VII. or $AC^2 : AD . DB :: Ca^2 : DE^2$,
and $AC^2 : AD . DB :: Ca'^2 : De^2$,
∴ $Ca^2 : Ca'^2 :: DE^2 : De^2$,
and $Ca : Ca' :: DE : De$.

Fig. 15. Cor. 4.—Let APGB be the rectangle of the Transverse axis, and Parameter BG, then DHGB is the rectangle of the absciss DB into the Parameter, and DIFB or the rectangle DI . DB is equal to the square of DE.

For (Cor. 2.) $AB : BG :: AD . DB : DE^2$,
and Sim. tri. $AB : BG :: AD : DI :: AD . DB : DI . DB$, 6. 1.
and $AD . DB : DE^2 :: AD . DB : DI . DB$;
$\therefore$ $DE^2 = DI . DB$.

Schol.—The square of the ordinate DE, therefore (which equals DIFB) is *less* than the rectangle of the absciss DB into the Parameter, or DHGB, by the rectangle IHGF similar and similarly situated to the whole rectangle PB It was on account of this *deficiency* in the square of the ordinate, compared with the rectangle of the absciss into the Parameter, that Apollonius named this section the *Ellipse*. The rectangle ABGP or AB . P was called the *Figure* of the section ; AB and BG its *latera*, and the *perpendicular* BG the *Latus rectum*.

PROPOSITION III.

As the square of the Conjugate axis,
To the square of the Transverse,
So is the rectangle of the abscisses of the Conjugate,
To the square of their ordinate.
That is $ab^2 : AB^2 :: ad . db : dE^2$.

Draw ED an ordinate to the Transverse AB, Fig. 2.

Then (II Cor. 1) $AC^2 : aC^2 :: CA^2 - CD^2 : DE^2$;
But $CD^2 = dE^2$, and $DE^2 = Cd^2$,
Therefore $AC^2 : aC^2 :: CA^2 - dE^2 : Cd^2$,
and $AC^2 : CA^2 - dE^2 :: aC^2 : Cd^2$,
" $AC^2 : dE^2 :: aC^2 : aC^2 - Cd^2$, 5. E.
" $AC^2 : aC^2 :: dE^2 : aC^2 - Cd^2$,
" $aC^2 : AC^2 :: aC^2 - Cd^2 : dE^2$;
But $aC^2 - Cd^2 = ad . db$, 2. 5.Cor.
Therefore $aC^2 : AC^2 :: ad . db : dE^2$.

Q. E. D.

Cor. 1.—From this Proposition it appears that the Conjugate axis, and the rectangles of its abscisses, have the same relations to their ordinates and to the Transverse, which is its conjugate (16), as those which have been demonstrated concerning the Transverse axis.

Hence 1st the Conjugate axis, also, bisects all its double ordinates.

2. Ordinates at equal distances from its vertices, or from the centre, are equal.

3. The Conjugate axis divides the Ellipse into two equal and similar parts.

4. The tangents at its vertices are perpendicular to it.

5. Ordinates from the opposite extremities of any diameter are equal and equally distant from the centre.

6. The square of the Conjugate axis, is to the square of the Transverse, as the difference between the squares of the semi-conjugate, and the distance from the centre to any ordinate, is to the square of that ordinate.

7. If two Ellipses have the same Conjugate axis, the corresponding ordinates to this axis, will be to each other as their Transverse axes.

8. The Conjugate axis is to its Parameter, as the rectangle of its abscisses, to the square of their ordinate.

9. The square of any ordinate is less than the rectangle of its absciss and the Parameter of the Conjugate, by a rectangle similar and similarly situated to the rectangle of the Conjugate and its Parameter. (See Prop. I and II.)

10. The squares of any two ordinates are as the rectangles of their abscisses ; or in other words, the rectangles of the segments of its double ordinates are as the rectangles of the segments into which they divide the Conjugate axis.

SCHOL. From PROP. III compared with PROP. II, it is evident, that if two unequal straight lines bisect each other at right angles, and either of them move parallel to itself on the other, varying in its length continually, so that the rectangles of its parts have always the same ratio to the rectangles of the parts into which it divides the other line,—the figure formed by the moving line will be an Ellipse.

COR. 2.—If two circles be described on the two axes of an Ellipse as diameters, the one being inscribed within the Ellipse, and the other circumscribed about it, then any ordinate in either circle is to the corresponding ordinate in the Ellipse as the axis of this ordinate is to the other axis.

That is, $CA : Ca :: DG : DE$.
and $Ca : CA :: dg : dE$.

For $AD \cdot DB = DG^2$,
Therefore (II) $CA^2 : Ca^2 :: DG^2 : DE^2$;
Or $CA : Ca :: DG : DE$.

In the same manner. $Ca : CA :: dg : dE$.

$DG : DE :: dE :: dg$.

Or the corresponding ordinates, in the circles and Ellipse are reciprocally proportional.

PROPOSITION IV.

The square of either axis, is the square of the other, as the rectangle of the segments of a line in the Ellipse parallel to the former, is to the rectangle of the corresponding segments of a line parallel to the latter.

That is $eh \cdot hg : Eh \cdot hG :: AB^2 : ab^2$. Fig. 3.
Or $Eh \cdot hG : eh \cdot hg :: ab^2 : AB^2$.

5. E. For (I) $ED^2 : eN^2 :: AD . DB : AN . NB$,
and $ED^2 : ED^2 - eN^2 :: AD . DB : AD . DB - AN . NB$.

But $ED^2 - eN^2 = ED^2 - hD^2 = Eh . hG$,

And $AD.DB - AN.NB = BD.\overline{DN+NA} - \overline{BD+DN}.NA$;

2. 1. And since, $BD . \overline{DN+NA} = BD . DN + BD . NA$,

And $\overline{BD+DN} . NA = BD . NA + DN . NA$,

Therefore, $AD.DB - AN . NB = DN . BD - NA . DN$
$= Dn . DN = hg . eh$;
$\therefore$ $ED^2 : Eh . hG :: AD . DB :: eh . hg$,
And $ED^2 : AD . DB :: Eh . hG : eh . hg$.
But (II.) $ED^2 : AD . DB :: ab^2 : AB^2$,

5. 11. $\therefore$ $Eh . hG : eh . hg :: ab^2 : AB^2$.

Q. E. D.

Cor. 1.—Hence by equality of ratios, $eh . hg : ek . kg :: Eh . hG : fk . kl$. Or, universally, the rectangles of the corresponding segments, of lines parallel to the two axes, mutually intersecting each other in the Ellipse, are to each other always in the same ratio; namely, the ratio of the squares of the two axes.

Schol.—As the square of either semi-axis, or of an ordinate, is the same as the rectangle of the two halves of the axis or double-ordinate. This proposition evidently includes Prop. II and III, as it demonstrates, that the relations there shown to exist between the two axes, extends to all lines in the section parallel to them.

For $AC^2 : ac^2 :: AD . DB : DE^2$ is the same as
$AC . CB : aC . Cb :: AD . DB : ED . DG$
which may be considered as included in this proposition.

If the cutting plane revolve until the section becomes a circle, we should infer, that the rectangles of lines, mutually intersecting each other at right angles, in the circle, have always the same ratio. This is demonstrated by Euclid, (III Book, 35 prop. See Schol. to Prop. I.) and also, that it is a ratio of equality.

If the plane be supposed to revolve in the opposite direction until the section becomes a Parabola, and the longer segments of the lines parallel to the transverse axis be considered equal.

Then the Prop. $AD . DB : eh . hg :: ED . DG : Eh . hG$
becomes $AD : eh :: ED . DG : Eh . hG$

The same as is demonstrated in the Parabola, Prop. II, Cor. 1.

PROPOSITION. V.

If lines parallel to the two axes, intersect each other, *without* the Ellipse, the rectangles of their corresponding segments are to each other as the squares of the axes to which they are parallel.

That is, $EH . HG : eH . Hg :: ab^2 : AB^2$, Fig. 4.
or $eH . Hg : EH . HG :: AB^2 : ab^2$.

For (I,) $eN^2 : ED^2 :: AN . NB : AD . DB$,
And $eN^2 : eN^2 - ED^2 :: AN.NB : AN . NB - AD.DB$.

But $eN^2 - ED^2 = HD^2 - ED^2 = EH . HG$,

Also $AN . NB = BN . \overline{AD+ND} = BN . AD + BN . DN$,

And $AD . DB = AD . \overline{BN+ND} = BN . AD + AD . DN$,
Therefore $AN . NB - AD . DB = BN . DN - AD . DN$,

$= DN . \overline{BN-AD} = DN . \overline{BN-Bd}$
$= DN . Nd = DN . Dn$.

Therefore $(eN^2 =) HD^2 : EH . HG :: AN . NB : (DN . Dn =) eH . Hg$,
And $HD^2 : AN . NB :: EH . HG : eH . Hg$;
But (II,) $HD^2 (= eN^2) : AN . NB :: ab^2 : AB^2$,
$\therefore$ $EH . HG : eH . Hg :: ab^2 : AB^2$.

Q. E. D.

Cor. 1.—Since this proposition is equally applicable to all lines parallel to the axes, it follows by equality of ratios, that, the rectangles of the segments of all lines parallel to the axes, intersecting each other wtthout the Ellipse, are to each other always in the same ratio; viz. that of the squares of the two axes.

That is $EH . HG : eH . Hg :: KI . IL : eI . Ig.$

Cor. 2.—If DH be supposed to move towards AP, the rectangle $EH . HG$ approaches continually to an equality with the square of AP, as its *limit*, and when D coincides with A, AP^2 may be substituted for $EH . HG$. In like manner, aS^2 may be subitituted for $eH . HG$.

That is, the rectangles of the segments of lines parallel to the axes, are to each other, as the squares of the intercepted parts of the tangents to the vertices of the axes.

Fig. 3. Fig. 4. Cor. 3.—Since (IV,) $Eh . hG : eh . hg :: ab^2 : AB^2$,
and from this $EH . HG : eH . Hg : ab^2 : AB^2$,
Therefore $Eh . hG : eh . hg :: EH . HG : eH . Hg.$

Schol.—When the Ellipse becomes a circle, the same relations exist betwen the rectangles of the corresponding segments, without the figure, and also between these rectangles and the squares of the tangents, as is demonstrated in Euclid 3 Book 36 and 37, but it is then a ratio of equality.

Also if the plane revolve in the contrary direction, until the section becomes a Parabola, and the longer segments are considered equal, then

$EH . HG : eH . Hg :: KI . IL : eI . Ig.$
becomes $EH . HG : eH :: KI . IL : eI$;
and $EH . HG : KI . IL :: eH . eI.$
As demonstrated of the Parabola II, Cor. 1.

The five preceding propositions may all be embraced in the following *general* proposition. *If lines parallel to one axis of the Ellipse intersect the other axis, or lines parallel to it, either within or without the section; the rectangles of the corresponding segments will always have to each other, the same ratio.*

PROPOSITION VI.

If a tangent and ordinate be drawn to any point of an Ellipse, meeting the transverse axis produced, the semi-transverse is a mean proportional between the distances of the two intersections from the centre,

That is, or $CM \cdot CT = CA^2$; $CM : CA : CT$, are continued proportionals. Fig. 5.

For from the point T draw any other line TEH to cut the curve in two points E and H; from which let fall the perpendiculars ED and HG, bisect DG in K,

Then (I) $AD \cdot DB : AG \cdot GB :: DE^2 : GH^2$,
And by sim. tri. $TD^2 : TG^2 :: DE^2 : GH^2$;
Therefore $AD \cdot DB : AG \cdot GB :: TD^2 : TG^2$.
But $DB = CB + CD = AC + CD = AG + DC - CG = 2CK + AG$,
And $GB = CB - CG = AC - CG = AD + DC - CG = 2CK + AD$.
$\therefore AD \cdot \overline{2CK + AD} \cdot AG : AG \cdot \overline{2CK + AG} \cdot AD :: TD^2 : TG^2$,
and $DG \cdot 2CK : (TG^2 - TD^2$ or$) \; DG \cdot 2TK :: AD \cdot \overline{2CK + AD} \cdot AG : TD^2$, 5. 17. 16 & A.
or $2CK : 2TK :: AD \cdot \overline{2CK + AD} \cdot AG : TD^2$
or $AD \cdot 2CK : AD \cdot 2TK :: AD \cdot \overline{2CK + AD} \cdot AG : TD^2$; 5. 19.
$\therefore AD \cdot 2CK : AD \cdot 2TK :: AD \cdot AG : TD^2 - AD \cdot 2TK$,
and $CK : TK :: AD \cdot AG : TD^2 - AD \cdot 2TK$, 5. A & 18.
and $CK : TC :: AD \cdot AG : TD^2 - AD \cdot \overline{TD + TA}$;
or $CK : CT :: AD \cdot AG : AT^2$.

But the *limit* of $AD \cdot AG$, when the line TH comes into the position of TL, is AM^2 (See Parab. Prop. IV,) and K, then coincides with M. The proposition therefore becomes, $CM : CT :: AM^2 : AT^2$;

That is $CM : CT :: \overline{CA - CM^2} : \overline{CT - CA^2}$,

5\. 19. or $CM : CT :: CM^2 + CA^2 : CA^2 + CT^2$,

and $CM : MT :: CM^2 + CA^2 : CT^2 - CM^2$,

or $CM : MT :: CM^2 + CA^2 : \overline{CT + CM} . MT$,

or $CM^2 : CM . MT :: CM^2 + CA^2 : CM . MT + CT . MT$;

Hence $CM^2 : CA^2 :: CM . MT : CT . MT$,

and $CM^2 : CA^2 :: CM : CT$,

6\.20. Cor. 2. Conversely. ∴ $CM : CA :: CA : CT$.

Q. E. D.

The converse of this poposition, may be very concisely demonstrated in the following manner, which as it does not depend on the preceding demonstration, may be made the principal demonstration, and the preceding inferred as the converse of it.

Fig. 5. On AB describe the circle AGB, produce ML to G, draw a tangent to the circle at G, intersecting the axis produced in T, join CG, then CGT is a right angle, 6. 8. and (CG=)CA, is a mean proportional between CM and CT.

Connect T and L, and the line TL is a tangent to the Ellipse at L. If not, it must cut it. Let it be supposed then to cut it at L and *e*, through *e* draw the ordinate *ed*, and produce it to cut the circle in *g*, and the tangent in *r*.

Then, sim. triang. $TM : Td :: ML : de$,

and " " $TM : Td :: MG : dr$,

∴ $ML : MG :: de : dr$,

but (II, Cor. 3.) $ML : MG :: de : dg$ ($Ca. : CA.$),

∴ $dr = dg$ — the less equals the greater, which is impossible. Therefore TL does not cut the Ellipse in L; but it meets it; it is therefore a tangent to the curve in L, and CT is a third proportional to CM and CA.

That is CM : CA : CT, are continued proportionals.

Q. E. D.

Cor. 1.—$CM : CT :: AM^2 : AT^2$, that is, the distances from the centre to the *two* intersections of the tangent and ordinate with the axis, are as the squares of the distances of the same intersections from the vertex.

Cor. 2.—If a tangent and ordinate be drawn from any point in the Ellipse, cutting the *conjugate axis* produced, the semi-conjugate is a mean proportional between the distances of the two intersections from the centre.

For the properties of the transverse, from which this proposition has been deduced, are the same in the conjugate. See Prop. II and III. Fig. 2.

The preceding demonstrations therefore, are applicable in every respect to the semi-conjugate axis.

That is $Cd : Ca : Ct$, are continued proportionals. Fig. 6.

Cor. 3.—As the demonstrations are applicable to a tangent to the curve at either extremity of the double ordinate, the two tangents drawn to the two extremities of any double ordinate to the axis, meet the axis in the same point, and at equal angles. Fig. 7.

And, conversely, a line drawn through the intersection of two such tangents, if it bisect the angle, will also bisect the double ordinate; and if it bisect the double ordinate, it will also bisect the angle, and in both cases it is the axis.

Cor. 4.—Also Kk, the tangent at the vertex is bisected by the axis,

For $TD : TA :: DE : AK$,
and $TD : TA :: DH : Ak$,
therefore $DE : AK :: DH : Ak$;
but $DE = DH \therefore AK = Ak$.
For a similar reason, $\therefore NF = NG$,
And as $NI = NM \therefore FI = MG$.

Cor. 5.—If any number of Ellipses have the same transverse axis, and an ordinate in one be continued to intersect the curves of all, the tangents from all the points of intersection, will meet the curve in the same point; for, since CD and CA, the two first of the continued proportionals, remain the same in all, the third CT will be the same.

It continues the same, if the Ellipse becomes a circle, as is evident from the first steps of the demonstration.

Cor. 6.—Two tangents to the curve at the extremities of any double ordinate, to the *conjugate axis*, cut it in the same point.

Cor. 7.—If any number of Ellipses have the same conjugate axis, the tangents drawn to each at the intersections of the same double ordinate produced, will cut the conjugate in the same point.

The same is true also of tangents to a circle described upon the conjugate axis as its diameter.

Fig. 6. Cor. 8.—The following list of proportionals is derived directly from this proposition, and its second Corollary, viz.

That CT : CA::CA : CD.

And Ct : Ca ::Ca : Cd.

Fig. 6. (1.) Then CT+CA : CT−CA::CA+CD : CA−CD

that is BT : AT::BD : AD.

or BT is harmonically divided.

In the same manner bt : at::bd : ad.

(2.) Also CT : CT+CA::CA : CA+CD.

that is CT : BT::CA : BD.

And Ct : bt::Ca : bd.

(3.) Again (2.) CT−CA : BT−BD::CT : BT,

That is AT : DT::CT : BT.

And, at : dt::Ct : bt.

(4.) Therefore sim. tri. AK : DE : : Ct : BH,
and " " TK : TE : : Tt : TH.

also ak : dE : : CT : bh,
and tk : tE : : tT : th.

PROPOSITION VII.

The rectangle of the Focal distances from the vertices, is equal to the rectangle of one fourth the parameter, and the transverse axis.

That is AF . FB$=\frac{1}{4}$P . AB. Fig. 8.

For (II Cor. 2.) AF . FB : FE^2 : : AB : P,
therefore AF . FB : FE^2 : : AB . AF : P . AF,
and AF . FB : AB . AF : : FE^2 : P . AF;
But (19) FE$=\frac{1}{2}$P.$\therefore$.FE$^2=\frac{1}{4}$P^2,
$\therefore$ AF . FB : AF . AB : : $\frac{1}{4}$P^2 : P . AF, 2.8.Cor.2.
$\therefore$ FB : AB : : $\frac{1}{4}$P : AF,
and FB . FA$=\frac{1}{4}$P . AB.

Q. E. D.

Cor. 1.—FB : AB : : $\frac{1}{4}$P : AF.

Schol.—If, as in the Parabola, FB and AB, be considered equal, then $\frac{1}{4}$P$=$AF as demonstrated in the V Prop. of Parabola.

Cor. 2.—Since FB.FA$=\frac{1}{4}$P.AB,and FB is less than AB,
$\therefore$ FA is greater than $\frac{1}{4}$P
In the Parabola, FA$=\frac{1}{4}$P.

Cor. 3.—AF . FB$=$Ca^2, or the rectangle of the Focal distances, equals the square of the semi-conjugate.

For since (19.) P . AB$=ab^2$.$\therefore$.$\frac{1}{4}$P . AB$=a$C$^2=$AF . FB,
or AF : aC : : aC : FB.

Cor. 4.—$AF.FB(=aC^2)=\frac{1}{4}ab^2$; or $\frac{1}{2}AB.\frac{1}{2}P=AC.FE$;
or $AF : FE :: AC : FB$.

Cor. 5.—(VI Cor. 1.) $aC^2=Ct \cdot Cd=Ct \cdot DE$.
but (VI. Cor. 7, 4.) $Ct \cdot DE=AK \cdot BH$,
$\therefore\ AK \cdot BH=aC^2=CA \cdot \frac{1}{2}P=AF \cdot FB$.

Cor. 6.—Therefore HFK, is a right angle.
for (Cor. 5.) $AK : AF :: FB : BH$,
∴ the triangles AKF and BHF are similar;
∴ — angle $BHF=AFK$ and $BFH=AKF$,
∴ also HFK is a right angle.

The same may be proved, in reference to the other Focus, therefore a circle described on HK, as a diameter, will pass through the two Foci.

PROPOSITION VIII.

The square of the distance of the Focus from the centre, is equal to the difference of the squares of the semi-axes.

Fig. 8. That is $CF^2=CA^2-Ca^2$,

For (III Cor. 1.) $CA^2 : Ca^2 :: CA^2-CF^2 : FE^2$,
6. 22. and (19) $CA^2 : Ca^2 :: Ca^2 : FE^2, (=\frac{1}{4}P^2,)$
Therefore $CA^2-CF^2=Ca^2$,
and $CF^2=CA^2-Ca^2$.

Q. E. D.

Fig. 9. Cor. 1.—Hence $Ff^2=AB^2-ab^2=\overline{AB+ab}\cdot\overline{AB-ab}$, or Ff is a mean proportional, between the sum, and difference of the two axes.

Cor. 2.—The two semi-axes, and the Focal distance from the centre, are the sides of a right angled triangle, in which the hypotenuse, or distance of the Focus from the vertex of the conjugate axis, is equal to the semi-transverse.

Fig. 8. For $Fa^2-Ca^2=CF^2$,
and $CA^2-Ca^2=CF^2, \therefore Fa=CA$.

COR. 3.—The square of the Focal distance from the centre, is equal to the rectangle of the semi-transverse, and the difference of the semi-transverse and the semi-parameter.

That is $FC^2 = CA \cdot \overline{CA - FE}$,

For $FC^2 = \overline{CA^2 - aC^2}$ and $aC^2 = (19)\ AC \cdot FE$,

Therefore $FC^2 = CA^2 - AC \cdot FE = AC \cdot \overline{AC - FE}$, 2. 1.

and $\therefore$ $AC : FC :: FC : \overline{AC - FE}$. 6. 17.

PROPOSITION IX.

The sum of two lines drawn from the Foci to meet at any point in the curve, is equal to the transverse axis.

That is $fE + fE = AB$. Fig. 9.

For, draw AG parallel and equal to Ca the semi-conjugate; and join CG meeting the ordinate DE in H; also take CI a fourth proportional to CA, CF, CD;

Then (II, COR. 1.) $CA^2 : AG^2 :: CA^2 - CD^2 : DE^2$, 6. 22. &
And, (sim. tri.) $CA^2 : AG^2 :: CA^2 - CD^2 : AG^2 - DH^2$, 5. E.
consequently $DE^2 = AG^2 - DH^2 = Ca^2 - DH^2$.

Also FD, is the difference between CF, and CD,
And $FD^2 = CF^2 - 2CF \cdot CD + CD^2$, 2. 4.
" $FE^2 = FD^2 + DE^2$,
Therefore $FE^2 = CF^2 + Ca^2 - 2CF \cdot CD + CD^2 - DH^2$.

But (VIII.) $CF^2 + Ca^2 = CA^2$;
and (Hyp.) $2CF \cdot CD = 2CA \cdot CI$;
Therefore $FE^2 = CA^2 - 2CA \cdot CI + CD^2 - DH^2$.

Again (Hyp.) $CA^2 : CD^2 :: CF^2 (\text{or } CA^2 - AG^2) : CI^2$;
And $CA^2 : CD^2 :: CA^2 - AG^2 : CD^2 - DH^2$,
$\therefore$ $CI^2 = CD^2 - DH^2$,
Consequently, $FE^2 = CA^2 - 2CA \,.\, CI + CI^2$.

2. 4. Cor. And the side of this square is $FE = CA - CI = AI$;
In the same manner it is found that, $fE = CA + CI = BI$;
$\therefore$ $FE + fE = AI + BI = AB$.

Q. E. D.

Cor. 1.—The difference between the semi-transverse, and the distance from the Focus to any point in the curve, is a fourth proportional to the semi-transverse, the distance from the centre to the Focus, and the distance from the centre to the ordinate to that point.

That is $CA : CF :: CD : CI$. (or $CA - FE$),

Cor. 2.—And $fE - FE = 2CI$. Or the *difference* between two lines, drawn from the Focus to any point in the curve is equal to twice the fourth proportional, to CA, CF, and CD. Therefore $CA : CF :: 2CD : fE - FE$.

Cor. 3.—Hence $CA \,.\, \overline{fE - FE} = 2CD \,.\, CF$
$= (2CF \text{ or}) fF \,.\, CD$.

Cor. 4.—Also $fE . FE = \overline{CA + CI} . \overline{CA - CI} = CA^2 - CI^2$.

Cor. 5.—From this proposition is derived the common method of describing this curve mechanically, by points, or with a thread thus :

Fig. 10. In the Transverse take the Foci F, *f* and any point I. Then with the radii AI, BI and centres F, *f*, describe arcs intersecting in *e*, which will be a point in the curve. In like manner, assuming other points I, as many other points will be found in the curve. Then with a steady hand, the curve line may be drawn through all the points of intersection e.

Or take a thread of the length AB of the Transverse axis, and fix its two ends in the Foci F, f by two pins. Then carry a pen or pencil round by the thread, keeping it always stretched and its point will trace out the curve line.

PROPOSITION X.

If there be any tangent, and two lines drawn from the Foci to the point of contact; these two lines will make equal angles with the tangent.

For draw the ordinate DE, and fe'; parallel to FE, Fig. 12. or Pl. XVI. 3.

Then (IX Cor. 1) CA : CD :: CF : CA−FE;

and (VI) CA : CD :: CT : CA,

therefore CT : CF :: CA : CA−FE,

and TF : Tf :: FE : 2CA−FE, (or fE)−(IX); 5. E.

But (sim. tri.) TF : TF :: FE : fe',

therefore fE $= fe'$ and the angle $e' =$ the angle fEe';

but because FE is parallel to fe', the angle $e' =$ the angle FET;

therefore The angle FET $=$ the angle fEe'.

Q. E. D.

Schol.—As opticians find that the angle of incidence is equal to the angle of reflection, it appears from this proposition, that rays of light issuing from one focus, and meeting the curve in every point, will be reflected into lines drawn from those points to the other focus, so the ray fE is reflected into FE, and this is the reason why the points F, f, are called the *foci* or burning points.

Appollonius called the Foci, of the Ellipse and Hyperbola *Puncta ex comparatione facta*, but he does not mention the Focus of the Parobala. We may hence conclude, perhaps, as well as from the term *Latus Rectum*, (see Prop. II, Schol.) that the latter was first fixed on, and that the

Foci, were discovered, or determined, from the relation of the ordinates to the Latus Rectum.

Newton and others called the Foci *Umbilici*. (Apollonii Conica—Hamilton's Conic Sections 102. Newton's Principia passim.)

Cor. 1.—Hence the *Normal*, EO, or the line drawn perpendicular to the tangent, from the point of contact, bisects the angle made by the two lines, drawn from the Foci. (Fig. 9.)

For since the *Normal*, makes equal angles with the tangent,

(ax. 3.) $\therefore$ OEP=OEF, or FEf is bisected by EO.

(6. 3. 5. E.) Cor. 2.—Consequently $FE : fE :: FO : fO$;

and $FE+fE : fE-FE :: FO+fO : fO-FO$;

That is (IX) $2CA : 2CI :: Ff : 2CO$,

therefore $CA : CI :: CF : CO$.

Cor. 3.—Therefore $CA^2 : CA . CI :: CF^2 : CF . CO$,

and $CA^2 : CF^2 :: (CA . CI=) CF . CD : CF . CO$,

$\therefore$ $CA^2 : CF^2 :: CD : CO$.

Cor. 4.—Therefore $CA^2-CF^2 : CF^2 :: CD-CO : CO$.

that is (VIII) $aC^2 : CF^2 :: DO : CO$.

Cor. 5.—$CA^2 : CA^2-CF^2 :: CD^2 : CD-CO$.

that is $CA^2 : Ca^2 :: CD : DO$.

(6. 20. Cor. 2.) Cor. 6.—Hence (19) $AB : P :: CD : DO$,

and $AC : \frac{1}{2}P :: CD : DO$,

$\therefore$ $AC . DO=\frac{1}{2}P . CD$.

And if, as in the Parabola, AC and CD, be supposed equal, then DO, the *sub-normal*, equals *half the Parameter.*

Cor. 7.—$2CA : 2CI :: fF : 2CO :: CF : CO$,

$\therefore$ $2CA : 2CA-2CI :: CF : CF-CO$.

that is (IX) $2CA : 2FE :: CF : OF$;

or $CA : FE :: CF : OF$.

PROPOSITION XI.

If a line be drawn from either Focus, perpendicular to a tangent to any point of the curve, the distance of their intersections from the centre will be equal to the Semi-transverse axis.

That is CP and C*p* each=CA or CB. Fig. 12.

For through, the point of contact E draw FE, and *f*E meeting FP produced in G, then the angle GEP=angle FEP, being each equal to the angle *fep*, and the angles at P being right and the side PE being common, the two triangles GEP, FEP, are equal in all respects, and so GE=FE, and GP=FP. Therefore, since FP=$\frac{1}{2}$FG, and FC=$\frac{1}{2}$F*f*, and the angle at F common, the side CP will be =$\frac{1}{2}$FG or $\frac{1}{2}$AB that is CP=CA or CB.
And in the same manner C *p*=CA or CB.

Q. E. D.

Cor. 1.—A circle described on the Transverse axis, will pass through the points of intersection P and *p*.

Cor. 2.—If from the intersections of a tangent with a circle, circumscribing an Ellipse, perpendiculars be drawn they will pass through the Foci.

Cor. 3.—The distances of the Foci from the point of contact, are to each other as the perpendiculars.
For the triangles EPE, *fp*E are similar;
Therefore FE : *f*E::FP : *fp*.

Cor. 4.—If PF and *p*C be produced, they will meet in the circumference of the circle in K, of which *p*K will be the semi-diameter, for *f*PK is a right angle.
also FKC, C*pf* are similar and equal triangles;
Therefore PF . *pf*=PF . FK=AF . FB=Ca^2. (VII. Cor. 5.)

PROPOSITION XII.

The distance of the Focus from any point in the curve, is equal to the ordinate to that point, continued until it meets the Focal tangent,

Fig. 11. that is $Fa=Ct$, $FA=AL$, $FB=Bz$, and generally $FM=DG$,

for (19) $FE=\frac{1}{2}P$,

and (VII Cor. 5) $FE . Ct=\frac{1}{2}PAC=FE . AC$,

therefore $Ct=AC=Fa$. (VIII)

Again (VI) $CF : AC :: AC : CT$,
and $AC : AF :: CT : AT$,
Sim. tri. $AT : TC :: AL : Ct, =CA$;
$\therefore$ $AF : AC :: AL : AC$,
therefore $AF=AL$.

That is, the distance from the vertex to the Focus, is equal to the tangent to the vertex, intercepted by the Focal tangent.

(VI Cor. 7,) $AF : FB :: AT : TB$,
$\therefore$Sim tri. $AT : TB :: AL (=AF) : BZ$,
therefore $AF : FB :: AF : BZ$,
$BZ=BF$.

(VI.) $AC^2=CF . CT=CF : \overline{CF+FT}=CF^2+CF . FT$,
$\therefore$ $AC^2-CF^2(=aC^2)=CF . FT$,
sim. tri. $TF : FE :: TC : Ct, =AC$,
but $TC : AC :: TC . CF : AC . CF$,
$\therefore$ $TC . CF (=AC^2) : AC . CF :: AC : CF$,
that is (I.) $TF : FE :: (AC : CF) :: TA : AL :: TD : DG$;
but (IX) $CT : AC :: CD : FM-CA$,
and $CT+CD : AC+FM-CA :: CT : AC$,
that is $TD : FM :: CT : AC$;
$\therefore$ $TD : DG :: TD : FM . \because . DG=FM$.

Q. E. D.

Cor. 1—If through T, a line be drawn perpendicular to TA, it is called the directrix.

The distance of any point in the curve, from the directrix as MN is equal to TD which has a constant ratio to DG or FM, the distance of the point from the Focus. In the Parabola it is a ratio of equality.

PROPOSITION XIII.

If tangents and ordinates be drawn from the extremities of any two conjugate diameters, meeting the Transverse axis produced, the distances of these intersections from the center are reciprocally proportional.

That is CD : Cd::Ct : CT. Fig. 14.

For (VI) CD : CA::CA : CT,
and ,, Cd : CA::CA : Ct,
∴ CD : Cd::Ct : CT.

Q. E. D.

Cor. 1.—Hence the distance from the center, to the ordinate, at the extremity of one diameter, is a mean proportional between the distances of the intersections of the tangent and ordinate drawn from the extremity of the other.

For since CD : Cd::Ct : CT,
and by sim. tri. Cd : DT::Ct : CT,
∴ CD : Cd::Cd: DT.

In like manner Cd : CD::CD : dt.

Cor. 2—The ordinates, also, are reciprocally as their distances from the center.
For sim. tri. DE : de::Cd : CD.

Cor. 3.—Therefore the ordinates are, as the distances of the intersections of their tangents from the center, or (Cor. 1,) DE : de::CT : Ct.

Cor. 4.—Also by (Cor. 2,) The rectangles $DE \cdot CD = de \cdot Cd$;
therefore, the triangle $CDE = Cde$.

PROPOSITION XIV.

If ordinates to the Transverse axis, be drawn from the extremities of any two conjugate diameters, the sum of their squares is equal to the square of the Semi-Conjugate axis, and the sum of the squares of two ordinates, drawn from the same extremities to the Conjugate axis, is equal to the square of the Semi-Transverse.

Fig. 14. That is, $De^2 + de^2 = Ca^2$.
And $D'e^2 + d'e^2 = CA^2$.

For (VI) $CD : CA :: CA : CT$,
$\therefore$ $CD : CA :: AD : AT$;
And $CD : DB :: AD : DT$,
$\therefore$ (XIII. 2) $CD \cdot DT\ (= Cd^2) = AD \cdot DB = CA^2 - CD^2$,
$\therefore$ $Cd^2 = CA^2 - CD^2$;
And $CD^2 + Cd^2 = CA^2$;
That is $D'E^2 + d'e^2 = CA^2$.
In the same manner $DE^2 + d'e^2 = Ca^2$.

Q. E. D.

Cor. 1.—Hence $AD \cdot DB = Cd^2$, and $Ad \cdot dB = CD^2$.

Cor. 2.—Hence $CD^2 + Cd^2 = CA^2$;
and $CD'^2 + Cd'^2 = Ca^2$;

Cor. 3.—Since $CA^2 : Ca^2 :: (AD \cdot DB =)\ Cd^2 : DE^2$,
$\therefore$ $CA : Ca :: Cd : DE$.
and $CA : Ca :: CD : de$.

PROPOSITION XV.

If from the extremity of any diameter a perpendicular be drawn to its conjugate, the rectangle of this perpendicular, and the part of it intercepted by the Transverse Axis, is equal to the square of the Semi-Conjugate axis.

That is $EP . EI = Ca^2$. Fig. 14.

Draw Cy parallel to EP, the triangle EDI, and Cyt are similar:

Therefore $Ct : Cy :: EI : ED$,
That is $Ct : EP :: EI : Cd'$,
Therefore $EP . EI = Ct . Cd = Ca^2$.

Q. E. D.

PROPOSITION XVI.

All parallelograms circumscribed about an Ellipse whose sides are parallel to two conjugate diameters, are equal to each other, being each equal to the rectangle of the two axes.

That is $KQNS = RYZX = AB . ab$. Fig. 13.

For (II & XIV) $CA^2 : Ca^2 :: (AD . DB$ or$) Cd^2 : DE^2$;
Therefore $CA : Ca :: Cd : DE$,
In like manner $CA : Ca :: CD : de$;
Or $CA : CD :: Ca : de$.
But (VI) $CA : CD :: CT : CA$,
$\therefore$ $Ca : de :: CT : Ca$,
Also, sim. tri. $Ce : de :: CT : Cy$,
Therefore $Ca : Ce :: Cy : CA$;
And $Ca . CA = Ce . Cy$ or $Gg . EP$.
$\therefore$ $4Ca . CA$, or $AB . ab = 4Ce . EP$, or QKSN.

Q. E. D.

PROPOSITION XVII.

The sum of the squares of every pair of conjugate diam ers, equal to the same constant quantity, viz. the sum of the squares of the two ..xes.

That is, $AB^2 + ab^2 = EG^2 + eg^2$.

Fig. 14. For $CE^2 + Ce^2 = CD^2 + Cd^2 + DE^2 + De^2$;
But (XIV) $CD^2 + Cd^2 = AC^2$; and $DE^2 + de^2 = Ca^2$;
therefore $CE^2 + Ce^2 = AC^2 + Ca^2$,
and $EG^2 + eg^2 = AB^2 + ab^2$.

Q. E. D.

PROPOSITION XVIII.

If the extremities of the transverse and conjugate axes be connected, the diameters which are parallel to the connecting line, are conjugate and equal to each other.

Fig. 16. That is, $EL = GK$, and they are conjugate to each other.

For $Aa = aB = Bb = bA \therefore AaBb$, is a parallelogram,
also the angles $GCB = aAB = bAB = LCB$,
and since $AB = 2CB \therefore Ba = 2BR$,
$\therefore$ aB is a double ordinate to GK,
and EL parallel to it, is conjugate to GK.
Also $CG = CL$ or $KG = EL$.

Q. E. D.

Corollary.—These are the only two congugate diameters in the Ellipse, which are equal.

This Corollary and the *last step in the demonstration* may easily be demonstrated by *reductio ad absurdum*.—See Emerson's Conic Sections—Ellipse—Prop. 35, 36, 37.

PROPOSITION XIX.

If from any point in the Ellipse, a line be drawn to the conjugate axis, equal to the semi-transvere axis, the part of the line intercepted between that point, and the transverse axis, is equal to the semi-conjugate.

That is, if $MG=AC$, then $MI=aC$. Fig. 8.

For, sim. tri. $MG : MR :: MI : ID$,
and $MG^2 : MR^2 :: MI^2 : ID^2$,
that is $AC^2 : CD^2 :: MI^2 : ID^2$,
and $AC^2 : AC^2-CD^2 :: MI^2 : MI^2-ID^2$,
" $AC^2 : AD . DB :: MI^2 : MD^2$,
" $AC^2 : MI^2 :: AD . DB : MD^2$;
but (II) $AC^2 : aC^2 :: AD . DB : MD^2$,
∴ $MI^2=aC^2$ and $MI=aC$.

Q. E. D.

Cor.—Conversely—If $MI=aC$ then $MG=AC$.

PROPOSITOIN XX.

If a tangent and ordinate be drawn from the extremities of the transverse axis, and of any other diameter, and be produced to meet that diameter and transverse axis, the triangles formed respectively by these lines, are equal.

That is, $CAE=CMT$ and $AGT=MGE$, Fig. 17.
also $CHN=CDM$ and $AHE=MDT$.

1. For CDM and CAE, also CHA and GMT are similar.
$CH : CM :: CA : CT$,
$:: CD : CA$,
$:: CM : CE$;
therefore DH, AM and TE are parallel;
∴ $AME=AMT$, and $CAE=CMT$.
2. Also taking CMGA from each, $AGT=MGE$. 6. 15.
3. ' adding MDA, to each, $MDT=AHE$.
4. therefore $CDM=CHA$.

Q. E. D.

Cor. 1.—$AME+AMD=MDAE=MDT$;
$AMH+AMT=AHMT=AEH$;
$MDAE=AHMT$.

Cor. 2. $QIR=AEFI$.

For sim.tri. $CA : AE :: CD : DM :: CD+CA : DM+AE$;
and $CA : AE :: CI : IF :: CI+CA : IF+AE$;

Again $AI . IB=AI . \overline{IC+CA}$;
and $AD . DB=AD . \overline{DC+CA}$;
therefore $AI . IB : AD . DB :: AI . \overline{IC+CA} : AD . \overline{DC+CA}$,
$:: AI . \overline{AE+IF} : AD . \overline{AE+DM}$,
$:: 2AEFI : 2AEMD$,
$:: AEFI : AEMD$.
But sim. tri. $QIR : MDT :: QI^2 : MD^2$,
$:: AI . IB : AD . DB$,
$:: AEFI : AEMD$;
But (Cor. 1.) $MDT=AEMD \therefore QIR=AEFI$.

Cor. 3. $QLF=LMTR$.

For take CLR from CAE and from CMT,
then $LRTM=LRAE=LRIF+AEFI=LRIF+RQI$
$=LQF$.

PROPOSITION XXI.

The square of any diameter,
Is to the square of its conjugate,
As the rectangle of its abscisses,
Is to the square of their ordinate.

Fig. 17. That is $NM^2 : VK^2 :: NL . LM : LQ^2$.

For (XX. Cor. 3.) $LQF=LRTM$;
and as LRQ moves parallel to itself, towards CK,
LRTM approaches to equality with CTM;
They may therefore be considered as ultimately equal;
and $CKO=CMT$.

But sim. tri. $CMT : CRL :: CM^2 : CL^2$, 6. 20.
and $CMT : CMT - CRL :: CM^2 : CM^2 - CL^2$,
that is $CMT : LRTM :: CM^2 : NL . LM$;
But sim. tri. $CK^2 : LQ^2 :: CKO : LQF$,
$:: CMT : LRTM$,
$:: CM^2 : NL . LM$,
therefore $CK^2 \; CM^2 :: LQ^2 : NL . LM$;
and $NM^2 : VK^2 :: NL . LM : LQ^2$.

Q. E. D.

Schol.—The preceding proposition shows, that any diameter of an Ellipse, and its rectangles, have the same relation to its conjugate, and ordinates, as have been demonstrated in the case of the Axes. All the properties therefore, which have been deduced from this relation, in the former case in reference to lines intersecting each other, parallel to the axes, and to tangents intersecting these lines and the curve, may be applied to any conjugate diameters. The first six propositions, therefore, of this chapter, with their corollaries, are *particular* truths which may safely be *generalized* by applying them to *any* two conjugate diameters, as well as to the two axes. The *general propositions*, however, will be stated, and the corresponding figures have been drawn and lettered in a similar manner, so that the demonstrations already given, for the axes, are applicable, with slight variations, to those propositions also, which will be arranged like the corresponding propositions, concerning the axes.

Cor. 1.—The squares of the ordinates of *any* diameter are to each other, as the rectangles of their abscisses.

That is $CK^2 : LQ^2 :: NC . CM : NL . LM$. (See I Prop.) Fig. 17.

Cor. 2.—The corresponding ordinates on each side of any diameter are equal.

Cor. 3.—Ordinates at equal distances from the center, or from the vertices of any diameter are equal; and, conversely, equal ordinates are at equal distances from the center and from the vertices.

Cor. 4.—The whole Ellipse is divided by every diameter into two equal parts, which may be placed so as to coincide in every respect.

Cor. 5.—Therefore the parts of the tangents, at the extremities of the diameter, would coincide, and the two tangents are parallel, as demonstrated, Prop. I and III.

Cor. 6.—*Any* diameter is to its parameter, as the rectangle of its abscisses, is to the square of their ordinates.

Cor. 7.—The rectangle of the absciss of any diameter and its parameter, exceeds the square of the ordinate, by a rectangle similar, and similarly situated to the rectangle of the diameter and its parameter.

That is $DB \cdot BG > DE^2$ by FIHG.

Cor. 8.—If a circle be described on any diameter,
The ordinate in the circle,
Is to the corresponding ordinate in the Ellipse,
As the given diameter,
To its conjugate.

Fig. 18. That is $CA : Ca :: DG : DE$.

PROPOSITION XXII.

The square of any diameter, is to the square of its conjugate, as the rectangle of the segments of a line in the Ellipse, parallel to the former, is to the corresponding rectangle of the segments of a line parallel to the latter.

Fig. 19. That is, $CA^2 : Ca^2$ or $AB^2 : ab^2 :: eh \cdot hg + Eh \cdot hG$. See Prop. IV, &c.

Cor. 1.—Therefore by equality of ratios:
The corresponding rectangles of the segments of lines intersecting each other, parallel to any diameter and its conjugate, have to each other, always the same ratio, viz. that of the squares of their diameters.

That is, $eh \cdot hg : ek \cdot kg :: Eh \cdot hG : ik \cdot kl$.

SCHOL.—If the Ellipse become a circle, then the same proposition holds true, as is demonstrated by Euclid, but the ratio is then, by alternating the terms, a ratio of equality, III Book, 35.

If hg, and kg, are considered equal, as in the Parabola, then the proposition is $eh : ek :: Eh . hG : ik . kl$.

As demonstrated of the Par. Prop. II.

PROPOSITION XXIII.

If lines in the Ellipse parallel to two conjugate axes, intersect each other *without* the Ellipse, the rectangles of corresponding segments, are to each other as the squares of the diameters to which they are parallel.

That is, $AB^2 : ab^2 :: eH . Hg : EH . HG$.

See Prop.V, &c.

COR. 1.—Hence the rectangles of the segments of all lines, parallel to two conjugate diameters, and intersecting each other without the Ellipse, have to each other always the same ratio, being the ratio of the squares of those two diameters.

That is, $EH . HG : eH . Hg :: KI . IL : EI . Il$.

COR. 2.—Also $EH . HG : ES . SG :: eH . Hg : Sa^2$.

COR. 3.—Also $ek . kn : Ek . kl :: EH . HG : eH . Hg$.

PROPOSITION XXIV.

The squares of *any two diameters*, are to each other, as the rectangles of the segments of one of those diameters, or of lines parallel to it, are to the corresponding rectangles of the segments of lines parallel to the other.

That is $AB^2 : EG^2 :: AD . DB : KD . DL$. Fig. 21.

There is of course no corresponding proposition, in the case of the two axes, as they have a definite position in regard to each other, but this more general proposition may be deduced from the preceding propositions by the ordinary changes of geometrical ratios.

Cor.—$ed \cdot dg : Kd \cdot dL :: CB^2 : CG^2 :: AB^2 : EG^2$, and $ed \cdot dg : eh \cdot hg :: Kd \cdot dL : Eh \cdot hG$.

That is, *the rectangles of the corresponding segments of all parallel lines, intersecting each other in the Ellipse, have to each other the same ratio.*

PROPOSITION XXV.

If any two lines, cutting the Ellipse, intersect each other, *without* it, the rectangles of their segments, are to each other, as the squares of the diameters, or semi-diameters to which they are parallel.

Fig. 23. That is $EH \cdot HG : eH \cdot Hg :: CG'^2 : Cg'^2$.

This is deduced in a manner similar to the last.

Cor. 1.—The squares of tangents intersecting each other, are to each other, as the squares of the diameters, or semi-diameters, to which they are parallel; or the tangents, are as the parallel diameters.

Fig. 22. That is $HM^2 : HN^2 :: Cm^2 : Cn^2$;
or $HM : HN :: Cm : Cn$.

Fig. 22. Cor. 2.—$IQ \cdot QS : ED \cdot DL :: DN^2 : QM^2$, and $IQ \cdot QS : ED \cdot DL :: IP \cdot PS : EP \cdot PL :: CG^2 : CK^2$.

Cor. 3.—$HN : HM :: TF : TM :: CG : CK$, and $LR \cdot RE : RF^2 :: ED \cdot DL : DN^2 :: CK^2 : CG^2$.

Schol.—In the two preceding propositions, with their Corollaries, the principle involved in the preceding *general propositions*, (and of course in the *particular propositions* relative to the axis,) is seen in its most general form. It may be expressed as follows. *If parallel lines in the*

Ellipse intersect another line, or intersect other parallel lines, either within or without the Ellipse, the rectangles of their corresponding segments, are always in the same ratio, and if any of the lines touch the Ellipse, the squares of the intercepted parts of those lines may be considered as rectangles, and are as the squares of diameters parallel to them.

It will be seen by trial, (reccollecting that the square of a semi-diameter, or of an ordinate is equal to the rectangle of the equal parts of the whole diameter, or double, ordinate) that this general proposition as now stated, is applicable to all the cases alluded to, and includes propositions **I, II, III, IV, V, XXI, XXII, XXIII, XXIV, XXV,** From it also may be very easily deduced, all the general propositions which are most useful, concerning diameters, tangents, &c.

PROPOSITION XXVI.

If a tangent and ordinate be drawn to any point in the curve, meeting any diameter, the half of that diameter is a mean proportional between the distances of the two intersections from the centre.

That is $CD \cdot CT = CA^2$,

or $CD : CA : CT$, are continued proportionals.

(*This may be demonstrated in the same manner as in Prop.* VI *but it is immediately deduced from the* General Pposition *just stated, as it is illustrated in Prop.* XXV, Cor. 3.)

For $HB : AK :: EH : EK :: BD : DA :: BT : AT$; Fig. 24.

and $BD - DA$ (or $2\,CD$) $: BD :: BT - AT$ (or BA) $: BT$,

$\therefore \quad CD : BD :: BC : BT$,

and $\quad CD : BD - CD :: BC : BT - BC$;

That is $CD : AC :: AC : CT$.

Q. E. D.

Cor. 1.—The two angents, ET, ET drawn at the two Fig. 25. extremities of the same double ordinate, intersect the diameter in the same point.

And, conversely, if a diameter pass through the point of intersection, it will bisect the line which connects the points of contact.

Fig. 24. Cor. 2.—HK and FG are bisected. Hence IF=MG.

Cor. 3.—DT is harmonically divided, or

DT : AT::BD : AD.

Cor. 4.—TA : TD::TC : TB.

also TK : TE::T*t* : TH.

and AK : DE::C*t*+BH.

Schol.—As the Transverse Axis, passes through the Foci, it has particular relations to those points, which no other diameter can have. In the relations which other diameters have to lines drawn through the Foci, there are however some analogies to those of the axis, as will be seen in the Notes.

OF THE HYPERBOLA.

PROPOSITION I.

THE squares of any two ordinates to the axis, are to each other, as the rectangles of their abscisses.

Let MVN be a triangular section through the axis of the cone (3), AGIH, another section perpendicular to the former, forming an Hyperbola; AH, the mutual intersection of the two, will be the transverse axis of the Hyperbola (11); let MIN, KGL be circular sections parallel to the base (4); FG, HI, the mutual intersections of these planes with that of the Hyperbola, will be ordinates, both in the Hyperbola, and in the circles, being perpendicular, both to MN and KL, and to AB. Fig. 1.

Then $FG^2 : HI^2 :: AF \cdot FB : AH \cdot HB$.

For sim. tri. $AF : AH :: FL : HN$,
And $FB : HB :: KF : MH$;
The rect. ∴ $AF \cdot FB : AH \cdot HB :: KF \cdot FL : MH \cdot HN$, 6. C.
But $KF \cdot FL = FG^2$, and $MH \cdot HN = HI^2$, 3.2.&6.8.
Therefore $AF \cdot FB : AH \cdot HB :: FG^2 : HI^2$.
Q. E. D. 5. 7.

SCHOL.—If then a straight line, placed perpendicularly upon another finite straight line produced, were to move parallel to itself, and to vary in its length continually, so that its squares should always have the same ratio to each other, as the rectangles of the parts between it and the extremities of the other line, the figure formed by the moving line, would be the portion of an Hyperbola between the axis and curve.

Cor. 1.—As the proposition and demonstration, are equally applicable to ordinates on opposite sides of the axis, having the same abscisses, those ordinates are equal or GS, is double of FG. Hence it is called a double ordinate.

Cor. 2.—The squares of the ordinates are the same as the rectangles of the equal segments of the double ordinates, therefore, the proposition may be thus expressed. *The rectangles of the segments of the double ordinates, are to each other as the rectangles of the abscisses, of the transverse produced.*

Cor. 3.—Ordinates at equal distances from either vertex, are equal; since the abscisses and consequently the rectangles of the abscisses are equal. And conversely, if the ordinates are equal, their distances from the vertices are equal.

Cor. 4.—Hence the Foci are equally distant from either vertex, for their ordinates are equal, being each equal to the semi-parameter, (20), their distances from the centre are therefore equal.

Cor. 5.—Hence also every diameter is bisected in the centre. For since at equal distances from the centre, the ordinates are equal, the third sides and the remaining angles of the right angled triangles, CDE, Cd′G are equal, therefore ECG, is a right line bisected in the centre.

Fig. 13.
1. 4.
1. 14.

And conversely, the ordinates from the ends of any diameter upon the transverse axis are equally distant from the centre.

Cor. 6.—Hence also, the whole Hyperbola, is by the transverse axis, divided into two equal parts, which if placed one upon the other will coincide in every respect. For if they should not coincide in any part, the ordinate in one, at that point, would be unequal to the corresponding ordinate in the other.

Cor. 7.—For similar reasons, the two parts of the tangent at the vertex, would coincide; that is, the tangent at the vertex makes equal angles with the axis, on each side of it. It is therefore perpendicular to the axis.

COR. 8.—The two opposite sections of a Hyperbola, are equal and similar, and if placed upon each other would coincide in every respect; for, as the ordinates at equal distances from either vertex are equal, the points in the curve which limit these ordinates would coincide.

COR. 9.—The two diameters, drawn from the extremities of any double ordinate to the axis, are equal.

COR. 10.—If a circular section as POC, pass through the centre of the Hyperbola, the semi-conjugate Ca, is a mean proportional between OC and CP. Fig. 1.

For the whole conjugate axis, is a mean proportional between BR and AQ, the diameters of the circular sections, passing through the two vertices of the Hyperbola.

(13.) or BR : ab : : ab : AQ,
but AB=2AC, ∴ BR=2CP, and QA=2OC,
therefore OC : Ca : : Ca : CP. 5. 15.

SCHOL.—It has been already remarked, that if the cutting plane be supposed to revolve about A, the axis AB will become more extended, the abscisses FB and HB, will approach to a ratio of equality, and the section itself will approach indefinitely near to a Parabola; and when it becomes parallel to the side of the cone, it will be a Parabola (5). If then the indefinite abscisses are considered equal, the preceding proposition will become, Fig. 2.

$$AF : AH :: FG^2 :: HI^2$$ 6. 1.

the same as the first proposition of the Parabola.

Again if the cutting plane, move parallel to itself, on the base MN, towards M, A and B will approach continually towards V, FB and HB will also approach to equality with FA and HA. When A and B coincide in V, the section will become a triangle (3) and FB, HB will be equal to FA, HA, and the rectangles AF . FB, AH . HB, will be equal respectively to AF^2, and AH^2, then the proposition becomes

$$AF^2 : AH^2 :: FG^2 : HI^2,$$
$$AF : AH :: FG : HI.$$ (See Euclid, 6, 1.)

PROPOSITION II.

As the square of the Transverse Axis,
Is to the square of the Conjugate Axis,
So is the rectangle of the abscisses of the Transverse,
To the square of their ordinate.

Fig 1. That is $AB^2 : ab^2 :: AD . DB : DE^2$.

For sim. tri. $BC : CO :: BF : FK$,
and „ „ $AC : CP :: AF : FL$,
∴ $AC . CB : CO . CP :: AF . FB : KF . FL$;
that is $AC^2 : Ca^2 :: AF . FB : FG^2$;
so $AC^2 : Ca^2 :: AD . DB : DE^2$,
5. 15. or $AB^2 : ab^2 :: AD . DB : DE^2$.

Q. E. D.

2. 6. Cor. 1.—Since the rectangle $AD . DB = CD^2 - CA^2$,

Therefore $AC^2 : aC^2 :: CD^2 - CA^2 : DE^2$;
or $AB^2 : ab^2 :: CD^2 - CA^2 : DE^2$.

Cor. 2.—(19) $AB : ab : ab :$ Parameter$=P$,
6. 20. Cor. 2. also $AB : P :: AB^2 : ab^2$,
5. 11. ∴ $AB : P :: AD . DB : DE^2$,
or The Transverse axis,
Is to its Parameter,
As the rectangle of its abscisses,
To the square of their ordinate.

Cor. 3.—If different Hyperbolas have the *same Transverse* axis the corresponding ordinates in each, will be to each other as their Conjugate axes.

For as the first and third terms in the proposition, will then be the same, the second will vary as the fourth.

Fig. 2. Cor. 4.—Let APGB be the rectangle of the Transverse axis, and Parameter BG, then D′HGB is the rectangle of the absciss D′B into the Parameter, and D′IFB or the rectangle D′I . D′B is equal to the square of DE.

For (Cor. 2.) $AB : BG :: AD' . D'B : DE^2$,
6. 1. and Sim. tri. $AB : BG :: AD' : D'I :: AD' . D'B : D'I' . D'B$,
and ∴ $AD' . D'B : D'E^2 :: AD' . D'B : D'I' . D'B'$;
$D'E^2 = D'I . D'B$.

SCHOL. 1.—The square of the ordinate D′E, therefore (which equals D′IFB) is *greater* than the rectangle of the absciss D′B into the Parameter, or D′HGB, by the rectangle IHGF similar and similarly situated to the whole rectangle PB. It was on account of this excess in the square of the ordinate, compared with the rectangle of the absciss into the Parameter, that Apollonius named this section the Hyperbola.

SCHOL. 2.—If the two axes are equal, that is if the section is an *Equilateral Hyperbola*, then the rectangles of the abscisses are *equal* to the square of the ordinate, the Equilateral Hyperbola, having the same relation to a circle, (which may be called an *Equilateral Ellipse*,) as any other Hyperbola, has to an Ellipse.

SCHOL. 3.—From the manner in which the Hyperbolic section is made, it is evident, that as the cones are more or less obtuse, the Conjugate axis, will be encreased or diminished, while the transverse remains the same ; and it appears from the 3d Cor. that the curve will also become more or less obtuse. A section therefore may have its Transverse axis, equal to the Conjugate of another section, and its Conjugate may be equal to the Transverse of the other ; and in this case they are *Conjugate Hyperbolas*, such are all those considered in the remaining propositions of this section, each pair of opposite Hyperbolas having the same properties.

PROPOSITION III.

As the square of the Transverse axis,
Is to the square of the Conjugate,
So is the sum of the squares of the semi-transverse, and of the distance from the centre to any ordinate,
To the square of that ordinate produced to the Conjugate Hyperbola.

That is, $AB^2 : ab^2 :: CA^2 + CD^2 : De^2$.

For (II Cor. 3 and Schol.) $Ca^2 : CA^2 :: Cd^2 - Ca^2 : de^2$;
But, $CD^2 = de^2$ and $De^2 = Cd^2$,
$\therefore$ $Ca^2 : CA^2 :: De^2 - Ca^2 : CD^2$,
and $Ca^2 : De^2 - Ca^2 :: CA^2 : CD^2$,
" $Ca^2 : De^2 :: CA^2 : CA^2 + CD^2$,
" $Ca^2 : CA^2 :: De^2 \; CA^2 + CD^2$,
" $CA^2 : Ca^2 :: CA^2 + CD^2 : De^2$;
or $AB^2 : ab^2 :: CA^2 + CD^2 : De^2$.

In like manner $AB^2 : ab^2 : CA^2 + CD'^2 : D'e'^2$,
also $ab^2 : AB^2 : Ca^2 + Cd^2 : dE'^2$.

Q. E. D.

PROPOSITION IV.

The square of either axis, is to the square of the other, as the rectangle of the segments of a line in the Hyperbola parallel to the former, is to the rectangle of the corresponding segments of a line parallel to the latter.

Fig 4. That is $eh \cdot hg : Eh \cdot hG :: AB^2 : ab^2$.
Or $Eh \cdot hG : eh \cdot hg :: ab^2 : AB^2$.

5. E. For (I) $ED^2 : eN^2 :: AD \cdot DB : AN \cdot NB$,
and $ED^2 : ED^2 - eN^2 :: AD.DB : AD.DB - AN.NB$.

But $ED^2 - eN^2 = ED^2 - hD^2 = Eh \cdot hG$,

And $AD.DB - AN.NB = BD.\overline{DN+NA} - \overline{BD-DN}.NA$;

2. 1. And since, $BD \cdot \overline{DN+NA} = BD \cdot DN + BD \cdot NA$,

And $\overline{BD-DN} \cdot NA = BD \cdot NA - DN \cdot NA$,
Therefore, $AD.DB - AN \cdot NB = DN \cdot BD + NA \cdot DN$
$= Dn \cdot DN = hg \cdot eh$;
$\therefore$ $ED^2 : Eh \cdot hG :: AD \cdot DB :: eh \cdot hg$,
And $ED^2 : AD \cdot DB :: Eh \cdot hG : eh \cdot hg$.
5. 11. But (II.) $ED^2 : AD \cdot DB :: ab^2 : AB^2$,
$\therefore$ $Eh \cdot hG : eh \cdot hg :: ab^2 : AB^2$.

Q. E. D.

Cor. 1.—Hence by equality of ratios, the rectangles of the corresponding segments, of lines parallel to the two axes, mutually intersecting each other in the Hyperbola, are to each other always in the same ratio; namely, the ratio of the squares of the two axes.

Schol.—As the square of either semi-axis, or of an ordinate, is the same as the rectangle of the two halves of the axis or double-ordinate; this proposition evidently includes Prop. II and III, as it demonstrates, that the relations there shown to exist between the two axes, extend to all lines in the section parallel to them.

For $AC^2 : ac^2 :: AD \cdot DB : DE^2$ is the same as
$AC \cdot CB : aC \cdot Cb :: AD \cdot DB : ED \cdot DG$
which may be considered as included in this proposition.

If the cutting plane move parallel to itself, until the section becomes a triangle, the preceding proposition should be applicable to the opposite triangles; which would also follow from similar triangles.

If the plane be supposed to revolve until the section becomes a Parabola, and the longer segments of the lines parallel to the transverse axis be considered equal.

Then the Prop. $AD \cdot DB : eh \cdot hg :: ED \cdot DG : Eh \cdot hG$,
becomes $AD : eh :: ED \cdot DG : Eh \cdot hG$,
The same as is demonstrated in the Parabola, Prop. II, Cor. 1.

PROPOSITION. V.

If lines parallel to the two axes, intersect each other, *without* the Hyperbola, the rectangles of their corresponding segments are to each other as the squares of the axes to which they are parallel.

Fig. 5. That is, $EH \cdot HG : eH \cdot Hg :: ab^2 : AB^2$,
or $eH \cdot Hg : EH \cdot HG :: AB^2 : ab^2$.

For (I,) $eN^2 : ED^2 :: AN \cdot NB : AD \cdot DB$,
And $eN^2 : eN^2 - ED^2 :: AN.NB : AN \cdot NB - AD.DB$.

But $eN^2 - ED^2 = HD^2 - ED^2 = EH \cdot HG$,
Also $AN \cdot NB = BN \cdot \overline{AD+ND} = BN \cdot AD + BN \cdot DN$,
And $AD \cdot DB = AD \cdot \overline{BN - ND} = BN \cdot AD - AD \cdot DN$,
Therefore $AN \cdot NB - AD \cdot DB = BN \cdot DN + AD \cdot DN$,
$= DN \cdot \overline{BN + AD} = DN \cdot \overline{BN + Bd}$
$= DN \cdot Nd = DN \cdot Dn$.

Therefore $(eN^2 =) HD^2 : EH \cdot HG :: AN \cdot NB : (DN \cdot Dn =) eH \cdot Hg$,
And $HD^2 : AN \cdot NB :: EH \cdot HG : eH \cdot Hg$;
But (II,) $HD^2 (= eN^2) : AN \cdot NB :: ab^2 : AB^2$,
$\therefore$ $EH \cdot HG : eH \cdot Hg :: ab^2 : AB^2$.
Q. E. D.

Cor. 1.—Since this proposition is equally applicable to all lines parallel to the axes, it follows, by equality of ratios, that, the rectangles of the segments of all lines parallel to the axes, intersecting each other without the Ellipse, are to each other always in the same ratio; viz. that of the squares of the two axes.

Cor. 2.—If DH be supposed to move towards AP, the rectangle EH . HG approaches continually to an equality with the square of AP, as its *limit*, and when D coincides with A, AP^2 may be substituted for EH . HG.

That is, the rectangles of the segments of lines parallel to the axes, are to each other, as the squares of the intercepted parts of the tangents to the vertices of the axes.

Fig. 3. Fig. 4. Cor. 3.—Since (IV,) $Eh \cdot hG : eh \cdot hg :: ab^2 : AB^2$,
and from this $EH \cdot HG : eH \cdot Hg : ab^2 : AB^2$,
Therefore $Eh \cdot hG : eh \cdot hg :: EH \cdot HG : eH \cdot Hg$.

SCHOL. If the Hyperbolas are *Equilateral*, the ratio will in all cases be that of equality.

That is EH . HG=eH . Hg.

Also if the plane revolve until the section becomes a Parabola, and the longer segments are considered equal, then the rectangles of the double ordinates produced, are as the external parts of the diameters.

That is EH . HG is as eH.

As demonstrated of the Parabola II, Cor. 1.

The five preceding propositions may all be embraced in the following *general* proposition. *If lines parallel to one axis of the Hyperbola intersect the other axis, or lines parallel to it, either within or without the section; the rectangles of the corresponding segments will always have to each other, the same ratio.*

PROPOSITION VI.

If a tangent and ordinate be drawn to any point of an Hyperbola, meeting the transverse axis produced, the semi-transverse is a mean proportional between the distances of the two intersections from the centre,

That is, or CM . CT=CA^2 ; CM : CA : CT, are continued proportionals. Fig. 6.

For from the point T draw any other line TEH to cut the curve in two points E and H; from which let fall the perpendiculars ED and HG, bisect DG in K,

Then (I) AD . DB : AG . GB::DE^2 : GH^2,
And by sim. tri. TD^2 : TG^2::DE^2 : GH^2;
Therefore AD . DB : AG . GB ::TD^2 : TG^2.
But DB=CB+CD=AC+CD=CG+DC−AG=2CK −AG,
And GB=CB+CG=AC+CG=CG+DC−AD=2CK −AD.

$\therefore AD \cdot 2CK - AD \cdot AG : AG \cdot 2CK - AG \cdot AD :: TD^2 : TG^2,$

5. 17. 16 & A. and $DG \cdot 2CK : (TG^2 - TD^2 \text{ or}) \; DG \cdot 2TK :: AD \cdot 2CK - AD \cdot AG : TD^2,$

or $2CK : 2TK :: AD \cdot 2CK - AD \cdot AG : TD^2.$

5. 19. or $AD \cdot 2CK : AD \cdot 2TK :: AD \cdot 2CK - AD \cdot AG : TD^2;$

$\therefore AD \cdot 2CK : AD \cdot 2TK :: AD \cdot AG : AD \cdot 2TK - TD^2,$

5.A & 18. and $CK : TK :: AD \cdot AG : AD \cdot 2TK - TD^2,$

2. 2. & 3. and $CK : TC :: AD \cdot AG : TD^2 - AD \cdot \overline{TD + TA};$

or $CK : CT :: AD \cdot AG : AT^2.$

But the *limit* of $AD \cdot AG$, when the line TH comes into the position of TL, is AM^2 (See Parab. Prop. IV,) and K, then coincides with M. The proposition therefore becomes, $CM : CT :: AM^2 : AT^2$;

That is $CM : CT :: \overline{CM - CA}^2 : \overline{CA - CT}^2,$

5. 19. or $CM : CT :: CM^2 + CA^2 : CA^2 + CT^2,$

and $CM : MT :: CM^2 + CA^2 : CM^2 - CT^2,$

or $CM : MT :: CM^2 + CA^2 : \overline{CT + CM} \cdot MT,$

or $CM^2 : CM \cdot MT :: CM^2 + CA^2 : CM \cdot MT + CT \cdot MT;$

Hence $CM^2 : CA^2 :: CM \cdot MT : CT \cdot MT,$

and $CM^2 : CA^2 :: CM : CT,$

6.20. Cor. 2. Conversely. $\therefore CM : CA :: CA : CT.$

Q. E. D.

Cor. 1.—$CM : CT :: AM^2 : AT^2$, that is, the distances from the centre to the *two* intersections of the tangent and ordinate with the axis, are as the squares of the distances of the same intersections from the vertex.

Cor. 2.—If a tangent and ordinate be drawn from any point in the Hyperbola, and the tangent be produced to cut the *conjugate axis* the semi-conjugate is a mean proportional between the distances of the two intersections from the centre.

For $CD : CL :: AC : CT$,
$\therefore$ $CD : CT :: CD^2 : CA^2$,
$\therefore$ (II, Cor.) $DT : CD :: CD^2 - CA^2 : CA^2 :: ED^2 : aC^2$;
But sim. tri. $DT : CT :: ED : Ct$,
$\therefore$ $ED : Ct :: ED^2 : Ca^2$,
That is $Cd : Ct :: Cd^2 : Ca^2$,
Therefore, $Cd : Ca :: Ca : Ct$. 6.19. Cor.

' That is $Cd : Ca : Ct$, are continued proportionals.

Cor. 3.—As the demonstration is applicable to a tangent to the curve at either extremity of the double ordinate, the two tangents drawn to the two extremities of any double ordinate to the axis, meet the axis in the same point, and at equal angles. Fig. 7.

And, conversely, a line drawn through the intersection of two such tangents, if it bisect the angle, will also bisect the double ordinate; and if it bisect the double ordinate, it will also bisect the angle, and in both cases it is the axis.

Cor. 4.—Also Kk, the tangent at the vertex is bisected by the axis,

For $TD : TA :: DE : AK$,
and $TD : TA :: DH : Ak$,
therefore $DE : AK :: DH : Ak$;
but $DE = DH \therefore AK = Ak$.

Cor. 5.—If any number of Hyperbolas have the same transverse axis, and an ordinate in one be continued to intersect the curves of all, the tangents from all the points of intersection, will meet the curve in the same point; for since CD, and CA, the two first of the continued proportionals, remain the same in all, the third CT will be the same.

Cor. 6.—Two tangents to the curve at the extremities of any double ordinate, cut the *conjugate axis*, at equal distances from the centre.

Fig. 8. Cor. 7.—If a perpendicular from either extremity of the transverse axis, be produced to intersect the Conjugate Hyperbola, a tangent to the curve at the point of intersection, will pass through the other extremity of the transverse axis.

Fig. 8. Cor. 8.—The following list of proportionals is derived directly from this proposition, and its second Corollary, viz.

That CT : CA::CA : CD.
And Ct : Ca ::Ca : Cd.

(1.) Then CT+CA : CA−CT::CA+CD : CD−CA
that is BT : AT::BD : AD.

In the same manner bt : at::bd : ad.

(2.) Also CT : CT+CA::CA : CA+CD.
that is CT : BT::CA : BD.

And Ct : bt::aC : bd.

(3.) Again (2.) CA−CT : BD−BT::CT : BT,
That is AT : DT::CT : BT.

And, at : dt::Ct : bt.

(4.) Therefore sim. tri. Ak : DE::Ct : BH,
and " " Tk : TE::Tt : TH.

Cor. 9.—When CD is indefinitely great, CT is indefinitely small, or the centre is the point to which the tangent approaches as a limit when the absciss recedes from it.

PROPOSITION VII.

The rectangle of the Focal distances from the vertices, is equal to the rectangle of one fourth the parameter, and the transverse axis.

That is AF . FB$=\frac{1}{4}$P . AB. Fig. 8.

For (II Cor. 2.) AF . FB : FE^2 : : AB : P,
therefore AF . FB : FE^2 : : AB . AF : P . AF,
and AF . FB : AB . AF : : FE^2 : P . AF;
But (19) FE$=\frac{1}{2}$P$\therefore$FE$^2=\frac{1}{4}$P^2,
$\therefore$ AF . FB : AF . AB : $\frac{1}{4}$P^2 : : P . AF,
$\therefore$ FB : AB : : $\frac{1}{4}$P : AF,
and FB . FA$=\frac{1}{4}$P : AB.

Q. E. D.

Cor. 1.—FB : AB : : $\frac{1}{4}$P : AF.

Schol.—If, as in the Parabola, FB and AB, be considered equal, then $\frac{1}{4}$P$=$AF as demonstrated in the V. Prop. of Parabola.

Cor. 2.—Since FB . FA$=\frac{1}{4}$P . AB, and FB is greater than AB,
$\therefore$ FA is less than $\frac{1}{4}$P.
In the Parabola, FA$=\frac{1}{4}$P, in the Ellipse FA$>\frac{1}{4}$P.

Cor. 3.—AF . FB$=$Ca^2, or the rectangle of the Focal distances, equals the square of the semi-conjugate.

For since (19.) P . AB$=ab^2\therefore\frac{1}{4}$P . AB$=aC^2=$AF . FB.
or AF : aC : : aC : FB.

Cor. 4.—AF.FB($=a$C^2)$=\frac{1}{4}ab^2$; or $\frac{1}{2}$AB . $\frac{1}{2}$P$=$AC.FE; Fig. 10.
or AF : FE : : AC : FB.

Fig. 8. Cor. 5.—(VI Cor. 1.) $aC^2 = Ct \cdot Cd = Ct \cdot DE$.
but (VI. Cor. 7, 4.) $Ct \cdot DE = Ak \cdot BH$,
$\therefore\ Ak \cdot BH = aC^2 = CA \cdot \frac{1}{2}P = AF \cdot FB$.

Cor. 6.—Therefore Hfk, is a right angle.
for (Cor. 5) $Ak : AF :: FB : BH$,
$\therefore$ the triangles Akf and BHF are similar;
$\therefore$ – angle $BHf = Afk$ and $BfH = Akf$,
$\therefore$ also Hfk is a right angle.

The same may be proved, in reference to the other Focus, therefore a circle described on Hk, as a diameter, will pass through the two Foci.

PROPOSITION VIII.

The square of the distance of the Focus from the centre, is equal to the sum of the squares of the semi-axes.

Fig. 9. That is $CF^2 = CA^2 + Ca^2$,

For (III Cor. 1.) $CA^2 : Ca^2 :: CF^2 - CA^2 : FE^2$,
6. 22. and (19) $CA^2 : Ca^2 :: Ca^2 : FE^2, (= \frac{1}{4}P^2,)$
Therefore $CA^2 - CF^2 = Ca^2$,
and $CF^2 = CA^2 + Ca^2$.

Q. E. D.

Fig. 9. Cor. 1.—Hence $FF^2 = AB^2 + ab^2$.

Cor. 2.—The two semi-axes, and the Focal distance from the centre, are the sides of a right angled triangle, in which the hypotenuse, is equal to the distance of the Focus from the centre.

For $CA^2 + Ca^2 = Aa^2$,

and $CA^2 + Ca^2 = CF^2, \therefore Aa = CF$.

Cor. 3.—Hence the distances of all the Foci, of any pair of Conjugate Hyperbolas are equal, being each equal to the sum of the squares of the semi-axes.

Cor. 4.—The square of the Focal distance from the centre, is equal to the rectangle of the semi-transverse, and the sum of the semi-transverse and the semi-parameter.

That is $FC^2 = CA \,.\, \overline{CA+FE}$,

For $FC^2 = \overline{CA^2 + aC^2}$ and $aC^2 = (19)\ AC \,.\, FE$,

Therefore $FC^2 = CA^2 + AC \,.\, FE = AC.\, \overline{AC+FE}$,

and $\therefore$ $AC : FC :: FC : \overline{AC+FE}$. 6. 17.

PROPOSITION IX.

The difference between two lines drawn from the Foci to meet at any point in the curve, is equal to the transverse axis.

That is $fE - FE = AB$. Fig. 10.

For, draw AG parallel and equal to Ca the semi-conjugate; and join CG meeting the ordinate DE in H; also take CI a fourth proportional to CA, CF, CD;

Then (II, Cor. 1.) $CA^2 : AG^2 :: CD^2 - CA^2 : DE^2$, 6. 22. &
And, (sim. tri.) $CA^2 : AG^2 :: CD^2 - CA^2 : DH^2 - AG^2$ 5. E.
consequently $DE^2 = DH^2 - AG^2 = DH^2 - Ca^2$.

Also FD, is the difference between CF, and CD,
And $FD^2 = CF^2 - 2CF \,.\, CD + CD^2$, 2. 4.
" $FE^2 = FD^2 + DE^2$,
Therefore $FE^2 = CF^2 - Ca^2 - 2CF \,.\, CD + CD^2 + DH^2$.

But (VIII.) $CF^2 - Ca^2 = CA^2$;
and (Hyp.) $2CF \,.\, CD = 2CA \,.\, CI$;
Therefore $FE^2 = CA^2 - 2CA \,.\, CI + CD^2 + DH^2$.

Again (Hyp.) $AC^2 : CD^2 :: CF^2 (\text{or } CA^2 + AG^2) : CI^2$;
And $CA^2 : CD^2 :: CA^2 + AG^2 : CD^2 + DH^2$,
$\therefore$ $CI^2 = CD^2 + DH^2 = CH^2$,
Consequently. $FE^2 = CA^2 - 2AC \cdot CI + CI^2$.

And the side of this square is $FE = CI - CA = AI$;

2. 4. Cor. In the same manner it is found that, $fE = CA + CI = BI$;
$\therefore$ $fE - FE = BI - AI = AB$.

Q. E. D.

Cor. 1.—The sum of the semi-transverse, and the distance from the Focus to any point in the curve, is a fourth proportional to the semi-transverse, the distance from the centre to the Focus, and the distance from the centre to the ordinate to that point.

That is $CA : CF :: CD : CI$. (or $CA + FE$).

Schol.—From this proposition is derived the common method of describing this curve mechanically, by points.

Fig. 9 and 10. In the Transverse take the Foci F, f and any point I. Then with the radii AI, BI and centres F, f, describe arcs intersecting in e, which will be a point in the curve. In like manner, assuming other points I, as many other points will be found in the curve. Then with a steady hand, the curve line may be drawn through all the points of intersection e.

PROPOSITION X.

If there be any tangent, and two lines drawn from the Foci to the point of contact; these two lines will make equal angles with the tangent.

Fig. 10. For draw the ordinate DE, and fe; parallel to FE,
Then (IX Cor. 1) $CA : CD :: CF : CA + FE$;
and (VI) $CA : CD :: CT : CA$,
therefore $CT : CF :: CA : CA + FE$,
2. 4. and $TF : Tf :: FE : 2CA + FE$, (or fE)--(IX);
But (sim. tri.) $TF : Tf :: FE : fe$,
therefore $fE = fe$ and the angle e = the angle fEe;
but because FE is parallel to fe, the angle e = the angle FET;
therefore The angle FET = the angle fEe.

SCHOL.—As opticians find that the angle of incidence is equal to the angle of reflection, it appears from this proposition, that rays of light issuing from one focus, and meeting the curve in every point, will be reflected into lines drawn through those points from the other focus, so the ray fE is reflected into FE, and this is the reason why the points F, f, are called the *foci* or burning points.

COR. 1.—Hence the *Normal*, EO, or the line drawn perpendicular to the tangent, from the point of contact, bisects the angle made by the two lines, drawn from the Foci.

For since the *Normal*, makes equal angles with the tangent,

$\therefore$ OEP=OEF, or FEf is bisected by EO. ax. 3.

COR. 2.—Consequently $FE : fE :: FO : fO$; 6. 3.
and $FE+fE : fE-FE :: FO+fO : fO-FO$; 5. E.
That is (IX) $2CI : 2CA :: 2CO : Ff$,
therefore $CA : CI :: CF : CO$.

COR. 3.—Therefore $CA^2 : CA . CI :: CF^2 : CF . CO$,
and $CA^2 : CF^2 :: (CA . CI=) CF . CD : CF . CO$,
$\therefore$ $CA^2 : CF^2 :: CD : CO$.

COR. 4.—Therefore $CF^2-CA^2 : CF^2 :: CO-CD : CO$.
that is (VIII) $Ca^2 : CF^2 :: DO : CO$.

COR. 5.—$CA^2 : CF^2-CA^2 :: CD : CO-CD$.
that is $CA^2 : Ca^2 :: CD : DO$.

SCHOL.—When the Hyperbola is Equilateral, CD=DO; and CE=EO; or (Fig. 13.) CE=EI.

COR. 6.—Hence (19) $AB : P :: CD : DO$, 6. 20.
and $AC : \frac{1}{2}P :: CD : DO$, Cor. 2.
$\therefore$ $AC . DO = {}^1P . CD$

And if, as in the Parabola, AC and DC, be supposed equal, then DO, the *sub-normal*, equals *half the Parameter*.

Cor. 7.—2CA : 2CI : : fF : 2CO : : CF : CO,
∴ 2CA : 2CI—2CA : : CF : CO—CF.
that is (IX) 2CA : 2FE : : CF : OF;
or CA : FE : : CF : OF.

PROPOSITION XI.

If a line be drawn from either Focus, perpendicular to a tangent to any point of the curve, the distance of their intersections from the centre will be equal to the Semi-transverse axis.

Fig. 11. That is CP and Cp each=CA or CB.

For through, the point of contact E draw FE, and fE meeting FP produced in G, then the angle GEP=angle FEP, and the angles at P being right and the side PE being common, the two triangles GEP, FEP, are equal in all respects, and so GE=FE, and GP=FP. Therefore, since FP=$\frac{1}{2}$FG, and FC=$\frac{1}{2}$Ff, and the angle at F common, the side CP will be =$\frac{1}{2}f$G or $\frac{1}{2}$AB, that is CP=CA or CB.

And in the same manner Cp=CA or CB.

Q. E. D.

Cor. 1.—A circle described on the Transverse axis, will pass through the points of intersection P and p.

Cor. 2.—If from the intersections of a tangent with a circle, described on the transverse axis, perpendiculars be drawn, they will pass through the Foci.

Cor. 3.—The distances of the Foci from the point of contact, are to each other as the perpendiculars.

For the triangles FPE, fpE are similar;
Therefore FE : fE : : FP : fp.

Cor. 4.—If PF and Cp be produced, they will meet in the circumference of the circle in K, of which pK will be the semi-diameter, for fPK is a right angle.

also FKC, Cpf are similar and equal triangles;
Therefore $PF \cdot pf = PF \cdot FK = AF \cdot FB = Ca^2$. (VII. Cor. 5.)

PROPOSITION XII.

The distance of the Focus from any point in the curve, is equal to the ordinate to that point, continued until it meets the Focal tangent,

that is $FA = AL$, $FB = Bz$, and generally $FM = DG$, Fig. 12.

for (19) $FE = \frac{1}{2}P$,

and (VI Cor. 2) $FE \cdot Ct = Ca^2 = FE \cdot AC$,

therefore $Ct = AC$.

Again (VI) $CF : AC :: CA : CT$,

and $AC : AF :: CT : AT$,

Sim. tri. $AT : TC :: AL : Ct, = CA$;

$\therefore$ $AF : AC :: AL : AC$,

therefore $AF = AL$.

That is, the distance from the vertex to the Focus, is equal to the tangent to the vertex, intercepted by the Focal tangent.

(VI Cor. 7,) $AF : FB :: AT : TB$,

$\therefore$ Sim. tri. $AT : TB :: AL (= AF) : BZ$,

therefore $AF : FB :: AF : BZ$,

$BZ = BF$.

(VI.) $AC^2 = CF \cdot CT = CF \cdot \overline{CF - FT} = CF^2 - CF \cdot FT$

$\therefore$ $CF^2 - AC^2 (= Ca^2) = CF \cdot FT$,

sim. tri. $TF : FE :: TC : Ct, = AC$,

but $TC : AC :: TC \cdot CF : AC \cdot CF$,

$\therefore$ $TC \cdot CF (= AC^2) : AC \cdot CF :: CA : CF$,

that is $TF : FE :: (AC : CF) :: TA : AL :: TD : DG$;

but (IX) $CT : AC :: CD : FM + CA$,

and $CD - CT : FM + CA - CA :: CT : CA$,

that is $TD : FM :: CT : AC$;

$\therefore$ $TD : DG :: TD : FM \therefore DG = FM$.

Q. E. D.

Cor.—If through T, a line be drawn perpendicular to TA, it is called the directrix.

The distance of any point in the curve, from the directrix as MN is equal to TD which has a constant ratio to DG or FM, the distance of the point from the Focus. In the Parabola it is a ratio of equality.

PROPOSITION XIII.

If tangents and ordinates be drawn from the extremities of any two conjugate diameters, meeting the Transverse axis produced, the distances of these intersections from the centre are reciprocally proportional.

Fig. 13. That is CD : Cd::Ct : CT.

For (VI) CD : CA::CA : CT,
and „ Cd : CA::CA : Ct,
∴ CD : Cd::Ct : CT.

Q. E. D.

Cor. 1.—Hence the distance from the centre, to the ordinate, at the extremity of one diameter, is a mean proportional between the distances of the intersections of the tangent and ordinate drawn from the extremity of the other.

For since CD : Cd::Ct : CT,
and by sim. tri. Cd : DT::Ct : CT,
∴ CD : Cd::Cd: DT.

In like manner Cd : CD::CD : dt.

Cor. 2.—The ordinates, also, are reciprocally as their distances from the centre.
For Cor. 1. and sim. tri. DE : de::Cd: CD.

Cor. 3.—Therefore the ordinates are, as the distances of the intersections of their tangents from the centre, or (Cor. 1,) $DE : de :: CT : Ct$.

Cor. 4.—Also by (Cor. 2,) The rectangles $DE \cdot CD = de \cdot Cd$;
therefore, the triangle $CDE = Cde$.

PROPOSITION XIV.

If ordinates to the Transverse axis, be drawn from the extremities of any two conjugate diameters, the difference between their squares is equal to the square of the Semi-Conjugate axis, and the difference between the squares of two ordinates, drawn from the same extremities to the Conjugate axis, is equal to the square of the Semi-Transverse.

Fig. 13.

That is, $de^2 - DE^2 = Ca^2$.
And $ED'^2 - ed'^2 = CA^2$.*
That is $CD^2 - Cd^2 = CA^2$.

For (VI) $CD : CA :: CA : CT$,
$\therefore$ $CD : CA :: AD : AT$;
And $CD : DB :: AD : DT$,
$\therefore$ (XIII. 1) $CD \cdot DT \; (= Cd^2) = AD \cdot DB = CD^2 - CA^2$,
$\therefore$ $Cd^2 = CD^2 - CA^2$;
And $CD^2 - Cd^2 = CA^2$;
That is $ED'^2 - e'd^2 = CA^2$.
In the same manner $de^2 - DE^2 = Ca^2$.

Q. E. D.

Cor. 1.—Hence $CD \cdot DT = AD \cdot DB = Cd^2 = CD^2 - CA^2$;

Cor. 2.—Hence $CD^2 - Cd^2 = CA^2$;
and $Cd'^2 - CD'^2 = (de^2 - DE^2)\; Ca^2$;

*The ordinates to the Conjugate axis, from E and e, are not drawn in the figure (13), but they are evidently equal to CD and Cd.

12

COR. 3.—Since $CA^2 : Ca^2 :: (AD . DB =) Cd^2 : DE^2$,

∴ $CA : Ca :: Cd : DE$.

and $CA : Ca :: CD : de$.

In Equilateral Hyperbolas $CD = de$ and $Cd = DE$.

PROPOSITION XV.

If from the extremity of any diameter a perpendicular be drawn to its conjugate, the rectangle of this perpendicular, and the part of it intercepted by the Transverse Axis, is equal to the square of the Semi-Conjugate axis.

Fig. 13. That is $EP . EI = Ca^2$.

Draw Cy parallel to EP, the triangle EDI, and CyT are similar:

Therefore $CT : Cy :: EI : ID$,

That is $CT : EP :: EI : ID$,

Therefore $EP . EI = CT . ID = Ct' . DE = Ca^2$.

Q. E. D.

PROPOSITION XVI.

All parallelograms described between four Conjugate Hyperbolas, whose sides are parallel to two conjugate diameters, are equal to each other, being each equal to the rectangle of the two axes.

Fig. 13. That is $KQNS = AB . ab$.

For (XIV. Cor. 3) $CA : Ca :: CD : de$;

Or $CA : CD :: Ca : de$.

But (VI) $CA : CD :: CT : CA$,

∴ $Ca : de :: CT : CA$,

Also, sim. tri. $Ce : de :: CT : Cy$,

Therefore $Ca : Ce :: Cy : CA$;

And $Ca . CA = Ce . Cy$ $(= EP.)$

∴ $4Ca . CA$, or $AB . ab = 4Ce . EP$, or QKSN.

Q. E. D.

PROPOSITION XVII.

The difference between the squares of every pair of conjugate diameters, is equal to the same constant quantity, viz. the difference between the squares of the two axes.

That is, $AB^2 - ab^2 = EG^2 - eg^2$. Fig. 13.

For $CE^2 - Ce^2 = CD^2 - Cd^2 + DE^2 - De^2$; 1. 47.
But (XIV) $CD^2 - Cd^2 = AC^2$; and $de^2 - DE^2 = Ca^2$;
therefore $CE^2 - Ce^2 = AC^2 - Ca^2$,
and $EG^2 - eg^2 = AB^2 - ab^2$.

Q. E. D.

SCHOL.—In Equilateral Hyperbolas, all conjugate diameters are equal to each other.

PROPOSITION XVIII.

If the extremities of the transverse and conjugate axes be connected, lines drawn through the centre, parallel to the connecting line, are the Assymptotes, and may be considered equal; and as the extremities of two conjugate diameters, recede from the vertex, the diameters approach to the Assymptotes, and to a ratio of equality.

For $Aa = aB = Bb = bA \therefore AaBb$, is a parallelogram, Pl. VIII. Fig. 15.
also the angles $GCB = aAB = bAB = LCB$,
$\therefore$ CG, CL, are parallel to Aa and Ab.

That *both* the Conjugate diameters increase in length, or approach to the Assymptotes, when either does, and also that they approach to a ratio of equality, is evident from the last Proposition.

Q. E. D.

PROPOSITION XIX.

If from any point of the Hype
and produced to cut both the axes, the part intercepted by the Conjugate shall be to that intercepted by the Transverse, as the square of the Transverse is to the square of the Conjugate.

Pl.XVI.8. That is EI : EO : : AB^2 : ab^2.

For AB^2 : ab^2 : : CD : DO;
: : EI : EO.

Q. E. D.

Cor. EF : E*f* : : OF : O*f*.

PROPOSITOIN XX.

If a tangent and ordinate be drawn from the extremities of the transverse axis, and of any other diameter, and be produced to meet that diameter and transverse axis, the triangles formed respectively by these lines, are equal.

Fig. 17. That is, CAE=CMT, and AGT=MGE,
also CHN=CDM, and AHE=MDT.

1. For CDM and CAE, also CHA and GMT are similar.
CH : CM : : CA : CT,
: : CD : CA,
: : CM : CE;
therefore DH, AM and TE are parallel;
∴ AME=AMT, and CAE=CMT.
6. 15. 2. Also taking CEGT from each, AGT=MGE.
3. " adding MDA=MHA and MGA, to each, MDT=[AHE.
4. therefore CDM=CHA.

Q. E. D.

Cor. 1.—AME+AMD=MDAE=MDT;
AMH+AMT=AHMT=AEH;
MDAE=AHMT.

COR. 2. $QID = AEFI$. Fig. 14.

For sim. tri. $CA : AE :: CD : DM :: CD+CA : DM+AE$;
and $CA : AE :: CI : IF :: CI+CA : IF+AE$;

Again $AI . IB = AI . \overline{IC+CA}$;

and $AD . DB = AD . \overline{DC+CA}$;

therefore $AI.IB : AD.DB :: AI . \overline{IC+CA} : AD . \overline{DC+CA}$.

$$:: AI . \overline{AE+IF} : AD . \overline{AE+DM},$$
$$:: 2AEFI : 2AEMD,$$
$$:: AEFI : AEMD.$$

But sim. tri. $QID : MDT :: QI^2 : MD^2$,
$$:: AI . IB : AD . DB,$$
$$:: AEFI : AEMD;$$

But (Cor. 1.) $MDT = AEMD \therefore QID = AEFI$.

COR. 3. $QLF = LMTD$.

For from CLD take CAE and CMT,
then $LDTM = LDAE = AEFI - LDIF = DQI - LDIF$
$= LQF$.

PROPOSITION XXI.

The square of any diameter,
Is to the square of its conjugate,
As the rectangle of its abscisses,
Is to the square of their ordinate.

That is $NM^2 : VK^2 :: NL . LM : LQ^2$. Fig. 14.

For (XX. Cor. 3.) $LQF = LDTM$;

Now suppose QLD to move parallel to itself, towards KC, it will first coincide with the tangent TM, and then both LQF and $LDTM$ will become equal to nothing. As it passes the tangent and comes into the situation KC, KCO corresponds to QLF, and TMC to $LTDM$;
therefore $KCO = TMC$.*

*The transfer of the relation of LQF and $LDTM$, to KCO and TMC, is founded on the analogy between Prop. II and III. The equality of KCO and TMC, may be rigorously deduced from Prop. III.—*See Emerson's Con. Sect.* 29 *and* 36.
Also $agC = TMC + Ckg$; $\therefore$ $akC = CTM$.

6. 20. But sim. tri. CMT : CDL::CM^2 : CL^2,
and CMT : CDL−CMT::CM^2 : CL^2-CM^2;
that is CMT : LDTM::CM^2 : NL . LM;
But sim. tri. CK^2 : LQ^2::CKO : LQF,
::CMT : LDTM,
::CM^2 : NL . LM,
therefore CK^2 : CM^2::LQ^2: NL . LM;
and NM^2 : VK^2::NL . LM : LQ^2.

Q. E. D.

Schol.—The preceding proposition shows, that any diameter of a Hyperbola; and its rectangles, have the same relation to its conjugate, and ordinates, as have been demonstrated in the case of the Axes. All the properties therefore, which have been deduced from this relation, in the former case in reference to lines intersecting each other, parallel to the axes, and to tangents intersecting these lines and the curve, may be applied to any conjugate diameters. The first six propositions, therefore, of this chapter, with their corollaries, are *particular* truths which may safely be *generalized* by applying them to *any* two conjugate diameters, as well as to the two axes. It is not thought necessary to state these *general* propositions, since they correspond to those concerning the axes, in precisely the same manner as in the Ellipse.

The relations of the Hyperbola to its Asymptotes, form a peculiar class of properties, to which there is nothing corresponding in the other sections;—a few of them will be stated.

PROPOSITION XXII.

If a double ordinate be produced to meet the Asymptotes, the rectangle of its segments will be equal to the square of the semi-axis to which it is parallel.

Fig. 15. That is HE . EK=He . eK=Ca^2.

and $he \cdot ek = hE \cdot Ek = CA^2$.

For $CA^2 : AL^2 = Ca^2 :: CD^2 : DH^2$,

And (II. Cor.) $CA^2 : Ca^2 :: CD^2 - CA^2 : DE^2$,

Therefore $CA^2 : Ca^2 :: CA^2 : DH^2 - DE^2 = HE \cdot EK$, 5. 19.

" $CA^2 = HE \cdot EK$.

Again $CA^2 : Ca^2 :: CD^2 : DH^2$,

And (III) $CA^2 : Ca^2 :: CA^2 + CD^2 : De^2$,

$\therefore$ $CA^2 : Ca^2 :: CA^2 : De^2 - DH^2 = He \cdot eK$,

Therefore $Ca^2 = He \cdot eK$.

In like manner, in the Conjugate Hyperbolas,
$he \cdot eh$ and $hE \cdot Ek = CA^2$.

Q. E. D.

SCHOL.—In Equilateral Hyperbolas $He \cdot eK = he \cdot ek$. &c.

COR 1.—Since $HE \cdot EK = He \cdot eK$,
$EH : eH :: eK : EK$.

COR. 2.—sim. tri. $Eh : EH :: Ek : EK$,
$\therefore$ $hE \cdot EK = HE \cdot Ek$.

COR. 3.—$HE = EK$, and $EG = Eg$, and $He = Ke$.

COR. 4.—Since $HE \cdot EK = LA^2 = LA \cdot AI$,
$\therefore$ $GE \cdot EK = PA \cdot AI$. 6. 14.

COR. 5.—EM, AS, EK, being parallel to HC, Fig. 14.
$GE \cdot EM = PA \cdot AS = NE \cdot EK$,
and $GEMC = PASC = NEKC$.

COR. 6.—Since all rectangles between the curve and Asymptotes, as CG, GE, are equal, therefore GE, is reciprocally as CG; and if CG be increased indefinitely GE, will be indefinitely diminished; that is the curve continually approaches the Asymptote, but never meets it. It is considered a tangent to the curve at an *infinite* distance.

COR. 7.—If $Cl : Cg : Cn$, be continued proportionals, $lK : gE : rm$, parallel to the other Asymptotes, are continued proportionals, decreasing; for they are reciprocally as Cl, Cg, Cn, being sides of equal, and equiangular parallelograms.

COR. 8.—CAE=PAEN=SAEK.

For CPAS=CNEK ∴ PI=IK,

therefore PAEN=SAEK.

Again from CAEK take CAS and AEK,

and CAE=SAEK=PAEN.

COR. 9.—If a cutting plane be supposed to pass through the vertex of the cone, parallel to the plane which forms the Hyperbolic section, forming the triangular section VCD; and if two planes VGC, VGD be supposed to *touch* the convex surface of the cone, in the sides of this triangle, then the intersections of these planes and the plane of the Hyperbola, will be the Asymptotes of the Hyperbola.

Pl. VIII. Fig. 19.

That is, *v*K, *v*L, are Asymptotes to the Hyperbolic section, H*hi*I; for HK, evidently equals LI, and *kh*=*li*, ∴ KH . HL=KI . IL. also *kh* . *hl*=*ki* . *il*.

3. 36. Also KH . HL=KC^2=kc^2=*kh* . *hl*. &c.

OF THE CURVATURE OF CONIC SECTIONS.

It is evident that the different parts of the curve of any Conic Section, have different degress of *curvature.* At the vertices of the Transverse Axis, for instance, they are more curved than in other parts, and the curvature evidently decreases with the distances from these vertices. A person, unacquainted with the doctrine of the *Curvature* of *Curves*, might naturally describe the difference between the curvature of an Ellipse, at the vertices of its Transverse and of its Conjugate axis, by saying that a circle, which should have the *same curvature* as the Ellipse, at the vertex of its Transverse Axis, or which should *coincide* with the curve at that point, would be smaller than the circle which would coincide with it, at the vertex of the Conjugate Axis. This obvious method of explaining the subject is substantially the same as that given by Mathematicians, and is capable of the precision of mathematical demonstration to a greater extent than might at first be supposed. Although no definite portion of a circle and Conic Section, can possibly coincide, since the nature of the curves and their equations are essentially different, yet, at their point of contact, they may have the *same* Curvature, and are often said to *coincide.**

Def. 1.—*The Curvature* of a line, is *its continued deviation* from a straight line; and the degree of its curvature is measured by the perpendicular distance of any point in the curve, from the straight line, at a given distance from their point of contact or concourse.

Thus, DE, DF, which are called subtenses of the angles DCE, DCF, being the distances of the points E and F in the curves, from the straight line AD, are *the measures* of *the curvature* of the two circles, or, *more accurate-* Fig. 2.

* Bridge's Conic Sections, pp. 75, 80

ly, if the line DEF, were to move parallel to itself towards C. until it should coincide with CG, the *limiting ratio*, of DE to DF, would be the ratio of the *curvatures* of the two circles, at that point. This *limiting ratio* of the subtenses, in many cases, is easily obtained.

Fig. 1 LEMMA.—If from C and E, any two points in the circumference of the circle CNE, straight lines be drawn to any other point, V, and a straight line CD, touching the circle in C; and if from E, ED and Ev be drawn parallel to CV and CD, also the chord CE,—then, if E be supposed to move in the curve towards C, the *tangent* CD; the chord CE, and the *ordinate* Ev, approach continually to a ratio of equality, as their *limiting ratio*; they are therefore ultimately equal.

3. 32 For, since the angles DCE,=EVC, and CV, DE, are paral. therefore by sim. tri. CE : CD : : CV : EV.

But as E approaches C, CV and EV, approach continually to a ratio of equality, therefore also, CE and CD approach to equality, and before E coincides with C, they differ from each other less than by any assignable difference; they are therefore *ultimately equal*. But Ev is equal to DC, therefore CD, CE, and Ev, are all ultimately equal.

Q. E. D.

PROPOSITION I.

The curvatures of different circles, are to each other inversely as their diameters, or radii.

Fig. 2. Let DCA, touch each of the circles, CEB, CFG, in the point C, and let Ee, Ff, be parallel to CD.

6. 8. then, sim, triang. Ce : CE : : CE : CB,

and " " Cf : CF : : CF : CG,

∴ (Ce=) DE . CB=CE^2, & DF . CG=CF^2,

∴ DE . CB : DF . CG : : CE^2 : CF^2.

But, if DEF, move towards CG, CE and CF, become ultimately equal, being each equal to CD,

then CE^2=CF^2 & ∴ DE . CB=DF . CG;

6. 16. therefore DE : DF : : CG : CB : : rad. of CFG : rad. of CEB; but the ratio of DE to DF, is the ratio of the curvatures of

the circles, CEB, and CFG at the point C; (1.) therefore their curvatures also, are as CG to CB, that is, they are inversely as the diameters or radii of the circles.

Q. E. D.

Cor. 1.—The curvature of the same circle is uniform, or remains the same in every part, since the preceding demonstration is applicable to *any* point of the circumference.

Cor. 2.—Also the curvatures of equal circles, are equal; since, they are inversely as their diameters, which are equal.

Cor. 3.—The curvature of different circles may differ indefinitely, and may approach indefinitely near to equality. Thus if B, move towards G, the diameters, and of course the curvatures of the two circles, approach continually to equality, until they differ from each other, less than by any assignable quantity, and when B and G coincide, then the circles, coincide, and the curvatures are the same.

Schol.—Hence, the curvature of *circles* is made the measure of the curvature of all curves, since their varying curvatures are easily compared with the uniform and definite curvature of the circle.

Def. 2.—If a circle touch a Conic Section,* *so that no other circle can be drawn between it and the curve*, it is called the *Circle of Curvature* to that point of the curve.

Cor..—If there can be one Circle of Curvature, to any point of a Conic Section, there can evidently be but one, since if there were two, one would pass between the other and the Section—contrary to the definition.

Schol.—The definition given by Euclid of a *straight line which touches a circle*, is not applicable to a circle which touches a curve, for this circle may cut the curve which it touches. It is sufficient that they both touch the same straight line, in the same point. There is however,

*Circles, and curves generally, are said to touch each other, when they touch the same straight line, in the same point.

a striking analogy between the Circle of Curvature, in relation to the curve which it touches, and a line touching a circle, for it is demonstrated concerning the latter (Euclid III. 16.) that no other straight line can be drawn between it and the circle.

PROPOSITION II.

If from the vertex of a Conic Section a part of the axis, be taken equal to its Parameter, a circle described on this line, will be the Circle of Curvature to the section at its vertex.

Fig. 4. & 6 That is, let RAS, be a Conic Section, AP=its Parameter, then AFP is the Circle of Curvature to the curve in A.

Fig. 4. 1. If CAG is a Parabola, draw AD=the Parameter, and DL parallel to the axis, join DP, and from any point in the diameter of the circle, draw EFGIH, perpendicular to

6. 4. it then because PA=AD, ∴ PE=EI,

then, $EF^2=AE \cdot EP=AE \cdot EI$.

But (19.) $EG^2=AE \cdot EH$,

therefore as EI is less than EH, EF^2 is less than EG^2; and EF is less than EG. But E is *any* point in the diameter AP. therefore every part of the circle AFP is *within* the curve CAG.

2. Let any line, A*p* be taken *greater* than the Parameter, and the circle AK *pg*, be described, whose centre is P, the part of this circle KA*g*, is *without* the curve CAG.

Take PO, =the Semi-Parameter, and draw OK, cutting the curve in C, and the circle in K, from C draw the tangent CT, and also AM a tangent to the vertex, join PM, PC is perpendicular to CT, (Par. VIII. Cor. 4.)

Also, (Par. VIII. Cor. 2.) CM=MT, & MT is less than MA,

Therefore CM is less than MA;

But $PC^2+CM^2=PM^2=MA^2+PA^2$,

And as CM^2 is less than MA^2, PA^2 is greater than PC^2,

Therefore PA=PK is greater than PC.

That is the circle is *without* the curve,

• If Ap, be nearer to equality with AP, or if p be supposed to move towards P, O will approach equally to A, and OK to AM, but PK, will always be greater than PC, until the moment that p and P, E and A, and the two circles coincide. If Ap therefore be in any degree *greater* than the Parameter, the circle described upon it, will fall *without* the curve, on each side of A, the point of contact but if it be *equal* to the Parameter it will fall wholly *within* the curve, therefore *no circle can be drawn between the circle* AFP described upon the Parameter, and the curve of the Parabola. In other words it is the Circle of Curvature. (def. 2.)

Q. E. D.

Schol.—The same may be demonstrated, in nearly the same manner, concerning the Ellipse and Hyperbola, if DL, instead of being drawn parallel to the axis, be drawn to its other vertex as DB. (Hamilton's Conic Sections, B.V. Prop. XVI, and XVII.) The demonstrations are omitted here, as the properties will be included, in some more general propositions which follow. Fig. 7.

It is evident from the preceding demonstration, that *any* circle, *greater* than the Circle of Curvature, is *without* the curve of the section, on each side of the vertex. And yet such a circle may approach the Circle of Curvarture, indefinitely; it may differ from it less than by any assignable difference, still the Parabola is within the circle, until the moment that the latter coincides with its Circle of Curvature. It is for this reason, that the curve is said to *coincide* with its Circle of Curvature, at the point of contact.

Cor.—The curve and its Circle of Curvature have *the same curvature*, at the point of contact.

For the curve has not a *greater* curvature, than its Circle of Curvature, because the latter is wholly *within* it. It has not a *less* curvature, because then a circle might be described between it and the curve. (I Cor. 3.) Since then, its curvature cannot be either greater or less than that of its Circle of Curvature, it must be the same, and their evanescent subtenses which measure their curvatures must be equal.

SCHOL.—This proposition and corollary, compared with Prop. I, illustrate the nature of Curvature, and its different degrees. In two unequal circles, whose curvatures consequently are unequal, DE and DF, which are the measures of these curvatures, do not *approach* to a ratio of equality, (when DF moves to CG) but remain ultimately *unequal*, in the definite ratio, of CG to CB; consequently the angles DCE, DCF, of which DE and DF are the subtenses, and to which they are nearly proportional, are ultimately unequal, in the same ratio. The curves EC, and FC, therefore *touch* the straight line AD, differently; since they approach the point of contact, with different inclinations.

Fig. 2.

Fig. 3.

But DE, and DF, the subtenses of the curve, and its Circle of Curvature do approach to a ratio of equality for the curves have both the *same curvature*, and DE and DF are therefore ultimately equal, that is, the corresponding point, in the curve and its Circle of Curvature, are ultimately at the same distance from the tangent, and the curves EC and FC, approach to perfect coincidence as the points F approach the point of contact; they meet the tangent with the same inclination, and have the same curvature.

DEF. 4.—The *Radius of Curvature*, the *Diameter of Curvature*, and the *Chord of Curvature*, of any curve, are the radius, the diameter, and the chord of the Circle of Curvature, which belongs to the curve in that point.

PROPOSITION III.

IN THE PARABOLA, *the Chord of Curvature which passes through the Focus*, is equal to the parameter of the diameter which passes through the point of contact.

Fig. 5.

That is CV=4CF=P.

Take any point in the curve, as E, draw ED parallel to CV, and Ev, parallel to CT, cutting CV, in x, forming the parallelogram, Dx; the triangle Cxv, being similar to CFT, (because xv is parallel to CT, and Cv to TF) and FT being equal to FC, (Par. VII, Cor.) Cx=Cv,∴Cv=ED. Now if D approaches to C, indefinitely, x and v approach to C also, and then Ev approaches indefinitely to equality with EC. Hence *ultimately*, Ev=EC.

Then, (in the cir. as Fig. 1.) $ED : EC :: EC : CV$,
And (bec. $Cv = ED$) $Cv : Ev :: Ev : CV$, 6. 8.
There'efore, $Cv \cdot CV = Ev^2$.
But (19.) $Cv \cdot P' = Ev^2$,
∴ (Par. XIV.) $P' = CV$, and $P' = 4FC$.

Q. E. D.

PROPOSITION IV.

A parallelopiped whose base is the square of the *Diameter of Curvature,* and its height is the distance of the Focus from the vertex of the Parabola, is equal to four times the cube of the distance of the Focus from the point of contact.

That is $CB^2 \cdot FA = 4FC^3$.

Let CB be drawn perpendicular to to the tanget CT, it is the *Diameter* of the Circle of Curvature, CVB. Also FG, perpendicular to CT, is parallel to CB. 3. 19.

Hence, sim. tri. $CB : CV :: CF : FG$,
And $CB^2 : CV^2 :: CF^2 : FG^2$.
But $FG^2 = FC \cdot FA$,
Therefore $CB^2 : CV^2 :: CF^2 : CF \cdot FA :: CF : FA$,
And $CB^2 \cdot FA = CV^2 \cdot CF = 4CF^2 \cdot CF = 4CF^3$.

Q. E. D.

Cor.—As FA, the height of the parallelopiped, is a constant quantity, the parallelopiped varies as its base : that is, as CB^2, which therefore varies as $4FC^3$. In other words, the *square* of the *Diameter* of *curvature*, varies as the *cube* of the distance of the Focus from the point of contact.* Sup. 3. 8.

As this point therefore becomes more distant, the Diameter, and Radius of Curvature increase, and the curvature itself diminishes. (See also Parabola, Prop. I. Schol.)

* Algebraically (Par. VIII. Cor. 5.) $FG^2 = FA \cdot FC$, and since FA constant.
∴ FG^2 varies as FC ∴ FC^3 varies as FG^5.
But CB^2 varies as FC^2 varies as FC^3 or as FG^5 ;
∴ CB varies as FG^3, or CN^3.
That is, the Diameter, or Radius of Curvature, varies as the cube of the *normal*.

SCHOL.—When the point of contact is the vertex of the Parabola, the diameter of Curvature coincides with the chord which passes through the Focus, and is equal to it. In that case therefore the Diameter of Curvature is equal to the Parameter of the axis, as was also demonstrated in Prop. II.

PROPOSITION V.

IN THE ELLIPSE, *the Chord of Curvature, which passes through the centre,* is equal to the parameter of the diameter which passes through the point of of contact.

That is $CL=P'$, or $CG : HK : CL$ are continued proportionals.

Fig. 6. For ED and Ev, being drawn parallel to CG and CD, then (Ell. XXI.) $Cv . vG : Ev^2 :: CG^2 : HK^2$.

Now as E approaches C, Ev approaches to equality with EC, and vG to equality with CG, as their limiting ratio. They are therefore ultimately equal.

Then, by substitution, $ED . CG : EC^2 :: CG^2 HK^2$,
but the chord $EC^2 = ED . CL$,
therefore $ED . CG : ED . CL :: CG^2 : HK^2$,
and $CG : CL :: CG^2 : HK^2$,
" $CL : CG :: HK^2 : CG^2$,
" $HK . CL : HK . CG :: HK^2 : CG^2$,
" $HK . CL : HK^2 :: HK . CG : CG^2$,
6. 1. therefore $CL : HK :: HK : CG$,
" $CL . CG = HK^2$; but $P' . CG = HK^2$,
" $CL = P'$: and $CG : HK : CL$, are continued proportionals.

Q. E. D.

PROPOSITION VI.

The diameter of curvature (CO) is a third proportional to twice CP and HK. That is 2CP : HK : CO, are continned proportionals, or $2CP \cdot CO = HK^2$.

For, sim. tri. $CO : CL :: Cc : CP :: 2Cc : 2CP$,
therefore (V) $CO \cdot 2CP = (2Cc)$ or $CG \cdot CL = HK^2$,

Q. E. D.

PROPOSITION VII.

The chord of curvature which passes through the Focus, is a third proportional to the transverse axis, and the diameter conjugate to that which passes through the point of contact.

That is AB : HK : CV, are continued proportionals.

For, sim. tri. $CV : CO :: CP : CN :: 2CP : 2CN$,
But (XI) $CN = Ac$, and $2CN = AB$,
∴ $CV : CO :: 2CP : AB$,
and (VI) $CV \cdot AB = CO \cdot 2CP = HK^2$,
or $AB : HK :: HK : CV$.

Q. E. D.

Cor. 1.—When the point of contact, is the vertex, the diameter of curvature, the chord which passes through the centre, and the chord which passes through the Focus, all coincide with each other, since they all coincide with the axis; they are therefore equal, and the diameter of Curvature is therefore a third proportional to the Transverse and Conjugate axes, that is, it is equal to the Parameter. (See Prop. II. Schol.)

Cor. 2.—When the point of contact is the extremity of the Conjugate axis, the Transverse axis is then conjugate to that which passes through the point of contact; and then the equation becomes

$$CV \cdot AB = AB^2, \text{ and } CV = AB.$$

That is, the chord of curvature passing through the Focus, then equals the Transverse axis.

COR. 3.—The chord of curvature which passes through the centre, is to that which passes through the Focus, as the Transverse axis, is to the diameter which passes through the point of contact.

For (demonstration) $CV : CO :: 2CP : AB$;
and (VI) $CL : CO :: 2CP : CG$,
$\therefore$ $CL : CV :: AB : CG$.

COR. 4.—The diameter of curvature, the chord which passes through the centre, and the chord which passes through the Focus, are inversely as the parts cut off from each, by the diameter conjugate to that which passes through the point of contact.

For (Cor. 2) $CL : CV :: AB : CG :: CN : Cc$,
and (VI) $CO : CL :: CG : 2CP :: Cc : CP$,
5.20. $\therefore$ $\left\{ \begin{array}{l} CO : CL : CV, \text{ are as} \\ CN : Cc : CP. \end{array} \right.$

COR. 5.—The diameter of Curvature, varies as the cube of the diameter, conjugate to that which passes through the point of contact.

For (Ell. XVI) $2CP \cdot HK = AB \cdot ab$,
but (VI) $2CP \cdot CO = HK^2$,
$\therefore$ $2CP \cdot HK : 2CP \cdot CO :: AB \cdot ab : HK^2$,
or $HK : CO :: AB \cdot ab : HK^2$,
$\therefore$ the parallelopiped $\overline{CO \cdot AB \cdot ab} = HK^3$.
But the base $AB \cdot ab$ is constant,
$\therefore$ CO varies as HK^3.

Consequently the *curvature itself* (II) at any point is *inversely* as the diameter conjugate to that which passes through that point, it is therefore least at the extremity of the Conjugate axis, and greatest at that of the Transverse.

COR. 6.—The diameter of Curvature, at the intermediate points, varies as the *cube* of the *normal*, CI.

For (Ell. XV) $CI \cdot CP = Ca^2$ $\therefore$ CP varies as CI,
and (Ell. XVI) $CP \cdot HC = Ac \cdot ca$ $\therefore$ CP varies as HC,
$\therefore$ CI varies as HC, and CI^3 as HC^3, as HK^3,
$\therefore$ CO which varies (Cor. 5) as HK^3, varies as CI^3.

PROPOSITION VIII.

IN THE HYPERBOLA, as in the Ellipse, the chord of curvature, which passes through the centre, is equal to the parameter of the diameter which passes through the point of contact.

That is CG : HK : CL, are continued proport's.

For $Cv \cdot vG : Ev^2 :: CG^2 : HK^2$,
which is ulti. $ED \cdot CG : ED \cdot CL :: CG^2 : HK^2$,
and $CG : CL :: CG^2 : HK^2$,
" $CL : CG :: HK^2 : CG^2$,
" $HK \cdot CL : HK \cdot CG :: HK^2 : CG^2$,
" $HK \cdot CL : HK^2 :: HK \cdot CG : CG^2$,
" $CL : HK :: HK : CG$.
But (19) $P' : HK :: HK : C'G$.
∴ $P' = CL$, and CG : HK : CL, contin'd. prop'ls.
Q. E. D.

PROPOSITION IX.

The Diameter of Curvature, is a third proportional to 2CP and HK; or 2CP : HK : Co, are continued prop's.

For sim. tri. $CO : CL :: Cc : CP :: 2Cc : 2CP$,
∴ $CO \cdot 2CP = CL \cdot CG = HK^2$,
Q. E. D.

PROPOSITION X.

The chord of curvature which passes through the Focus, is a third proportional to the transverse-axis, and the diameter conjugate to that which passes through the point of contact.

That is AB : HK : CV, are continued proportionals,

In sim. tri. $CV : CO :: CP : CN :: 2CP : 2CN = AB$
∴ $CV \cdot AB = CO \cdot 2CP = HK^2$,
or AB : HK : CV, are continued proportionals.
Q. E. D.

Cor. 1.—At the vertex of the Hyperbolas, the diameter of curvature, coincides with the chord which passes through the centre, and the chord which passes through the Focus, since they all coincide with the axis; it is therefore equal to the Parameter. (See Schol. Prop. II.)

The diameter of curvature varies as the cube of the *normal* (CI), and therefore the curvature itself is inversely as CI^3.

Schol.—The analogy between the curvature of the Ellipse and Hyperbola is complete, since the corresponding propositions are capable of being expressed in the same words. The terms, however, as was remarked in reference to the general properties of the two sections, have somewhat different significations; and in particular the *normal*, the cube of which is proportional to the radius of curvature, increases more rapidly in the Hyperbola than in the Ellipse, and never decreases again, after reaching a *maximum*, as it does in the Ellipse. The curvatures of the two sections therefore are widely different. That of the Hyperbola, constantly decreases, and the curve never returns into itself. When the two conjugate axes, are unequal, the curvature is greatest at the vertex of the greater axis, as in the Ellipse, and if the two axes of a Hyperbola are equal to the two axes of an Ellipse, each to each, the *curvature* at their vertices, in one section is the same as in the other.

If the two axes of a Hyperbola are equal to each other, (that is if the Hyperbolas are Equilateral,) in which case the section corresponds to the *circle* into which the Ellipse passes as its two axes become equal, the curvature at the four vertices of the Conjugate Equilateral Hyperbolas is the same, and is equal to that of a circle, whose diameter is equal to the axis; and the diameter of curvature to any other point, is as the distance of that point from the centre. The curvature itself therefore of Equilateral Hyperbolas is the same, in any point, as that of a circle, whose circumference is at an equal distance from its centre. In other words, in both figures, the curvature is always inversely as the distance of the curve at that point, from the centre. In this respect, therefore, there is a striking analogy between an Equilateral Hyperbola, and the circle, which may be considered as an *Equilateral Ellipse.* The difference however, between the two curves is great, arising from the fact, that in one the curve is convex towards the centre, and in the other it is concave.

In a general comparison of the curvature of the circle, with that of the other Conic Sections, the distinguishing peculiarity of the former is, that its curvature is *uniform*, remaining the same in every point of its circumference, while that of the three other sections *changes* continually, and is never the same at two successive points. In all of them, the curvature is greatest at the vertex of the sections, and decreases according to a given law, as any point is removed from the vertex.

This continual *Variation of Curvature*, as it is termed, is the reason why no definite portion of any Conic Section, coincides with its Circle of Curvature. If the curvature of the section, remained the same, or was uniform for any distance, *that portion* would be a part of a circle, and would consequently coincide with its Circle of Curvature, since it has been demonstrated that the latter has the *same curvature* as the curve of the section at the point of contact. Since however the curvature of a Conic Section, as has been shown, never continues the same, in two successive points, but *varies* continually, the curve immediately deviates from the tangent more or less than the Circle of Curvature does, and therefore passes in the former case *within*, and in the latter *without* the circle. As the curvature in all the sections, is greater at the vertex, than in any other point, the whole section passes *without* the Circle of Curvature to that point; for a similar reason, the Ellipse is wholly *within* the Circle of Curvature, to the curve at the extremity of its Conjugate axis: and in all the sections, that part of the curve which is towards the vertex from any point, being more curved than at the point, passes *within* the Circle of Curvature to that point, while the part more remote, being less curved, passes *without* it.

Fig 7.

APPENDIX TO CONIC SECTIONS.

I. Similar and Sub-contrary Sections.

Def. 1.—If a point be taken above the plane of a circle; and one extremity of a straight line, remain in that point, while the line moves around in the circumference of the circle, the figure described is called an *Oblique Cone*.

Pl. VII. Fig. 8. as VLK.

2.—If either of the Conic Sections be supposed to revolve around its axis, which remains fixed, the solid thus generated, may be called a *Conoid.**

Conoids receive specific names, from the different figures by which they are described; as Paraboloids, Ellipsoids and Hyperboloids.

Fig. 4. 5. and 6. As GAH, AHBG, and GAHO.

PROPOSITION I.

All the sections of a cone, made by parallel planes, are *similar* figures.

Fig. 1. Parabolic sections, AD : *ad* : : DG : *dg*.

Fig. 2. Ellipses, AB : *ab* : : A'B' : *a'b'*.

* Perhaps this use of the term is not authorized : but some *general* name was needed in the following propositions, and it seemed most easy and proper to *extend* the application of a term already in use. Archimedes gives the name *Conoids* to *Paraboloids* and *Hyperboloids*, (probably because they obscurely resemble cones,) and for a similar reason to *Ellipsoids*, he gives the name of *Spheroids*.

Hyperbolic Sections, AB : conj. of AB::*ab* : conj. of *ab*. Fig. 3.

For, the conj. of AB=a mean proportional between AQ and BR,
and the conjugate of *ab*=a mean proportional between *aq* and *br*.
therefore AB : its conjugate ::*ab* : its conjugate.

Cor. 1.—If a plane perpendicular to the triangular section, pass through the vertex of the cone, cutting the base, it will cut off *similar segments*, from similar parallel sections;

that is, in the Parabola, AD : DG::*ad* : *dg*. Fig. 1.

" " " Ellipse, AD : DE::A'*d* : *de*. " 2.
and " BD : DE::B'*d* : *de*.
" " AD : A'*d*::AB : A'B'.
∴ DE : *de*::*ab* : *a'b'*.

" Hyperbola AD : DG::*ad* : *dg*. " 3.
" BD : DG::*bd* : *dg*,
" AD : *ad*::AB : *ab*,
∴ DG : *dg*::conjugate of AB : conjugate of *ab*.

Cor. 2.—Similar segments may be cut from all Parabolas, that is, all Parabolas are similar sections. Plate XVI, 2.
For, if there be taken $DG^2 : DG'^2$ ::AD : AH,
then AD : DG::AH : HK.

Cor. 3.—Similar Polygons may be inscribed in similar segments of similar sections.

PROPOSITION II.

If any Conoid, be cut by a plane parallel to its base, or to the plane of its revolving axis, the section will be a circle. Fig. 4. 5. and 6.

Thus MZN and HS*g*G are circular sections.

PROPOSITION III.

If any Conoid, be cut by a plane, coinciding with its axis, the section will be a figure, equal and similar to that, by the revolution of which the Conoid was generated.

Fig. 4. 5. and 6. For it coincides with the revolving figure, in one of its positions.

Thus GAHO, may be considered a section of the Conoids.

PROPOSITION IV.

If a Conoid be cut by a plane *parallel* to its fixed axis, the section will be of *the same kind* with that by the revolution of which, the Conoid was generated, and *similar* to it.

Fig. 4. 5. and 6. Suppose a plane GAHO to coincide with the axis producing the section by which the conoid was described, and cutting the given plane at right angles. Let there be two circular sections MZN and HSgG perpendicular to the fixed axis, and therefore at right angles to each of the two other planes. The mutual intersection of these planes RZ and DG will be ordinates, both in the circular sections, and in the first sections, and it will be easy to prove,

Fig. 4. 1. In the Parabola, that the squares of the ordinates are as their abscisses (See Par. Prop. 1.)

That is, $aR : ad :: RZ^2 : dg^2$.

And that the section, is similar to that which coincides with the axis, or to that by which the Conoid was generated.

That is, $ad : dg :: AD : DG$.

Fig. 5. 3. In the Ellipse, that the squares of the ordin s a e as the rectangles of their abscisses (Ellipse Prop.

That is $aR \,.\, RB : ad \,.\, db :: RZ^2 : dg^2$.

and that $AD : DG :: ad : dg$,

and $AB : ab :: HG : 2dg$.

3. in the Hyperbola, if a plane parallel to that in which the axes of four Conjugate Hyperboloids are placed, cut them, it may, in a similar manner, be proved, that the two opposite sections, are Hyperbolas, and in every respect equal and similar to each other, and that the four sections are Conjngate Hyperbolas, similar to the four by which the Hyperboloids were generated.

That is $cR^2 - ca^2 : cd^2 - ca^2 :: RZ^2 : dg^2$, and $AB : ab :: A'B' : a'b'$. Fig. 6. " 7.

COR.—If a plane cut two opposite Hyperboloids, being perpendicular to the plane in which their axes are placed, and passing through their centre (and therefore coinciding with a diameter of the Hyperbolas, by which the Hyperboloids were generated,) the sections will be two opposite Hyperbolas equal to each other in every respect, but *not similar* to the generating Hyperbolas.

That is OIQ is a Hyperbola, or $Im \cdot mH : Is \cdot sH : mi^2 : sv^2$, and the opposite section will be equal and similar.

If then another plane cut the two Conjugate Hyperboloids in the same manner, but coinciding with KB″ the diameter which is conjugate to the other, this section will also form two opposite Hyperbolas which will be conjugate to the other two.

and $HI^2 : KB''^2 :: Im \cdot mH : mi^2$. Fig. 7.

PROPOSITION V.

If in *any* Conoid, a plane pass so as to intersect all sides of it, the section in every case, will be an *Ellipse*.

In addition to the planes supposed in the last figure, let a plane coincide with RZ (the mutual intersection of the circular section MZN, and the section parallel to the axis, $aZgd$) and let it cut the axis of the Conoid and all its sides; it can easily be shewn that this oblique section has in every case the properties of an Ellipse. (See Ellipse, Prop. I.) Fig. 4. 5. and 6.

That is $XR \cdot RO : XP \cdot PO :: RZ^2 : PS^2$.

COR.—All sections parallel to any oblique section are Ellipses and are similar.

PROPOSITION VI.

Every section of an oblique cone, which is parallel to its base, is a circle. Every section which is *sub-contrary* to the base is a circle.

Every other section is an Ellipse, and all the Elliptic sections which are *between* the sections parallel and sub-contrary to the base, may be called *Ellipses sub-contrary* to all other Elliptic sections of the same cone, that is their Transverse and Conjugate axis, are placed in contrary positions.

1. Sections parallel to the base, are proved to be circles, by shewing that all lines drawn in them, from the axis of the cone to its circumference are equal.

Fig. 8. that is $cB=BA$.

2 If a plane, cut the cone so as to make *the same angles* with the axis and with the opposite sides of the cone, that are made by a section parallel to its base, but in a *contrary order*, it is called a *sub-contrary* section, and is proved to be a circle, as in case first.

That is, if $VBA=VAB$ and cd be the mutual intersection of the circular section, parallel to the base $B'cR'$, and the given sub-contrary section, BA, then $Bc \, . \, cA=cd^2$.

3. That all other sections are Ellipses, is proved as by Prop. I Ellipse.

4. Of any section between the parallel and sub-contrary sections as AB', AB' is the conjugate, or smaller axis.

Fig. 8. SCHOL.—Those circular sections which are parallel to the base, and those which are sub-contrary, in an oblique cone, are *limits* between sub-contrary Elliptic sections, and hold the place of transition from one of those Elliptic sections to the other.

Thus, the sections ÂB, AB are *limits* between AB' and AB'', two Elliptic sections, one on each side, but having their axes in a contrary position.

PROPOSITION VII.

Every section of a Cylinder, parallel to its base is a circle, every other section is an Ellipse.

The first case is evident.—The second is proved in a manner similar to Prop. I, of the Ellipse. Fig 9.

That is, BD . DA : BS . SA : : DE^2 : SZ^2.

SCHOL.—The demonstrations of the preceding propositions have not been given in full, on the supposition that to persons well acquainted with the propositions on Conic Sections, the truth of them will be almost immediately evident; to others, it may afford a useful and agreeable exercise, to make out the demonstrations, by deducing them from the properties of the curves previously demonstrated.

II. COMPARISON OF CONIC SECTIONS.

PROPOSITION I.

A Pa   la is equal to four thirds of a Triangle, having the sam   and altitude. Or it is two thirds of a circumscribing parallelogram.

Pl. VIII. Fig. 10.

That is *Par* BAC$=\frac{4}{3}$ Tri. BAC.

Let the base of the Parabola BAC, be bisected continually, dividing it into the equal parts, BR, RG, GS, SD, DX, XE, EY, YC, and let lines be drawn through the intersections parallel to the axis AD, meeting the curve. Let CA be drawn parallel to the axis, and made of such length that C*a* : DA::4 : 3.

Connect *a*B also AB, AH, HB, AC, AF, FC, forming the inscribed polygon BHAFC. Then RP, passing through that division of the base, which is next to B, and meeting B*a* in *d* and BH, in *l*; and GH, meeting *a*B, AB, in *c* and L; HQ also being drawn parallel to BC, and P*q* parallel to BA, and DO, HO′ perpendicular to AB,

Then tri. B*a*C : BAC::C*a* : DA::4 : 3,
and sim. tri. B*a*C=4B*b*D, or B*b*D$=\frac{1}{4}$B*a*C,
∴ B*a*C : *ba*CD::4 : 3,
therefore *ba*CD=BAC.

Again tri. BHA : BDA::HO′ : DO::HL : AD,
BHA+AFC : BAC::HL (=AQ) : AD,
::HQ^2 : BD^2,
:: 1 : 4.
And DG*cb* : CD*ba* (=BAC) :: 1 : 4.
Therefore BHA+AEC=DG*cb*.

like manner $BPH+HMA+ANF+FNC$:
$BHA+AFC$, $::Pl:HL$,
$::Pq^2:AL^2$,
$::RG^2:BG^2$,
$::1:4$,
But $GRdc:GcbD(=BHA+AFC)::1:4$;
Therefore $BPH+HMA+ANF+FVC=GRdc$.

In the like manner if the parts of the base be bisected continually, and the sides of the inscribed polygon multiplied in the same proportion, the polygon will *in all cases*, be equal to the trapezium cut from the triangle BaC, by the parallel to the axis, nearest to B. But the polygon, thus approaches continually and indefinitely towards equality with the Parabola, which is its *limit;* also the trapezium approaches indefinitely towards the triangle BaC, which is evidently its *limit;* and since the polygon is always equal to the trapezium, the *limit* of the polygon, that is, the *Parabola is equal* to the *limit* of the trapezium, or *the triangle* BaC.

If this conclusion is denied, it can be shown that the denial leads to absurdity; for if the triangle BaC, for instance were supposed to be greater than Parabola, then the process could be carried on, until the trapezium taken from the triangle, should be also greater than the Parabola; but the trapezium is always *equal to the polygon inscribed in the Parabola*, and therefore always *less* than the Parabola.

A like absurdity follows the supposition that the Parabola is greater than the triangle.

Therefore the triangle BaC, is equal to the Parabola.

But the triangle $BAC=\frac{3}{4}BaC$,

∴ also the inscribed triangle $BAC=\frac{3}{4}$ Parabola,

Or the Parabola $=\frac{4}{3}$ Triangle BAC.

Q. E. D.

Cor. 1.—Hence if two Parabolas have equal altitudes, they are to each other as their bases.

For this is true of the circumscribing parallelograms, or the inscribed triangles. 6. 1. 5. 15.

Cor. 2.—And if two Parabolas have equal bases, they are to each other as their altitudes.

COR. 3.—Any two Parabolas are to each other, in the ratio compounded of their bases and altitudes.

COR. 4.—Hence *similar* Parabolic Sections have to each other, the duplicate ratio of their bases, or altitudes.

PROPOSITION II.

The Ellipse has to the circle which is described upon its Transverse Axis, the same ratio, which the Coujugate axis has to the Transverse;—and to the circle described upon its Conjugate axis, the same ratio which the Transverse has to the Conjugate.

Pl. VIII. Fig. 12.

That is Ellip. A*a* B*b* : Cir. AGB : : *ab* : AB.
And Ellip. A*a* B*b* : Cir. *agb* : : AB : *ab*.

Let AGB. be a semi-circle, described upon AB, the Transverse Axis, of the Ellipse AKB, let NHF, MKG be any two ordinates in the Ellipse, produced until they meet the circle, join GF and KH,

Then (Ell. III Cor. 2.) MK : MG : : NH : NF ;
Therefore GF. KH, produced will meet in R, in MN produced. Also, if GF be bisected in D, and the ordinate DEI be drawn, then the tangents to the Circle and Ellipse at the points D and E, will meet CA produced in the same point T. (Ellip. VI. Conversely.)

6. 1. Then. tri. MGR : MKR : : MG : MK,
Also " NFR : NHR : : NP : NH : : MG : MK,
5. 19. ∴ trap MGFN : MKHN : : MG : MK.
also tri. NEA : NHA : : NF : NH : : MG : MK,
and " MGB : MKB : : MG : MK,
∴ PolAFDGB : AHEKB : : MG : MK : : CA : C*a*.
6. 4. But, DT is parallel to GR,
∴ ER is parallel to KR and the triangles GDF, KEH, are each greater than half their segments GDF, KEH, therefore the inscribed polygons, as the sides are multiplied, will approach indefinitely to equality with the semi-circle and the semi-ellipse, as their *limits*.
∴ Cir : Ellipse : : Pol. AGB : Pol. AKB : : CA : C*a*.
Pl. III. Fig. 2. In like manner Ellipse A*a*B : Cir. D*ab* : : CA : C*a*.

Q. E. D.

COR. 1.—Any Ellipse is a mean proportional, between the circles described upon its two axes.

That is, Cir. AGB : Ell. $AaBb$ :: Ell. $AaBb$: Cir. agb.

COR. 2.—An Ellipse is equal to a circle whose diameter is a mean proportional between its two axes.

For let Z be such a circle, Fig. 13.

Then cir. AGB : Z :: AB^2 : $A'B'^2$

Also, cir. AGB : Ellip. :: AC : Ca,

:: AC^2 : $AC \cdot Ca$,

:: AB^2 : $2AC \cdot 2Ca = A'B'^2$.

COR. 3.—Hence any circle is to the square of its diameter. as the inscribed Ellipse, is to the rectangle of two axes.

COR. 4.—The areas of any two Ellipses, are to each other as the rectangles of their Transverse and Conjugate axes.

That is, Ellip. $A'aB'b$: Ellip. $Aa'Bb'$ :: $A'B' \cdot ab$: $AB\ a'b'$. Fig. 14.

SCHOL.—The two preceding propositions are demonstrated after the manner of the ancient Geometers. See Archimedes "De Quadraturæ Parabolæ" and "De Conoidibus et Spheroidibus."

It is called the method of *Exhaustions*, and is much admired for its ingenuity and rigorous exactness.

"*Though few things,* says Playfair, *more ingenious than this method have been devised, and though nothing could be more conclusive than the demonstrations resulting from it,* yet it laboured under two very considerable defects. In the first place the process by which the demonstrations were obtained was long and difficult, and in the second place it was indirect, giving no insight into the principle on which the investigation was founded."

This method has been much abridged by the moderns, especially by Cavaleri, in the method of *Indivisibles*, a method not only more concise, than that of the ancients, but much more easily and extensivly applicable to a great variety of propositions, and at the same time no less rigorously exact, when properly managed.

An application of it to the second of the preceding propositions will illustrate its nature, and prepare the student for the demonstrations which follow.

Fig. 11. Any trapezium as KMGD, is to the corresponding trapezium in the Ellipse KLED, as the sum of the parallel sides KM and DG, is to the sum of KL, and DE, for each trapezium may be divided into two triangles, which will be to each other, respectively as these lines.

Consequently, *all* the trapeziums taken together, or the polygon in the semi-circle, will be to the polygon in the Ellipse, as the *sum* of the ordinates in the circle, is to the sum of the ordinates in the Ellipse, that is (Euclid V. 12. as any one ordinate in the circle, to the corresponding ordinate in the Ellipse; in other words, as the Transverse to the Conjugate axes. But by increasing the number of ordinates, and consequently the number of sides of the polygons, the latter approach indefinitely to equality with the circle and Ellipse; therefore the circle is to the Ellipse, in the same ratio as the polygon in the circle, to the polygon in the Ellipse, that is, as the sum of the ordinates in one, to the sum of the corresponding ordinate in the other, or as the Transverse to the Conjugate axes.

This method of *Indivisibles*, or of supposing ordinates to be encreased indefinitely in number while their distances are diminished in the same proportion, often appears unsatisfactory, not only on account of the abridged mode of expression, by which important steps in the demonstration are omitted, but because it is common to say, that the spaces compared are *made up of an infinite number* of parallel ordinates, whereas it is evident from the definitions of Geometry, that no possible number of lines can "make up" a space, or form any thing but a line; not to mention the inconsistency of an "infinite number" of lines

The truth is, as has just been shown, that the *rectilineal spaces* compared, have to each other *the same ratio* which the sums of corresponding ordinates have, and *curvilineal spaces*, which are their *limits*, have exactly the same ratio. The exact truth of this Proposition, may be rigorously demonstrated in all cases by a *reductio ad absurdum*.

The method of Indivisibles then, when properly considered and applied, is an abridged form of the method of Exhaustions; the ungeometrical language, so often used, and before alluded to, will be avoided in the following propositions.

PROPOSITION III.

Two Ellipses having the same, or equal Transverse axes, are to each other as their Conjugate axes.

That is Ellip. $Aa\ Bb$: Ellip. $Aa'\ Bb'::ab : a'b'$,

For any ordinate DE : De::CA : Ca, — Pl. VIII. Fig. 14.
Therefore by indivis. CAa : CAa'::Ca : Ca';
Or Ellipse AaBb : Aa'Bb'::ab : $'ab'$.

Q. E. D.

Cor. 1.—This reasoning is equally applicable to two Ellipses, having the same or equal *conjugate* axes. They are to each other as their transverse axes.

That is, Ellip. A$'a$B$'b$: AaBb::A$'$B$'$: AB.

Cor. 2.—Any Ellipses have to each other the ratio compounded of the ratios of their two axes.

For Ellipse Aa'Bb' : Ellip. AaBb::$a'b'$: ab,
and Ellip. AaBb : Ellip. A$'a$B$'b$::A$'$B$'$: AB,
∴ Ellip. Aa'Bb' : Ellip. A$'a$B$'b$ in a ratio compounded of the ratios of $a'b'$ to ab, and of A$'$B$'$ to AB. — 5. G.*

Cor. 3.—*Similar Ellipses* have to each other the duplicate ratio which either of their axes have,

That is *Ellipses* Aa'Bb': A$'a$B$'b$::AB^2 : A$'$B$'^2$::$a'b'^2$:ab^2,

For (22) AB : $a'b'$::A$'$B$'$: ab,
and AB^2 : $a'b'$. AB::A$'$B$'^2$: ab . A$'$B$'$,
∴ AB^2 : A$'$B$'^2$::AB . $a'b'$: A$'$B$'$. ab::Ellip. Aa'Bb'Ellip. A$'a$B$'b$.

* Simpson's Euclid.—See also Book VI, Prop. XXIII.

PROPOSITION IV.

Two Hyperbolas, having the same or equal Transverse axes and equal abscisses, are to each other as their Conjugate axes.

Fig. 15. That is, ADE : ADE′::C*a* : C*a*′.

This is evident from the method of indivisibles, (See Hyp. II, Cor. 3.)

PROPOSITION V.

Similar segments of similar Conic Sections, are to each other in the ratio compounded of the ratios of their abscisses and ordinates.

Pl. VII.

Fig. 1. ADG : *adg*::AD . DG : *ad* . *dg*.
Fig. 2. ADE : A′*de*::AD . DE : A*d* . *de*.
Fig. 3. ADG : *adg*::AD . DG : *ad* . *dg*.

For their abscisses, being divided into an equal number of parts, these parts will be to each other as the whole abscisses, also the corresponding trapeziums in the inscribed polygons, will be to each other in the ratio compounded of their bases and altitudes; and the whole polygons, and consequently the Conic Sections which are their *limits*, will be to each other as the *sums* of the *bases and altitudes*, or as the abscisses and ordinates.

Cor. 1.—Similar segments are to each other as the rectangles of the Transverse and Conjugate axes of their sections.

Cor. 2.–Similar segments, are to each other, in the duplicate ratio of either of their axes.

Schol.—This proposition and corollaries, follow from Prop. I, Cor. 3, Append. I.

PROPOSITION VI.

Any diameter of any Conic Section, bisects any segment included between a double ordinate or the conjugate diameter, and the curve.

1. In the Parabola HA and HV, bisect the segments KAI and KVI. Fig. 12. and 18.

Or KAH=HAI, and KVH=HVI. Fig. 3.

2. In the Ellipses AC, bisects aAb and EAG,
Or ACa=ACb, and ADE=ADG. Fig. 13.
also CGg=CEg=CEe=CGe.
The same reasoning evidently applies to the circle.

In the Hyperbola AD bisects the segment EAG, or ADE=ADG. Fig. 4.

Also in a triangle, whether isosceles or scalene, a line which bisects the base, bisects the triangle.

For in each case the diameter bisects the double ordinate which limits the segment, and all double ordinates within the segment, therefore by the method of indivisibles, it bisects the space in which they are drawn.

Cor. 1.—The spaces included between the curve and two tangents drawn from the extremities of any double ordinate, are bisected by the diameter of that double ordinate.

That is CAT=TAD, TNC=TNL and TAE=TAH. Fig. 16. and 24.

Also the space between the curve, the tangent and double ordinates produced are equal. Fig. 7. & of Ell. & Hyp.

That is COP=LIR, FEI=GHM, also AKE=AKh.

Cor. 2.—The spaces between the curve of an Ellipse, and a circumscribing parallelogram, whose sides are parallel to two coujugate axes, are all equal.

That is, EKG=gSG=EQe=Gte. Fig. 13.

This is equally applicable to the circle.

Cor. 3.—The spaces between the curve of a Hyperbola and its asymptotes, and bounded by a double ordinate, are bisected by the diameter of that double ordinate.

That is CAEH=CAE*h*.

PROPOSITION VIII.

Similar Conoids, are to each other, in the ratio compounded of the ratios of their bases and altitudes, or, of the squares of the diameters of their bases and altitudes.

Similar Segments, being supposed to revolve on their axes, generate similar Conoids, and if the inscribed polygons, of which they are *limits*, be supposed to revolve with them, it will generate a solid, of which the Conoid is the *limit*. Hence, by *Indivisibles* or Exhaustions, the proposition may be deduced.

Cor. 1.—Similar Conoids are to each other, in the ratio compounded of the ratios of the squares of the diameters of their bases and their altitudes, or of the squares of their revolving axes or ordinates, and their fixed axes or abscisses.

Cor. 2.—Similar Conoids are to each other, in the triplicate ratio of their altitudes, or the diameters of their bases.

PROPOSITION IX.

1. An *OblateSpheroid*, or one described about the Conjugate axis, is to a *Prolate Spheroid*, or one about the Transverse of the same Ellipse, as the Transverse axis is to the Conjugate axis.

2. The circumscribing sphere, is to the oblate Ellipsoid, as the Transverse axis is to the Conjugate.

3. The Prolate Ellipsoid is to the inscribed sphere, as the Transverse axis, is to the Conjugate.

4. The circumscribing sphere, is to the oblate Ellipsoid, as the prolate Ellipsoid is to the inscribed sphere. That is, the four solids are Geometrical Proportionals.

5. They are *continued* proportionals, each having to the following, the ratio of the Transverse axis to the Conjugate; therefore the circumscribing sphere has to the inscribed sphere, the triplicate ratio of the Transverse axis to the Conjugate, as is evident also from Euclid, Sup. III, 11–21.

PROPOSITION X.

Every Paraboloid, is equal to one half its circumscribing cylinder.

This is proved, in a manner very similar to that in which the corresponding proposition concerning the sphere is demonstrated. (Plate XVI. 9)

PROPOSITION XI.

Every Ellipsoid, Prolate or Oblate, is $\frac{2}{3}$ of its circumscribing cylinder.

The Ellipsoid and the sphere, have evidently the same relations to the circumscribing cylinders. Pl. VIII. Fig. 12.

SPHERICAL GEOMETRY.

DEFINITIONS AND SECTIONS OF THE SPHERE.

Def. 1.—*A Sphere* is a solid, generated by the revolution of a semi-circle about its diameter, which remains fixed. Eucl. Sup. 3. Def. 7.

2.—The *Axis* of the sphere, is the fixed diameter about which the semi-circle is supposed to revolve. "Def. 8.

3.—The *Centre* of a sphere is the same as that of the semi-circle, by which it is described. "Def. 9.

4.—The *Radius* of the sphere, is any straight line, drawn from the centre to the circumference of the sphere.

Cor. 1.—The Radii of the sphere are all equal; each being equal to the radius of the generating semi-circle, In other words, every point on the surface of the sphere is equally distant from its centre.

Cor. 2.—Hence if two portions of the surface of the same sphere or of equal spheres, be placed upon each other *they will coincide* in one superfices.

Def. 5.—The *Diameter* of the sphere, is any straight line which passes through the centre, and is terminated both ways by the surface of the sphere. Sup. 3. Def. 10.

Cor.—The diameters of a sphere are all equal; being each equal to twice the radius, or to the diameter of the generating semi-circle.

Any diameter, therefore, may be assumed, as the axis of the sphere.

6.—The *Poles* of the sphere, are the extremities of its axis.

Cor.—The extremities of any diameter may be assumed as the poles of the sphere. (Def. 5, Cor.)

7.—*A plane touches* the sphere when it meets the surface of the sphere, but does not cut it.

8.—*A straight line touches* the sphere when it meets the surface, but does not cut it.

PROPOSITION I.

If a plane cut the sphere passing through its centre, the mutual intersection of the sphere and plane, will be a *circle*, which is equal to that by which the sphere was described.

This is evident because every point of the surface of the sphere, (Def. 4, Cor. 1.) is equally distant from the centre; and this distance is equal to the radius of the generating semi-circle.

1 Def.12

Cor. 1.—All circles on the sphere whose planes pass through the centre, are equal to each other.

Scholium.—Circles on the sphere whose planes pass through the centre, are called *Great Circles*.

Such upon the Armillary sphere, are the *Equator*, the *Horizon*, the *Ecliptic* and the *Meridians*.

Cor. 2.—*A great circle* may pass through any two points in the surface of the sphere.

For a plane may be supposed to pass through *any three given points.* As the plane of a great circle passes through the centre of the sphere, it may also pass through any two points in its surface.

Cor. 3.—The straight line which connects any two points, in the surface of the sphere, falls wholly within the sphere.

For it is a chord of the great circle which passes through these two points.

Cor. 4.—Any straight line which touches the sphere, touches the great circle whose plane coincides with the line. (Def. 8.) 3 Def. 1.

Cor. 5.—The straight line which passes through the point of contact and the centre of the sphere, is perpendicular to a line touching it.

In other words, any diameter of the sphere, is perpendicular to all lines, which meet its extremity and touch the sphere. 3. 16.

Cor. 6.—An indefinite number of great circles may pass through any given point in the surface of the sphere.

For planes may pass through the centre, through a given point in the surface, and through *any other point* in the surface. And the intersection of each of them with the sphere will be a great circle. (Cor. 2.)

Schol.—Thus there are an indefinite number of Meridians, all passing through the poles of the terrestial sphere.

Cor. 7.—The mutual intersections of a plane which touches a sphere, and the planes of any number of great circles passing through the point of contact, will be straight lines touching the sphere, (Def. 7.) and touching those great circles respectively. (Cor. 4.)

Schol.—Thus if a plane touch the globe at one of its poles, the mutual intersections of this plane with the planes of the Meridians, will all be lines touching those Meridians, and also touching the sphere.

Cor. 8.—The radius or diameter of the sphere, which passes through the point of contact, is perpendicular to the *plane touching* the sphere.

Sup. 2. Def. 1. For it is perpendicular to every straight line which it meets in that plane. (Cor. 5, and 7.)

Schol.—Thus the axis of the Earth, is perpendicular to planes touching it, at its poles.

Cor. 9.—The mutual intersection of two great circles, is a diameter of the sphere.

For as the plane of each great circle passes through the centre of the sphere, the line of their mutual intersection therefore passes through the centre of the sphere. It is therefore a diameter. (Def. 5.)

Cor. 10.—Hence all great circles mutually bisect each other: that is, they divide each other into two equal parts, or semi-circles.

Schol.—Thus the Ecliptic and the Equator, bisect each other. All the Meridians bisect each other, &c.

PROPOSITION II.

If any plane cut a sphere, but do not pass through the centre, the section will be a circle, whose radius is equal to that ordinate in the generating semi-circle which coincides with the plane.

For in describing the sphere, the revolutions of that ordinate will describe the circle.

Cor. 1.—The circles formed by the section of planes which do not pass through the centre of the sphere, since they are to each other as the squares of their diameters or radii,—that is, the squares of the ordinates in the semi-circle, are as the rectangles of the segments into which their planes divide the axis.

That is, the cir. $EcQ : HDI :: PC \cdot Cp : PD \cdot Dp$. Fig. 1.

Schol.—As these circles are smaller than the generating circle of the sphere, and therefore smaller than the circular sections which pass through the centre, they are called *Small Circles* of the sphere. Such are all the parallels of latitude on the terrestial sphere.

Cor. 2.—Small circles whose planes are equally distant from the centre, are equal.

And conversely.—If they are equal, they are equally distant from the centre.

For then, the corresponding rectangles, of the axis, will be equal. (Cor. 1.)

Cor. 3.—If two small circles are unequal, the less is more distant from the centre, and conversely.

Schol.—Thus the two Tropics, or the Polar Circles, are respectively equal, and equally distant from the centre; but the Polar Circles are more distant from the centre than the Tropics, and smaller.

Cor. 4.—A small circle may pass through any three points, in the surface of the sphere, if those points are not all in the same great circle.

For a plane may pass through *any three points*, and if it do not pass also through the centre of the sphere, the section will be a small circle.

Cor. 5.—If any two circles meet, but do not cut each other, on the surface of the sphere, the straight line which touches one of the circles, touches the other also.

For it meets the other circle, but does not cut it. If the line did cut the second circle, it would fall wholly within the sphere (I, Cor. 2.) and therefore would not touch the first circle, contrary to the supposition.

And conversely, Circles which touch the same straight line touch each other.

Cor. 6.—Hence the mutual *intersection* of the planes of two circles which touch each other on the sphere, is a straight line which touches both of them.

For the line is in the planes of both the circles, and it touches each of them.

Schol.—As illustrations: The Tropics touch the Ecliptic. Each of them therefore, and the Ecliptic touch the same straight line, at the point of contact, and that line is the common intersection of their planes.

Def. 9.—The extremities of the axis, are the *poles* of *all circles* which are perpendicular to the axis.

Cor. 1.—Hence two *great circles* cannot have the same poles, not having a common axis.

Cor. 2.—There may be an indefinite number of small circles, all parallel to a great circle, and having the same poles with it, being all perpendicular to the same axis.

Schol.—Thus all the parallels of latitude have the same poles as the Equator.

Cor. 3.—Circles which have the same poles are parallel to each other, and the axis passes through the centre of each of them.

The axis therefore passes through the poles of any circle, through its centre, and through the centre of the sphere.

Sup.2.17 Cor. 4.—The plane of any great circle, is at right angles to all the circles through whose poles it passes. For it coincides with the axis, which (Def. 9.) is perpendicular to all their planes.

Schol.—Thus the Meridians are at right angles to the Equator and to all its parallels.

Cor. 5.—And conversely, All the great circles, which are at right angles to a given great circle, intersect each other in its poles.

For they all pass through its poles, and must therefore intersect each other there.

Schol.—For example, all the Meridians are at right angles to the Equator, and they all intersect each other in its poles.

Cor. 6.—So also, if two great circles are at right angles to each other, the poles of each are in the circumference of the other: and if any number of great circles pass through the poles of another great circle, the poles of all the former, are in the circumference of the latter.

Schol.—Thus the poles of all Meridians, are in the Equator.

Cor. 7.—If three great circles are at right angles to each other, the pole of each is in the circumference of *both* the others, and therefore in the point of their intersection.

Cor. 8.—A circle whose plane coincides with the axis, and therefore passes through the poles, bisects all the circles which are perpendicular to the axis,

For it passes through their centres. (Cor. 3.)

Cor. 9.—The angle made at the centre of the sphere, by the axes of two great circles, is equal to the inclination of their planes.

That is $PCP' = QCL$,

For $PCQ = P'CL =$ a right angle, therefore taking away the common part $P'CQ$, and $PCP' = QCL$. Fig. 3.

Cor. 10.—If two circles touch each other, on the surface of the sphere, the plane which passes through the point of contact, and is perpendicular to the line, touching both of them, (II Cor. 5.) that is, to the line of their mu-

Sup.2.17. tual intersection, (II Cor. 6.) is perpendicular also to their planes. It therefore passes through their poles, (Cor. 5.) and through the centre of the sphere, (Cor. 3.)

Conversely, The plane of a great circle which passes Sup.2.18. through the poles of two other circles which touch each other, passes also through the point of contact and is perpendicular to the line of their mutual intersection.

Schol.—For example. The solstitial Colure, passes through the points of contact, of the Ecliptic and Tropics, and through their poles. It is perpendicular also to their planes, and to the lines of their mutual intersection, each of which lines touches both the Ecliptic and Tropic in the points of their contact.

Cor. 11.—Any circle on the sphere cuts off equal arcs from all the great circles which pass through its Fig. 2. poles.

That is, PH=PF and PE=PG.

For in the description of the sphere, the arc of the generating semi-circle between the poles, and any plane perpendicular to the axis, will coincide successively with the arcs of all great circles passing through the pole.

Schol.—Thus the Equator, or any of its parallels, cuts off equal arcs from all the Meridians.

Cor. 12.—Hence any two parallel circles, intercept equal arcs of all great circles passing through their poles.

That is EH=GF.

This is inferred from the last, by taking equal arcs from those which are equal.

SCHOL.—Thus arcs of Meridians, intercepted between two parallels of Latitude, or between any parallel and the Equator, are equal.

PROPOSITION III.

If two circles on the sphere, pass through the remotest poles of two great circles, they will cut off equal arcs from those circles.

That is $Ee=Gg$, Fig. 3.

For the arc $PE=pG$; $\therefore$ $PG=pE$,
Therefore the chord $PG=pE$, and $PD=pD$. 3. 36.
Again the chord $Pe=pg$, $\therefore$ arc $Pe=pg$.
Therefore the arc $Pg=pe$ and chord, $Pg=pe$, and $Pd=pd$;
Therefore, the triangle $DPd=Dpd$ in all respects; 1. 8.
$\therefore$ also $GPg=Epe$, and chord $gG=Ee$; and arc $Gg=Ee$.
Q. E. D.

PROPOSITION IV.

If from any point which is not the pole of a great circle, there be described arcs of great circles to that circle, the greatest is that which passes through its pole. And the other arc of the same circle is the least; and of the others, that which is nearer to the pole, is greater than any other which is more remote.

Let the common section of the planes of the great circles ACB, AD B at right angles to each other, be AB; and from C, draw CG perpendicular to AB, which will also be perpendicular to the plane ADB, Fig. 7. Sup. 2. Def. 2

Join GD, GE, GF, CD, CE, CF, CA, CB.

Of all the straight lines drawn from G, to the circumference ADB, GA is the greatest and GB the least; and GD which is nearer to GA is greater than GE, which is more remote. The triangles CGA, CGD are right angled at G, and they have the common side CG; therefore the squares of CG, GA together, that is the square of CA, is greater than the squares of CG, CD together, that is, the square of CD: and CA is greater than 3. 7

3. 28. CD, and therefore the arc CA is greater than CD. In the same manner, since GD is greater than GE, and GE than GF, &c. it is shown that CD is greater than CE, and CE than CF &c. and consequently, the arc CD, greater than the arc CE, and the arc CE greater than the arc CF, &c. And since GA is the greatest, and GB the least of all the straight lines drawn from G to the circumference ADB, it is manifest that CA is the greatest, and CB the least of all the straight lines drawn from C to the circumference: and therefore the arc CA is the greatest, and CB the least of all the circles drawn through C, meeting ADB.

Q. E. D.

SPHERICAL GEOMETRY.

II. *OF SPHERICAL TRIANGLES.*

DEF. 1—*A spherical angle* is the angle made by two circles which intersect each other, on the surface of the sphere. Or *it is the angle made by the straight lines which touch those circles at the point of their intersection.*

SCHOL.—The first of the preceding definitions, is one frequently given by writers on spherics. "*The angle made by two circles,*" however is two indefinite an expression. Cagnoli endeavors to render it more precise, by adding that *it is the angle made by the arcs,* "*considered in the points immediately contiguous to that in which they meet each other.*" (*consideree dans les points immediatement contigus a celui dans lequel ils se rencontrent,* (279.) and this angle he infers is the same as that made by the lines touching them in that point. But not to mention the want of mathematical precision in the phrase "points *immediately contiguous* to that &c," the equality of this angle and that of the touching lines, cannot be rigorously demonstrated except by the method of *ultimate* or *limiting ratios*, a method far too refined and too far removed from the principles of Elementary Geometry to be involved in a simple definition. To avoid this difficulty the second of the preceding definitions is given, which will be referred to in the following demonstrations.

COR. 1.—Hence if two circles cut each other on the surface of the sphere, the adjacent angles are either two right angles, or are together equal to two right angles. 1. 13.

That is BCA+BCD=ECA+ECD=2 right angles. Fig. 10.

For this is true of the angles made by the straight lines touching them at the point of intersection.

Cor. 2.—For the same reason, the angles made by any two circles, or any number of circles intersecting each other in one point, on the surface of the sphere, are together equal to four right angles.

Cor. 3.—A spherical angle is always less than two right angles.

1. 15. Cor. 4.—If two circles cut each other on the sphere, the vertical or opposite angles are equal.

Fig. 11. That is BCA=ECF.

Cor. 5.—The angle which two great circles make upon the surface of the sphere, is the same as the inclination of the planes of these circles.

Fig. 2. That is GPE=GCE.

For the lines touching the circles in the point of intersection, are perpendicular to that diameter of the sphere, which is the line of the mutual intersection of their planes; (1 Cor. 5.) the angle formed by the touching lines therefore, is the inclination of their planes. It is also the angle made by the circles. (Def. 1.)

Sup. 2. Def. 4.

Cor. 6.—Hence the angles made by any two great circles, on the opposite side of the sphere are equal.
That is GPE=G*p*E.

Cor. 7.—The angles made by two great circles are equal to that made by two of their radii, which are in the plane of another great circle whose pole is the angular point.

For these two radii are parallel to the two lines touching the circles in the angular point. (See Cor. 5.)

Cor. 8.—Hence the angles which great circles make with each other on the surface of the sphere, are proportional to the arcs, intercepted by them, of a great circle whose pole is the angular point.

For EG is the measure of GCE, therefore, it is also the measure of GPE. (Cor. 7.) Fig 2.

Schol.—Thus any arc of the Equator intercepted between two Meridians, is the *measure* of the angle which those meridians make at its poles. Whenever spherical angles are compared with arcs of great circles, the *measures* of these angles, are really compared with those arcs.

Cor. 9.—The arcs of parallel circles, intercepted by great circles which pass through their common poles, (9 Cor. 2.) are to each other as their circumferences, (since they subtend equal angles at their centres,) that is as their diameters or radii. Therefore the arcs of *any* circle may be made the *measures* of the angles formed by great circles at its poles, since they are always proportional to those angles.

That is FH is the measure of FDH=GCE=GPE.

Def. 2.—A *Spherical Triangle*, is a figure formed on the surface of the sphere, by the arcs of three great circles, which intersect each other.

Cor. 1.—The side of a spherical triangle is always less than a semi-circle.

For any two great circles intersect each other in points diametrically opposite, dividing each other into semi-circles. (I, Cor. 10.) Therefore if a third circle intersect each of these, it must intersect the semi-circles into which they divide each other, and consequently each intercepted arc will be less than a semi-circle.

That is, PH*p* and PE*p* are semi-circles; therefore if any arc, as HF intersect them, forming the triangle *p*HF, *p*H, or *p*F, are each less than *p*HP, or *p*FP, and therefore less than a semi-circle. Fig. 2.

DEF. 3.—The different kinds of Spherical Triangles.—*right-angled*, *acute-angled*, *obtuse-angled*, and *equi-angular; also equi-lateral*, *isoceles*, and *scalene*, are distinguished and defined in the same manner as in plane triangles. (See Euclid, Book I, Def. 20 to 25.)

DEF. 4.—A *rectilateral*, or *quadrantal*, spherical triangle, has one side or quadrant.

DEF. 5.—An *oblique angled triangle*, is a general term applied to all triangles in which there is not a right-angle, nor a quadrantal side.

PROPOSITION I. PROB.

On a given arc, to describe (on the surface of the sphere) an Equilateral Spherical Triangle.

Let AB be the given arc, on which it is required to describe an Equilateral Spherical Triangle. Fig. 5.

Let one extremity of the arc as A, be taken as a pole, and through the other extremity, as B, let a small circle be described whose pole is A, (Def. 9, Cor. 2, also II, 2,) then let B be taken as a pole, and a small circle described about it passing through A, they will intersect each other in C. Then let two great circles pass through A and C, and through B and C (I, Cor. 2,) ABC is the Triangle required.

For AC=AB (II. Cor. 11,) and BC=BA.·.AC=BC=AB.

COR.—Therefore, from any given point on the surface of the sphere, an arc of a great circle may be described equal to a given arc, in a manner analogous to that followed in the second proposition of the first book of Euclid; for a small circle cuts off equal arcs of great circles passing through its poles; as in a plane, a circle cuts off equal segments from lines passing through its centre.

PROPOSITION II.

If two spherical triangles have two sides of the one, equal to two sides of the other each to each, and the contained angles equal, the triangles are equal in all respects; that is, their areas are equal, the remaining sides in each are equal, and the angles opposite the equal sides are equal.

If the two equal sides in one, *correspond* to those in the other in position, as in the triangles CAE and CAB, their equality may be proved in the method followed by Euclid Prop. IV, by super-position, and coincidence. (4. Def. Cor. 2.) Fig. 10. and 12.

If they do not correspond in position, as CGE and CGF, supposing that CE=CF, and GE=GF it is evident they cannot be made to coincide; it is then perhaps a *sufficient reason* for believing them equal, that there is no reason why they should differ. Their equality however may be rigorously demonstrated. * Fig. 11.

PROPOSITION III.

If two spherical triangles, have one side in each equal, and the angles adjacent to the equal sides equal, each to each, the two triangles shall be equal in all respects. Fig. 10. and 12.

That is if ACE and ACB have the side AE in one, equal to AB in the other, and CAE, CEA, in one, equal CAB, CBA, in the other, then AC=AC and CE=CB and ACE=ACB.

This proposition may be demonstrated by super-position, as Prop. II; or it may be demonstrated by a *reductio ad absurdum* founded on Prop. II, as Euclid has done in regard to plane triangles. (Book I, 26, first part.)

PROPOSITION IV.

The angles at the base of an isosceles spherical triangle are equal.

That is, if CE=CF, then CEF=CFE. Fig 11.

* See Notes.

This proposition may be demonstrated from Prop. II, in a manner similar to that in which the corresponding proposition is demonstrated in plane triangles. (See Euclid I Prop. 5.)

1.5. Cor. Cor. 1.—Hence an Equilateral Spherical Triangle, is also Equiangular.

1. 6. Cor. 2.—And conversely, if the angles at the base of a spherical triangle are equal, the opposite sides are also equal.

This is derived from the Proposition and may be demonstrated by *reductio ad absurdum* in the manner of Euclid Prop. VI.

1.6. Cor. Cor. 3.—Hence, every Equiangular Spherical Triangle, is also Equilateral.

PROPOSITION V.

If two Spherical Triangles have the three sides of one equal to the three sides of the other, the two triangles will be equal in all respects.

I. 7. This proposition may be demonstrated by proving that upon the same base, and upon the same side of it, two spherical triangles cannot be described, which have their corresponding sides equal, as is done in plane triangles.

Schol.—The Problems, for *bisecting* a *spherical angle*; —for *bisecting* a *given arc* of a great circle, for describing an arc which shall be at *right angles* to another arc, either from a point within the arc, or from a point above it, depend, like the similar Problems in plane triangles, upon describing an Equilateral Triangle, upon a given base, and drawing another arc through two given points.

Prop. I, describes an Equilateral Spherical Triangle, and an arc of a great circle may be described through *any two points* on the *surface of a sphere*, (I, Cor. 2.) and therefore, through the vertices of two triangles, in the first and second cases, and through the vertex of a triangle, and a point in the base, in the two latter cases.

Also, by the same means, upon a given base, a spherical triangle may be described, whose sides shall be equal to those of a given spherical triangle. And consequently, a spherical angle may be described, equal to any given spherical angle. (See Euclid I, 22, 23.) Figures 4, and 5, may be used for the illustration and demonstration of all the propositions referred to.

PROPOSITION VI.

Any two sides of a spherical triangle are greater than the third side.

That is AB+BC>AC, and AB+AC>BC, and BC+AC >AB. Fig. 8.

For the planes of the three circles all pass through D, the centre of the sphere, forming a solid angle at D, any two angles of which are greater than the third ; therefore the same is true of the arcs AB, BC and AC which have to each other the same ratios as the angles. Sup. 2. 20. 6. 33.

Q. E. D.

PROPOSITION VII.

The three sides of a spherical triangle, are together less than a circle.

For any two great circles, as ACD, ABD, bisect each other, therefore ACD and ABC, are together equal to a circle. Fig. 10.

Let the arc CB, intersect both these circles.

In the triangle CAB the two sides CA and AB, are together greater than the third side CB (VI,) therefore CD DB and BC, are together less than CD, DB, BA and AC, that is less than a circle.

Q. E. D.

COR. – Each of the three sides of a spherical triangle may be greater than a quadrant.

Also each of them, evidently, may be less than a quadrant.

PROPOSITION VIII.

In any spherical triangle, the greatest side is opposite the greatest angle; and conversely.

Fig. 10. That is, if the angle ACE is greater than AEC then also AE is greater than AC.

For, let the angle ACB be made equal to BAC, then BA=BC (III) and therefore CB+BE=AE; but CB+BE>CE, therefore AE>CE, that is the greater side, AE in the triangle ACE, is opposite to the greater angle C.

Q. E. D.

PROPOSITION IX.

If the three angles of any spherical triangle are made the poles of three great circles, the intersection of those circles will be a triangle, whose sides will be supplemental to the measures of the angles of the given triangle, and the measures of its angles will be supplemental to the sides of the given triangle.

Fig 9. That is DF, FE, and ED are supplemental of the angles C, B and A, and the angles at E, D and F, are supplemental of the arcs DF, FE and ED.

Since A is the pole of FE and the circle AC passes through A, EF will pass through the pole of AC (9, Cor. 6.) and since C, is the pole of FD, FD also will pass through the pole of AC; therefore the pole of AC is in the point F, in which the arc DF, EF intersect each other. In the same manner D is the pole of BC, and E the pole of AB.

And since F, E, are the poles of AL and AM, FL and EM are quadrants, and FL, EM together, that is FE and ML together, are equal to a semi-circle. But since A is the pole of ML, ML is the measure of the angle BAC.(II, 1 Cor. 8.)

In the same manner, ED, DF are the supplements of the measures of the angles ABC, BCA.

Since likewise CN, BH are quadrants, CN, BH together, that is, NH, BC together are equal to a semi-circle; and since D is the pole of NH; NH is the measure of the angle FDE; therefore the measure of the angle FDE is the supplement of the side BC. In the same manner, it is shown that the measures of the angles DEF, EFD are the supplements of the sides AB, AC, in the triangle ABC.

Q. E. D.

SCHOL.—The triangle DEF, is called the *Supplemental* or *Polar* triangle of ACB.

PROPOSITION X.

If two triangles have the three angles of one equal to the three angles of the other each to each, the two triangles shall be equal in every respect.

For the *sides* of their *Polar* Triangles are then equal, each to each (IX,) therefore the two *Polar* Triangles shall be equal to each other *in every respect*,(V.) Therefore the sides of the *given triangles*, which are supplemental to the *angles*, respectively, of the *polar triangles*, shall be equal to each other, each to each, therefore the two given triangles are equal in all respects. (V.)

Q. E. D.

PROPOSITION XI.

The three angles of any spherical triangle are together greater than two right angles, and less than six right angles.

That is $A+B+C >$ 2 right angles and $<$ 6 right angles. Fig. 9.

First they are greater than two right angles, for the sides of its polar triangle DF, FE and ED, which are supplemental to their angles, are together *less* than a circle (VII) which is the measure of four right angles, therefore their three supplemental angles C, A, B are *greater* than *two* right angles.

Second, As any spherical angle (II Def. 1, Cor. 3.) is less than two right angles, the three angles of any spherical triangle, are together less than six right angles.
Q. E. D.

SCHOL.—In the property now demonstrated, there is a striking difference between spherical and plane triangles, (see Euclid 1, 32.) which is the foundation of corresponding differences in all the relations deducible from these different properties.

COR. 1.—If a side of a spherical triangle be produced, the exterior angle, is *less* than the sum of the two interior and opposite angles.

Fig. 10. That is, CED is less than ECA+EAC.

For this exterior angle CED together with its adjacent interior angle CEA is equal to two right angles, but the angle CEA together with the two opposite interior angles, is greater than two right angles.

COR. 2.—Also if the sides of a spherical triangle be produced, the exterior angles are together less than four right angles.

COR. 3.—The three angles of a spherical triangle, may be each greater than a right angle.

PROPOSITION XII.

In any spherical triangle, if the sum of the two sides, be equal to a semi-circle, the external angle at the base, will be equal to the interior and opposite angle; and therefore the sum of the two angles at the base, will be equal to two right angles.

Fig. 10. That is, if AC+CE=AD, then CED=CAE.
and CAE+CEA=two right angles.

For then CE=CD.·.(IV) CED=CDE=CAE.
And CAE+CEA=CEA+CED=two right angles.
Q. E. D.

Cor. 1.—Conversely, if the angles at the base, are equal to two right angles, the sides are together equal to a semi-circle; and if the angles are *unequal*, that opposite the greater angle, is greater than a quadrant, (VIII) and that opposite the less, is less than the quadrant: and conversely.

But if the angles are equal, the angles at the base are right angles.

That is, if $CAE+CEA=2$ right angles, then $AC+CE=AD$.

And if $CAE=CEA$, then CE and CA, each equal a quadrant.

If $CAE>CEA$ " CE is greater, and CA less than a quadrant.

Cor. 2.—If the two sides are together greater than a semi-circle, the interior angle will be greater than the exterior and opposite angle, and the sum of the angles at the base, will be greater than two right angles, and the greater side will be greater than a quadrant. Conversely, if the interior is greater than the exterior angle, then the two interior angles are together greater than two right angles, and the two opposite sides are together greater than a semi-circle.

That is, If $CA+CE<AD$,

Then $CAE>CEA$, and $CAE+CEA>2$ right angles, and conversely, if $CAE>CED$, then $CAE+CEA>2$ right angles, and $CA+CE>AD$.

Also if the two sides be less than a semi-circle, the interior angle will be less than the exterior and opposite angle, and the two angles at the base, will be less than two right angles. Also the less side will be less than a quadrant.

That is. if $CA+CE<AD$, then $CAE<CED$;

And $CAE+CEA<2$ right angles, and conversely.

Also the less side is less than a quadrant.

Cor. 3.—Hence if each of the sides of any spherical triangle, is greater than a quadrant, as *any two* of them are greater than a semi-circle, the sum of any two angles is greater than two right angles, and therefore each of the angles of the triangle, is greater than a right angle; and conversely.

For a similar reason, if each side is less than a quadrant, each angle will be less than a right angle, and conversely. And if each side is equal to a quadrant, each angle will be equal to a right angle, and conversely.

PROPOSITION XIII.

If from the extremities of the base of a spherical triangle arcs of great circles be described, meeting in a point within the triangle, these arcs shall together be less than the two sides of the triangle; *and if two sides of the triangle are together not greater than a semi-circle,* the arcs shall make a greater angle, than that made by the sides of the triangle.

The former part of this proposition follows from Prop. VI, and the latter from Prop. XII, in the manner of plain triangles. (*See Euclid* 1, 21.)

The necessity of the *condition* in the latter part is owing to the difference between spherical and plain triangles before noted. (*See Prop.* XI, *also Euclid* 1, 16, 21.)

PROPOSITION XIV.

In any right angled spherical triangle, the sides are of the *same affection* as the opposite angles; that is, *if either side is greater or less than a quadrant, the opposite angle is greater or less than a right angle.*

Fig. 10. Let ABC, be a spherical triangle, right angled at A, either side as AB will be of the same affection with the opposite angle ACB.

Fig. 12. Let AB be less than a quadrant, let AE be a quadrant, and let EC be a great circle passing through E, C. Since A is a right angle, and AE a quadrant, E is the pole of the great circle CA, and ECA is a right angle. But ECA is greater than BCA, therefore BCA is less than a right angle.

Let AB be greater than a quadrant, make AE a quadrant, and let a great circle pass through C, E; ECA is a right angle as before, and BCA is greater than ECA that is greater than a right angle. Fig. 12. or Pl. XV. Fig. 6.

Q. E. D.

PROPOSITOIN XV.

If the two sides of a right angled spherical triangle, be of the same affection, the hypotenuse will be less than a quadrant. if they be of different affection, the hypotenuse will be greater than a quadrant; and if each side be equal to a quadrant, the hypotenuse will be a quadrant.

Let ABC be a right angled spherical triangle, if the two sides AB, AC be of the same or different affection, the hypotenuse BC will be less or greater than a quadrant. Pl. XV. Fig. 5.

1.—Let AB, AC be each less than a quadrant. Let AE, AG be quadrants, G will be the pole of AB, and E the pole of AG, and EC a quadrant; but (I. Prop. IV.) CE is greater than CB, since CB is farther off from CGD, than CE. In the same manner, it is shown that CB, in the triangle CBD, where the two sides CD, BD are each greater than a quadrant, is less than CE, that is less than a quadrant.

2.—Let AC be less, and AB greater than a quadrant, then the hypotenuse BC will be greater than a quadrant, for let AE be a quadrant, then E is the pole of CA, and EC is a quadrant. But CB is greater than CE (I. Prop. IV.) since AC passes through the pole of ABD. Fig. 6.

Q. E. D.

Cor. 1.—Conversely, accordingly as the hypotenuse, is greater, less, or equal to a quadrant, the sides will be of different, or of the same affection, or will be quadrants. Also the oblique angles will be of different or of the same, affection, or right angles. (XIV.)

COR. 2.—When an angle and the side adjacent are of the same affection, the hypotenuse is less than a quadrant, and conversely.

COR. 3.—If in a right angled spherical triangle, one side be less than a quadrant, it is less than the hypotenuse; for its opposite angle, is less than a right angle, and therefore the side is less than the hypotenuse which is opposite the right angle.

For a similar reason, if one side be greater than a quadrant, it is greater than the hypotenuse. And if it be equal to a quadrant, it is equal to the hypotenuse.

COR. 4.—In a right angled spherical triangle, the *measure* of an acute angle is not *less* than the opposite side; and the *measure* of an obtuse angle is not *greater* than the opposite side.

PROPOSITION XVI.

In any spherical triangle, if the perpendicular from one of its angles upon the base, fall *within* the triangle, the angles at the base will be of the same affection, but if it fall *without* the triangle, the angles will be of different affection.

Fig. 14 Let ABC be a spherical triangle, and let the arc be drawn from C perpendicular to the base AB.

1. Let CD fall within the triangle; then, since ADC, BDC are right angled spherical triangles, the angles A, B must each be of the same affection with CD. (XIV.)

Fig. 15. 2. Let CD fall without the triangle; then the angle B is of the same affection with CD; and the angle CAD is of the same affection with CD; therefore the angle CAD and B are of the same affection, and the angle CAB and B are therefore of different affections.

Q. E. D.

COR.—Conversely, if the angles at the base of any triangle, be of the same affection, the perpendicular will fall *within* the triangle, but if the angles be of different affection, the perpendicular will fall *without* the triangle.

PROPOSITION XVII.

A perpendicular being drawn to the base of a spherical triangle, if the sum of the sides is *less* than a semi-circle, then the *least* segment of the base is adjacent to the *least* side of the triangle; but if the sum of the sides be *greater* than a semi-circle, the *least* segment is adjacent to the *greatest* side.

Let ABEF be a great circle of a sphere, H its pole, and GHD any circle passing through H, which therefore is perpendicular to the circle ABEF. Let A and B be two points in the circle ABEF, on opposite sides of the point D, and let D be nearer to A, than to B, and let C be any point in the circle GHD between H and D. Through the points A and C, B and C, let the arcs AC and BC be drawn, and let them be produced, till they meet the circle CBEF, in the points E and F, then the arcs ACE, BCF are semi-circles. Fig. 11.

Also, ACB, ACF, CFE, ECB are four spherical triangles, contained by arcs of the same circles, and having the same perpendiculars CD and CG.

1. Now because CE is nearer to the arc CHG than CB is, CE is greater than CB, and therefore CE and CA are greater than CB and CA; wherefore CB and CA are less than a semi-circle; but because AD is by supposition less than DB, AC is also less than CB. (I. Prop. IV.) and therefore in this case, viz. when the perpendicular falls within the triangle, and when the sum of the sides is less than a semi-circle, the least segment is adjacent to the least side.

2. Again, in the triangle FCA the two sides FC and CA are less than a semi-circle; for since AC is less than CB, AC and CF are less than BC and CF.

Also AC is less than CF, because it is more remote from CHG than CF is, therefore in this case, also.viz, when the perpendicular falls without the triangle, and when the sum of their sides is less than a semi-circle, the least segment of the base AD is adjacent to the least side.

3. But in the triangle FCE, the two sides FC and CE are greater than a semi-circle; for since FC is greater than CA, FC and CE, are greater than AC and CE. And because AC is less than CB, EC is greater than CF, and EC is therefore nearer to the perpendicular CHG than CF is, wherefore EG is the least segment of the base, and is adjacent to the greatest side.

4. In the triangle ECB, the two sides EC, CB are greater than a semi-circle, for since by supposition CB is great-greater than CA, EC and BC are greater than EC and CA.

Also, EC is greater than CB; wherefore in this case also, the least segment of the base FG is adjacent to the greatest side of the triangle. Wherefore, when the sum of the sides is greater than a semi-circle, the least segment of the base is adjacent to the greatest side, whether the perpendicular fall within or without the triangle: and it has been shewn, that when the sum of the sides is less than a semi-circle, the least segment of the base is adjacent to the least of the sides, whether the perpendicular fall within or without the triangle.*

Q. E. D.

PROPOSITION XVIII.

The sums of the opposite angles of a quadrilateral figure, on the sphere, which can be inscribed in a small circle, are equal

Pl. XVI. Fig. 7.

That is $A+C=B+D$.

* When the perpendiculars fall *without* the triangle, it is that which is nearest to the triangle to which the demonstration is applicable.

For if arcs of great circles pass through the Pole of the small circle and each angle of the quadrilateral, they will divide it into isosceles triangles, (Def. 9. Cor 7.) whose angles at the base in each will be equal, (IV.) and consequently the sums of equals will be equal.

That is PBA+PBC+PDA+PDC=PAB+PAD+PCD+PCB.

Or ABC+BDC=BAD+BCD.

Q. E. D.

PROPOSITION XIX.

If two spherical triangles, have two right angles, *in other words*, If their sides intersect each other in the Pole of their base, then the two triangles are to each other as their bases.

If a great circle should pass through P, and bisect the base, EG, of the triangle PEG, it would also bisect the triangle, (V) for the two triangles into which it would divide PEG would be equal. Also if any *multiple* of the base should be taken, the triangle formed by a great circle passing through its extremity and the Pole (P) would be the same *multiple* of the triangle PEG. Fig. 2.

Therefore the demonstration of Euclid (VI, 33.) concerning the angles at the centre of a circle, and the circular sectors, is applicable in every respect, to the angles at the Pole of a great circle on the sphere, and to the triangles formed between it, as a vertex, and the arcs of that circle as bases.

That is, the angles at the Pole, and also the triangles themselves, have to each other respectively, the same ratio, which their bases have.

Q. E. D.

Cor. 1.—In such triangles, the area, varies as the base, or as the angle at the Pole, of which the base is a measure; that is the base is a measure of the area of the triangle.

Cor. 2.—Hence, since the triangle E*p*G, equals EPG in every respect, therefore, the arc EG which is the measure of the angle at the Pole, is also the measure of the *lune* PE*p*G ; which is composed of the two triangles.

Thus, when the *lune* is right angled, or EG= a quadrant, it is one half of a hemisphere, and as EG approaches to a semi circle, the *lune* approaches to a Hemisphere.

Also a spherical triangle which has its three angles right, being the half of a right-angled *lune*, is equal to one fourth of a Hemisphere.

PROPOSITION XX.

As Four right angles, is to the excess of the angles of any spherical triangle above two right angles ; called the *Spherical Excess* ; so is the area of the Hemisphere, to the area of the triangle.

Let ABC be a spherical triangle, BCEF, be the circle, of which BC is an arc ; let ABDCA, be the *lune* formed by the two semi-circles, of which AB and AC are arcs, and the parts CAF, BAE, two semi-circles, meeting the circle BCEF, (I, Prop. I Cor. 10,) then the triangle EFA =BDC on the opposite Hemisphere (II) therefore the lune ABDC, which equals ABC+DBC, equals also ABC+AFE.

Pl. XVI. Fig. 6.

Also ABC+ABF= lune CAFB,
and ABC+ACE= lune BAEC,
and ABC+AFE+ABF+ACE= the Hemisphere.

Now these lunes are to each other respectively as their angles (XIX, Cor 2.), that is, as the angles of the spherical triangle, and each of them, is to the Hemisphere as its angle, is to two right angles,

That is, 2. right angles : angle A : : Hem. : ABC+AFE,
and 2. " " : " B : : Hem. : ABC+ACE,
" 2. " " : " C : : Hem. : ABC+ABF,
∴ two right angles are to the angles at A, B, C, taken together, as a Hemisphere is to 3 ABC+AFE+ACE+ABF ; therefore also two right angles, are to the *Excess* of A+B+C, above two right angles, as a Hemisphere is

5. 12.

5. 16. and D.

to the excess of 3ABC+AFE+ACE+ABF, above a Hemisphere, that is to 2 ABC.

Or 4 right angles : Spherical Excess :: 2 Hem. : 2ABC,
:: Hem. : ABC.

Q. E. D.

Cor. – As the first and third terms are constant, therefore, by equality of ratios the *Spherical Excess* is the *measure* of the area, of a spherical triangle.

III. INTERSECTIONS

IN THE

PLANE OF A GREAT CIRCLE OF THE SPHERE.

PROPOSITION I.

If a line be drawn from any point of the Sphere, to the pole of a great circle, the distance from the centre of that circle, to the point where the line meets its plane, is equal to the tangent * of half the distance of the given point, from the other pole of the circle.

Fig. 18. That is $Cf=pt$, where $pO=\frac{1}{2}pE$.

For the triangles $CPf=pCt$, are similar and equal in all respects.

Q. E. D.

PROPOSITION II.

If one extremity of a straight line, remain in the pole of a great circle, while the line is carried around in the circumference of any parallel circle, it will describe the convex surface of a right cone; and the intersection of this surface, with the plane of the great circle, is a circle whose centre will coincide with that of the Primitive circle.†

* The word *tangent* is here used, as in Trigonometry, for *that part of a line* touching a circle, which is intercepted between the point of contact, and any diameter of the circle produced.

Also, *distance* on the sphere, is used for *the arc of a great circle, which passes through any two points*; for this arc on the sphere, like a straight line, in a plane, is the least line which can be drawn between two points.

† The term *Primitive* is here applied to that great circle, in whose plane the intersections are supposed to be made.

1. The Figure FPI described by the motion of the line, is evidently a right cone. (See Euclid, Sup. III, def. 11.) Fig. 17.

2. Its intersection with the plane of the Primitive is a circle. (Con. Sect. def. 4.)

3. The centre will coincide with that of the Primitive, for the axis of the cone, PH, is the axis of the sphere and of the great circle EQ, through the centre of which, and of all circles parallel to it, it passes. (Def. 9, Cor. 3.)

Q. E. D.

Cor.—If the vertex of the cone, or P, be the *more distant* pole, the circular intersection will be *within* the Primitive, as *fhi*, and consequently less than the Primitive. But if the vertex be the *nearer* pole, the intersection will be *without* the Primitive, and larger as *f'i'*.

PROPOSITION III.

The same being supposed as in the last proposition, but the circle oblique to the Primitive, the figure described will be an oblique cone, and its intersection with the Primitive will be a circle.

1. The figure described FPI is an *oblique cone*. (Con. Sect. Appendix Def. 1.) Fig. 16.

2. Its intersection with the Primitive, *fhi*, is a circle, for it is a *sub-contrary* section, since IPx = Iif, and also = IFP. (Con. Sec. App. VI.) 1. 29. 3. 32.

Q. E. D.

Cor. 1.—If F be between E and P, a part of the intersection will be in the plane of the Primitive, and the remainder without it, and if the oblique circle is a *Great Circle*, the corresponding circular intersection, will cut the primitive in two points diametrically opposite. (Prop. I. Cor. 10.)

The preceding proposition and demonstration, are applicable to any small circle which is at right angles to the

Fig. 19. primitive as EFI, whose corresponding circular intersection is fhi.

PROPOSITION IV.

The centres of the circular intersections in the plane of the Primitive arc, in all cases, are in the line of mutual intersection of the Primitive and a plane passing through the centre of the sphere, and through the poles of the Primitive, and of the given circle, upon its surface.

Fig. 16. 17. & 19. That is, the centre of fhi is in QE, which is the intersection of the Primitive, and of IpE, or a great circle passing through the poles of the Primitive, and of the circle FHI.

For the circle IpDE, is perpendicular to the Primitive, and to the circle FHI, (Def. 9. Cor. 4.) and coincides with the axes of both, and consequently passes through the centres of both, bisecting both the circle FHI, on the sphere, and the corresponding circular intersection in the plane of the primitive.

Q. E. D.

Cor. 1.—If the circle on the sphere be an oblique great circle, the centre of the corresponding circular intersection will be in that diameter of the Primitive produced, which is perpendicular to its intersection with the oblique circle.

Fig. 23. 3. 1. That is, in QEi, which is perpendicular to FI.

Cor. 2.—Hence if any number of oblique great circles, cut the Primitive in the points F and I, the centres of all their circular intersections, will be in the same line. Viz. in QEi produced.

Fig. 20. Cor. 3.—In like manner if a small circle be at right angles at the Primitive, the centre of the corresponding circular intersection, will be that diameter of the Primitive, which is perpendicular to its intersection with the small circle.

And the centres of all such circles, will be in the same straight line, Ei produced.

Cor. 4.—If any number of great circles oblique to the Primitive, intersect each other in one point on the Sphere, their corresponding circular sections, will intersect each other in one point in the plane of the Primitive, and their centres will be in a straight line KL, perpendicular to the diameter of the Primitive which passes through the point of their mutual intersection, and passes through, h, the centre of the circular section whose chord, at its intersections with the Primitive, is parallel to KL. Fig. 21.

PROPOSITION V.

The distance between the centres of the Primitive and of any circular intersection of a great circle in its plane, is equal to the tangent of the arc which measures the inclination of the circles on the sphere, I. Prop. Cor. I.

That is $Ph=Qt$. Eig. 23.

For APQ, being the inclination of the circles on the sphere, and Qt, the tangent of QA, which measures their inclination,

Then QPA=FPK; & FPK+FPA=right angle.

Also $AIP+PIh+HIi=PIh+PhI$=right angle, 3. 31. & 1. 32.

$\therefore$ $AIP+hIi=PhI=hIi+hiI=2hIi$;

$\therefore$ $AIP=hIi$; & 2AIP, or $APF=2hIi=PhI$,

$\therefore$ $APF+PIh=FPK+APF$,

$\therefore$ $PIh=FPK=APQ$; & $Ph=Qt$.

Q. E. D.

Cor. 1.—Hence Ih the radius of the circular intersection $FfIi$, is equal to Pt, the *secant* of the arc QA, which measures the inclination of the cicles.

Cor. 2.—Pk, the distance from the centre of the Primitive to the point, where a line connecting the pole of the oblique circle with that of the Primitive, meets the plane of the latter, is equal to the tangent of half the the angle of their inclination.

For $PIk=\frac{1}{2}FPK=\frac{1}{2}APQ$.

PROPOSITION VI.

If a small circle be at right angles to the Primitive, the radius of its circular intersection, will be the tangent of its distance from the nearer pole.

Fig. 20. That is, Fh is the tangent of FE.

For xDF=DBF=PFB,
And xDi=hiF=hFi,

3. 31. ∴ PFB=hFi, ∴ PFh=fFi=right angle.
∴ Fh is the tangent of FE.

Cor.—Ph the distance of the centre of the circle Ff Ii, from that of the Primitive, is the *Secant* of FE, the distance of the small circle from its nearest pole.

PROPOSITION VII.

The angle made by two circles which cut each other on the sphere, is equal to the angle made by the corresponding circular intersections in the plane of the Primtive.

Fig. 22. That is, DFK, or D,FK′=dfk or D′fK′.

For suppose planes to coinoide with FK and FD, the *tangents* to the circles, at the point of their intersection, and to pass through P; these planes will *touch* the convex surfaces of the cones, whose bases are the two circles. Also let another plane pass through P, and cut those tangents, in K and D, forming a pyramid PKDG, which is cut by a plane KDG, parallel to the primitive. Let FI be parallel to EQ.

3. 32. Then PIF=PFK′=GFK, also PIF=PFI=FGK;
Therefore FK=KG.
In like manner FD=DG.

Therefore $KFD=KGD=kfd$, in a plane parallel to KGD. But since the planes PFK, PFD *touch* the cones, their intersection with the primitive, will be straight lines *touching* the circular intersections, in the same plane. And the angle made by two circles which cut each other, in a plane, is here considered as equal, or is the same as the angles made by lines touching them at the point of intersection. 1. 8.

In most cases, the demonstration will be much more concise and simple, by taking the vertical angle of the tangents, viz: $D'FK'$, and proving its equality with $D'fK'$ in the plane of the Primitive.

For $D'FP=FIP=D'fF$, therefore $D'F=D'f$.

In like manner $K'F=K'f$, and therefore, $D'FK'$ $D'fK'$.

Q. E. D.

PROPOSITION VIII.

If on the sphere, any number of small circles pass through one pole of the Primitive, and the farther pole of any oblique great circle, their circular intersections will all pass through the centre of the Primitive, and also through the point, where a line drawn from the other pole of the Primitive through the pole of the oblique circle, meets the plane of the Primitive. Pl. XIV. Fig. 4&5.

That is, the circular intersections, will all pass through P and p'.

This is evident from the demonstrations of the preceding propositions.

Cor. 1.—MO, ON are respectively equal to the arcs *mo*, *on*, of the oblique circle on the sphere.

Cor. 2.—The intersections of the *planes* of the small circles, with the Primitive, are straight lines which intersect each other, in the point (p'') where a line drawn from the given pole of the oblique circle, to that of the Primitive, meets the plane of the latter. Fig. 4.

Pl. XI. Fig. 12.

PB, PN, PX, are the intersections of the pimitive, and of the planes of small circles, passing through its pole, and the farther pole of the oblique circle **CRSOD.**

Cor. 3.—**RS** and **SO** correspond to arcs of the oblique circle on the sphere, respectively equal to **XN** and **NB.**

APPENDIX TO SPHERICAL GEOMETRY.

PROJECTIONS.

Def. 1.—To *Project an object*, is to represent every point of it, on a plane, as it appears to the eye in a certain position.

2.—The *Plane of Projection*, is that on which the object is represented.

3.—The *Projecting Point*, is that point where the eye is supposed to be placed.

4.—The *Orthographic Projection* of the sphere is that in which a great circle is assumed as the plane of projection, and a point at an infinite distance in the axis produced, as the projecting point.

5.—The *Stereographic Projection* of the sphere is that, in which a great circle is assumed as the plane of projection, and one of its poles, as the projecting point.

6.—The *Gnomonic Projection* of the sphere, is that in which the plane of projection touches the sphere, and the centre is the projecting point.

Schol.—In the Theory of Projections, the rays of light are supposed to move in straight lines, from any point of an object to the eye, and the projection of each point, is the intersection of such a line, and the plane of projection.

7.—*A direct circle* is parallel to the plane of projection.

8.—*An oblique circle* is oblique to the plane of projection.

9.—*A right circle*, is that whose plane coincides with the axis of the eye.

Cor.—A Great Circle which is *right*, is perpendicular to the plane of projection.

The following results or Laws of Projection will follow as Corollaries, from the preceding definitions and Scholium.

I. *Of Orthographic Projection.*

1.—The Rays of light being supposed to come from an indefinite distance, may be considered as *parallel* to each other, and *perpendicular* to the plane of projection.

2.—A straight line, perpendicular to the axis of the eye, is projected into a point.

3.—A straight line parallel to the plane of Projection, is projected into a line equal to itself.

4.—A straight line oblique to the plane of Projection is projected into a line less than itself, in the ratio of the sine of the angle which it makes with any ray of light to radius.

6.—So a plane, perpendicular to the plane of Projection, is projected into a straight line, and a plane parallel to the plane of Projection, is projected into a plane equal and similar to itself.

Pl. VIII. Fig. 11. 7.—A circle oblique to the plane of Projection is projected into an Ellipse.

For let AMB represent the circle to be projected, and ALB the figure into which it is projected. Any ordinates in the circle will be projected into lines less than themselves, in the same ratio. (4.)

That is, KM : DG : : KL : DE ;
which is the property of an Ellipse.

As Orthographic Projection, is not necessarily limited in its application, to the sphere, having no particular relation to it, and is principally used in a more general application to mathematical figures, it is sufficient here, to give a mere sketch of its general properties.

II. *Stereographic Projection*, is confined to the sphere, and on many accounts, is the most convenient method of representing on a plane, figures on the surface of the sphere.

By comparing the preceding definitions with the propositions, in the third part of Spherical Geometry, it will be evident,

1.—That all circles, on the surface of the sphere, are projected into straight lines, or into circles. This renders the practical operations in this method of projection very easy.

2.—That these projected circles, in all cases, make the same angle, on the plane of Projection, which the circles to be projected make on the sphere.

3.—That the *centres* of projected circles and their projected *poles* are accurately and easily found; by geometrical operations.

III. *Gnomonic Projection*, has a particular reference to the sphere, but is of a very limited application ; being used to explain the theory of Dialing, or the Geometrical construction of Dials.

If the Earth were supposed transparent, and its axis capable of casting a shadow, this shadow from the Earth's rotation would coincide successively with the planes of different Meridians ; and as the rotation of the Earth is supposed to be uniform, the shadow, would in equal times, coincide with Meridians at equal distances from each other, or which make with each other equal angles.

If successive meridians were taken, making angles of 15° with each other, the shadow of the axis would move from one to the other is one hour, or $\frac{1}{24}$ part of an entire revolution. Such Meridians are called *hour circles*.

If the planes of these Meridians or *hour circles*, were produced to intersect a plane touching the sphere, their mutual intersections would be straight lines, which to an eye placed at the centre of the sphere, would coincide with the planes of the Meridians. They would therefore be the *Projections* of those planes, or of their *circumferences* on the surface of the Sphere. Hence in every kind of Dial, the *edge* or *part of the Gnomon*, which casts the shadow must be placed *parallel* to, or *coinciding* with the Earth's axis, and supposing a sphere to surround it, the *hour lines* are the intersections of the hour circles with the dial plate, which is the plane of projection.

Pl. XV. Fig. 1. Thus in the common Horizontal Dial, CAB, is a triangular plate, of which the edge, which casts the shadow, is CA, and the angle CAB being made equal to the latitude of the place, i. e. to the elevation of the pole, CA will be parallel to the axis of the earth. Suppose a sphere to surround CA, as its axis, whose centre is D, then DF perpendicular to CA, will be the intersection of its equator with the plane CAB, and HFG will be the intersection with the Dial plate.

Making FI and *fi* each equal to FD, the quadrants, FK, *fk* will represent half the Equator, or a circle concentric with it, equally divided by I*m*, I*m* &c. *in*, *in*, &c. which represent meridians, or hour circles; F*m*, F*m*, &c. also *fn*, *fn*, &c. are the intersections of these circles, with the line HG, in the plane of the Dial.

Hour lines drawn from A and *a*, through *m*, *m*, &c. *n*, *n*, &c. will intersect the circumference of the Dial plate, in the divisions which mark the hours.

The same principles, will serve for the construction and explanation of all Dials.

APPENDIX TO SPHERICAL GEOMETRY.

PROBLEMS

IN

STEREOGRAPHIC PROJECTION.

PROBLEM I.

To project a great Circle, of which a given point is the projected Pole.

1. *The given point being the centre of the Primitive.*

Then the Primitive itself is the circle required.

2. *The given point being in the circumference of the Primitive.*

Let the given point be A. Draw through the centre of the Primitive the diameter ACE, and a perpendicular diameter BCD, which is the projected circle, whose poles are A and E. Fig. 1.

Also B and D are the poles of the projected circle ACE.

3. *The given point, being in the plane of the Primitive.*

Let the given point be p. Draw the diameter BpD, and its perpendicular AC. From A reduce p to G, in the circumference of the Primitive; make GF equal to a quadrant, project F in E; a circle through A, E and C, is the projected circle, whose pole is p. Fig. 2.

COR.—By reversing the process in each case, the projected pole of a given projected circle may be found.

PROBLEM II.

To Project a Small Circle, parallel to a given projected Great Circle, and at a given distance from it.

1. *The given Great Circle being the Primitive.*

Fig. 15. Draw any diameters as BCD, and its perpendicular. From B set the given distance to O; project O in X; then the circle XFE, is the circle required.

2. *The given Circle being a right Circle.*

Fig. 16. Let the given projected circle, be BD. Set the given distance from B to F, project F in G on the perpendicular diameter AY; a tangent to the primitive, at F, will intersect AY produced in (K) the centre of the required circle, which will pass through F and G.

3. *The given Circle being oblique.*

Fig. 17. Let the given oblique circle, be AED. Find its Pole (Prob. I. Cor.) P; reduce P to N, and from N set off on each side, the complement of the given distance of the small circle, to R and Q; project R and Q in X and Y, XY is the diameter of the small circle (BXY) required.

COR.—By reversing the process, the pole of a given small circle, parallel to a large circle, may be found, and also the distance of the small circle from its large circle.

PROBLEM III.

To project a *great circle, through two given points* in the *Primitive.*

1. *One of the points being in the centre of the Primitive.*

Let the given points be C and A, or C and R. Then the diameter ACE or YRB will be the circle required. Fig. 7.

2. *One of the points being in the circumference of the Primitive.*

Let the given points be C and K. Draw the diameter CXD, the circle required will pass through C, K and D. Fig. 8.

3. *Neither of the points being in the centre or circumference of the Primitive.*

Let the given points be P and Y. Draw the diameter BPA and its perpendicular CD, reduce P to G; the required circle will pass through, YP and G. Fig. 9.

PROBLEM IV.

Through any given point, to project two circles which shall make a given angle.

1. *When the angle is at the centre of the Primitive.*

Draw any diameter as ACE, set the measure of the given angle from A to H, draw HC, then AC and HC are projected circles making the required angle at C. Fig. 1.

2. When the angular point is in the circumference of the Primitive.

Fig. 3. Let A be the angular point. Draw the diameter AC, and the perpendicular DB. Set the measure of the given angle from D to E; project E to F, AFC, is a projected circle which makes the given angle with the Primitive at A.

3. When the angular point is in the plane of the Primitive.

Fig. 5. Let O be the given point. Draw the perpendicular diameters, AOB and CD. From C reduce O to I; make DK equal to AI; project K in P, CPD is the projected circle, of which O is the projected pole. (Prob. I.) Set the given angle from C to L, let a line from L to O (which will be the projection of a small right circle) intersect the circle CPD in Q, from Q as a pole, project the great circle EOF, (Prob. I.) which makes the given angle EOA, with the right circle AOB.

4. When the circles required are both oblique.

Fig. 14. Let E be the given point. Through E project any oblique great circle, as AED. Find its pole, P. Let the straight line EP intersect the Primitive in T; set off TV the measure of the given angle; draw the diameter VY, and its perpendicular FN, NEA, is the required angle, NEF, AED, the required circles.

Cor.—By reversing the process, in each case the angle made by two given projected great circles can be measured.

PROBLEM V.

Through a given point to project a great circle perpendicular to a given great circle.

1. When the given circle is the Primitive.

Let Y or X be the given point. The diameter YCB, or AXCE, will be the circle required. Fig. 7.

2. When the given circle is right, and the given point the centre of the Primitive.

Let AB be the given projected circle, the perpendicular CXD is the circle required. Fig. 8.

3. When the given circle is right, and the given point not in the centre of the Primitive.

Let the given projected circle be AB, and K, in it, the given point. Draw the perpendicular diameter CD, then the required circle will pass through C, K, and D.

4. When the given circle is oblique, and the given point in the middle.

Let DCK be the given oblique circle, and K the given point. The diameter AKB, is the required projected circle. Fig. 8.

5. When the given circle is oblique, and the given point not in the middle of it.

Let Y be the given point in CYD. Find the poles P, *p* of CYD, and PY*p*, is the circle required. Fig. 9.

PROBLEM VI.

To project a great circle, through a given point which shall make a given angle with the Primitive.

Let O be the given point. Describe the circle CPD, whose pole is O, as in Prob. IV. 3. Make BL equal to Fig. 5.

the given angle, project L in M; and describe the small circle EQN, which is parallel to the Primitive and whose distance from the centre is equal to the given angle, intersecting CPD, in Q. From Q as a projected pole, describe the circle EOF, which makes the required angle OEA.

Cor.—Conversely BL is the measure of the angle, AEO.

PROBLEM VII.

To project a great circle, which shall make given angles with two given circles.

1. *When one of the given angles is a right-angle.*

Fig. 3. Let the Primitive and the right circle DFB, be the given circles. It is required to project a great circle which shall make right angles with DB, and any given angle with the Primitive. Draw AC perpendicular to DB. Set off the given angle from D to E, project E in F, and CFA, is the circle required, F and A the required angles.

2. *When neither of the angles are right.*

Fig. 32. Let the Primitive and AB*d* be the given circles.

Find *q* the pole of AB*d*, around it describe a small circle at the distance of the angle B. (Prob. II.) Set the angle C from *d* to *b*, project *b* in *a*, around the centre or pole of the Primitive, describe a small circle at the distance of *a*, intersecting the other small circle in *s*, which is the pole of the circle required, viz, CB*e* which describe by Prob. I.

PROBLEM VIII.

"*To set off any number of degrees on a projected circle.*" Or, *to cut off from a projected circle an arc. which corresponds to an arc of the circle to be projected equal to a given arc.*

1. *When the projected circle is the Primitive, or one of its parallels.*

Make DG, equal to the given arc, it is the arc required. Fig. 15.
EF on a parallel circle, is an arc of the same "number of degrees," or it is the same portion of the circle EFX, that DG is of DAB.

2. *When the projected circle is a right circle.*

Let the projected circle be ACB; make AE or DH, RN or RX, equal to the given arc; then AF or CG, OH or OG will be the required arcs. Fig. 10 & 11.

If TU be made equal to the given arc, then OB will be the required arc, on the great circle, and XY, a proportional arc, on the parallel small circle MN. Fig. 18.

3. *When the projected circle is oblique.*

Let the projected circle be CSD. Find its pole P. (Prob. I. Cor.) Draw PSN, from N, cut off NB or NX equal to the given arc; draw PB and PX, intersecting CSD in R and O; SR or SO, are the arcs required. Fig. 12.
Here PX and PB, represent a small circle passing through the remotest poles of the Primitive, and of the oblique circle CSD. (Sp. Geom. III. Prop. VIII. Cor. 2.)

If CPA, represent the given oblique circle, and NL the given arc, then OQ will be the required arc on the great circle, and XY the corresponding arc on a small circle parallel to it. Fig. 19.

Cor.—Conversely, if the arc of a projected circle be given, the corresponding arc, on the circle to be projected may be found and measured.
Thus AE and DH, RN and RX are the measures of AF and CG, OH and OG. Fig. 10 & 11.
Also TU, measures OB and XY, and LN measures OQ and XY. 18 & 19.

PROBLEM IX.

To project a Hemisphere on the plane of the Equator

Pl. XIV. Let the Primitive ENWS, represent the Equator.

Draw the perpendicular diameters EW, NS.

Divide each quadrant into nine equal parts, then the lines drawn from their division to the centre, will be the projection of Meridians.

Project parallels of Latitude, corresponding to the divisions of the Primitive, (Prob. II.) also the Tropic and Polar circles, at the proper distances from the centre or pole of the Primitive; The Ecliptic may be projected by Prob. IV.

PROBLEM X.

To project a Hemisphere on the plane of the Meridian.

Describe the Primitive and divide it as before. Let it represent the solstitial colure.

WE is the projected Equator. Project parallels to it, with the Tropics and Polar circles, by Prob. II.

Tangents to the Primitive, at the several divisions, will intersect NS produced in the centres of their parallel circles, as in y, y.

The meridians may be projected by Prob. IV. (their centres are all in WE produced.)

The projection of the Ecliptic is obvious.

SPHERICAL TRIGONOMETRY.

PART I.

Geometrical Principles of Spherical Trigonometry; or the mutual relations of the Trigonometrical lines, corresponding to the arcs and angles of Spherical Triangles.

PROPOSITION I.

In right angled spherical triangles, the sine of either of the sides about the right angle, is to the radius of the sphere, as the tangent of the remaining side is to the tangent of the angle opposite to that side.

Let ABC be a triangle having the right angle at A. Then, sine of AB : rad : : tang. AC : tang. ABC.

Pl. X. Fig. 24.

Let D be the centre of the sphere; join AB, AD, AC, and let AF be drawn perpendicular to BD, which therefore will be the sine of the arc AB; and from the point F, let there be drawn in the plane BDC the straight line FE at right angles to BD, meeting DC in E, and let AE be joined. Since therefore the straight line DF is at right angles to both FA and FE, it will also be at right angles to the plane AEF, wherefore the plane ABD, which passes through DF is perpendicular to the plane AEF, and the plane AEF, is perpendicular to ABD: But the plane ACD or AED, is also perpendicular to the same ABD because the spherical angle BAC is a right angle: Therefore AE, the common section of the planes AED, AEF is it at right angles to the plane ABD and EAF, EAD are right angles.

Sup. 2. 4.
" " 17.
Sup. 2. 18.

Therefore AE is the tangent of the arc AC and in the rectilineal triangle AFE having a right angle at A,

AF : rad.::AE : tang. AFE,

but AF=sine of AB, and AE=tang. of AC, and AFE= [ABC.

∴ sin AB : rad.::tang. AC : ABC.

Q. E. D.

Cor.—Since sin AB : rad::tang. AC : tang. ABC, and (Pl. Trig. 93.) R : cot. ABC::tang. ABC : rad, ∴ Sin. AB : cot. ABC::tang. AC : rad.

PROPOSITION II.

In right angled Spherical Triangles, the sine of the Hypotenuse is to the radius, as the sine of either side is to the sine of the angle opposite to that side.

Fig. 25. That is, sin BC : rad::sin AC : sin ABC.

Let D be the centre of the sphere, and let CE be drawn perpendicular to DB, which will therefore be the sine of the Hypotenuse BC: and from the point E let there be drawn in the plane ABD the straight line EF, perpendicular to BD, and let CF be joined: then CF will be at right angles with the plane ABD. Sup.2.18.

Wherefore CFD, CFE are right angles, and CF is the sine of the arc AC: and in the triangle CFE, having the right angle CFE,

CE : rad::CF : sin CEF,

But CEF=ABC, (Spher. Geom. II Def. 1. cor. 5.)

∴ sin. BC : rad::sin AC : sin ABC.

Q. E. D.

PROPOSITION III.

In right-angled spherical triangles, the cosine of the Hypotenuse is to the radius, as the cotangent of either of the angles is to the tangent of the remaining angle.

That is, cos. BC : rad.::cot. ABC : tan. ACB. Fig. 26.

Describe the circle DE, of which B is the pole, and let it meet AC in F, and the circle BC in E; and since the circle BD passes through the pole B, of the circle DF, DF must pass through the pole of BD.

And since AC is perpendicular to BD, (Sp. Geom. Def. 9, cor. 5.) therefore AC must also pass through the pole of BAD; wherefore the pole of the circle BAD is in the point F, where the circles AC, DE intersect.

The arcs FA, FD are therefore quadrants, and likewise the arcs BD, BE. Therefore, in the triangle CEF right-angled at the point E, CE is the complement of BC, the Hypotenuse of the triangle ABC; EF is the complement of the arc ED, the measure of the angle ABC; and FC, the Hypotenuse of the triangle CEF, is the complement of AC; and the arc AD, which is the measure of the angle CFE, is the complement of AB.

But (I) Sin. CE : R::tan. EF : tan. ECF,

That is, cos. BC : R::cot. ABC : tan. ACB.

Q. E. D.

Cor.—Since cot. ACB : R::R : tan. ACB, Pl. Trig. 93.

$\therefore$ cot. ACB : cos. BC::R : cot. ABC.

PROPOSITION IV.

In right-angled spherical triangles, the cosine of an angle is to the radius, as the tangent of the side adjacent to that angle, is to the tangent of the Hypotenuse.

That is, cos. ABC : rad.::tang. AB : tang. BC. Fig. 26.

For (I) sine. FE : rad.::tang. CE : tan. CFE,
But sine EF=cos. ABC, tang. CE=cot. BC, and tang. CFE=cot. AB.
∴ cos. ABC : rad.::cot. BC : cot. AB.

Again cot. BC : rad.::rad. : tang. BC,
And cot. AB : rad.::rad. : tang. AB,
∴ cot. BC : cot. AB::tang. AB. : tang. BC;
Therefore, cos. ABC : rad.::tang. AB : tang. BC.

Q. E. D.

Cor. 1.—From the demonstration it is manifest, that the tangents of any two arcs AB, BC are reciprocally proportional to their cotangents.

Cor. 2—Cos. ABC : cot. BC::tan. AB : rad.

PROPOSITION V.

In right-angled spherical triangles, tne cosine of either of the sides is to the radius, as the cosine of the Hypotenuse, is to the cosine of the other side.

That is, cos. CA : rad. : cos. BC : cos. AB.

For (II) Sin. CF : rad.::sin. CE : sin. CFE,
But Sin. CF=cos. CA, sin. CE=cos. BC, and sin. CFE=[cos. AB.
∴ cos. A : rad.::cos. BC : cos. AB.

Q. E. D.

PROPOSITION VI.

In right-angled spherical triangles, the cosine of either of the sides is to the radius, as the cosine of the angle opposite to that side, is to the sine of the other angle.

That is, cosine CA : rad::cosine ABC : sin. BCA.

For (II) sin. CF : rad. :: sin. EF : sin. ECF,
But " sin. CF = cos. CA, sin. EF = cos. ABC, and sin. ECF = sin BCA.
∴ cos. CA : rad. :: cos. ABC : sin. BCA.

Q. E. D.

PROPOSITION VII.

In spherical triangles, whether right-angled, or oblique-angled, the sines of their sides are proportional to the sines of the angles opposite to them.

That is, sin. AC : sin. B :: sin. AB : sin. C. Fig. 13.

1. For (II) sin. BC : rad. (= sin. A.) :: sin. AC : sin. B,
Also " sin. BC : sin. A :: sin. AB : sin. C,
∴ " sin. AC : sin. B :: sin. AB : sin. C.

2. Then, sin. BC : sin AC :: sin A : sin B, Fig. 14. & 15.
Through C, draw a perpendicular CD to the opposite side, then
(II) sin. BC : rad. :: sin. CD : sin. B,
" sin. AC : rad. :: sin. CD : sin. A.
∴ sin. BC : sin. AC :: sin. A : sin. B.

In like manner, sin. BC : sin. AB :: sin. A : sin. C.
Q. E. D.

PROPOSITION VIII.

In oblique angled spherical triangles, a perpendicular arc being drawn from any of the angles on the opposite side, the cosine of the angles at the base, are proportional to the sines of the segments of the vertical angle. Fig. 14.

Let ABC be a triangle, and the arc CD perpendicular to the base BA.
Then, cos. B : cos. A :: sin. BCD : sin. ACD.

For (VI) cos. CD : rad. :: cos. B : sin. DCB,
and " cos. CD : rad. :: cos. A : sin. ACD,
∴ cos. B : cos. A :: sin. DCB : sin ACD.

Q. E. D.

PROPOSITION IX.

The same things remaining, the cosines of the sides are proportional to the cosines of the segments of the base.

Fig. 14. That is, cos. BC : cos. AC :: cos. BD : cos. AD.

For (V) cos. BC : cos. BD :: cos. DC : rad.
And " cos. AC : cos. AD :: cos. DC : rad.
∴ cos. BC : cos. BD :: cos. AC : cos. AD,
And cos. BC : cos. AC :: cos. BD : cos. AD.

Q. E. D.

PROPOSITION X.

The same construction remaining, the sines of the segments of the base are reciprocally proportional to the tangents of the angles at the base.

That is, sin. BD : sin. AD :: tan. A : tan. B.

For (I) sin. BD : rad. :: tan. DC : tan. B,
and " sin. AD : rad. :: tan. BC : tan. A,
∴ " sin. BD : sin. AD :: tan. A : tan B.

Q. E. D.

PROPOSITION XI.

The same construction remaining, the cosines of the segments of the vertical angle, are reciprocally proportional to the tangents of the sides.

That is, cos. BCD : cos. ACD : : tan. AC : tan. BC.

For (IV) cos. BCD : R : : tan. CD : tan. BC,
And " cos. ACD : R : : tan. CD : tan. AC,
∴ cos. BCD : cos. ACD : : tan. AC : tan. BC.

Q. E. D.

PROPOSITION XII.

The tangent of half the sum of the segments of the base, is to the tangent of half the sum of the sides, as the tangent of half the difference of the sides, is to the tangent of half the difference of the segments of the base.

Then, $\tan. \frac{1}{2}\overline{BD+AD} : \tan. \frac{1}{2}\overline{BC+AC} :: \tan. \frac{1}{2}\overline{BC-AC} : \tan. \frac{1}{2}\overline{BD-AD}$.

For (IX.) cos. BC : cos. AC : : cos. BD+cos. AD,
∴ cos BC+cosAC : cos. BC − cos. AC : : cos. BD+cos. AD : cos. BD − cos. AD.
But $\cos. BC+\cos. AC : \cos. BC-\cos. AC :: \cot. \frac{1}{2}\overline{BC+AC} : \tan. \frac{1}{1}\overline{BC-AC}$,*
Also, $\cos. BD+\cos. AD : \cos. BD-\cos. AD :: \cot. \frac{1}{2}\overline{BD+AD} : \tan. \frac{1}{2}\overline{BD-AD}$.

∴ $\cot. \frac{1}{2}\overline{BB+AC} : \tan. \frac{1}{2}\overline{BC-AC} :: \cot. \frac{1}{2}\overline{BD+AD} : \tan. \frac{1}{2}\overline{BD-AD}$.

But, $\tan. \frac{1}{2}\overline{BC+AC}\times\cot. \frac{1}{2}\overline{BC+AC} : \tan. \frac{1}{2}\overline{BC+AC}\times\tan. \frac{1}{2}\overline{BC-AC} :: \tan. \frac{1}{2}\overline{BD+AD}\times\cot. \frac{1}{2}\overline{BD+AD} : \tan. \frac{1}{2}\overline{BD+AD}\times\tan. \frac{1}{2}\overline{BD-AD}$, 6. 1.

But, $\tan. \frac{1}{2}\overline{BC+AC}\times\cot. \frac{1}{2}\overline{BC+AC} = \text{rad.} = \tan. \frac{1}{2}\overline{BD+AD}\ \cot. \frac{1}{2}\overline{BD+AD}$,

∴ $\tan. \frac{1}{2}\overline{BC+AC}\times\tan. \frac{1}{2}\overline{BC-AC} = \tan. \frac{1}{2}\overline{BD+AD}\times\tan. \frac{1}{2}\overline{BD-AD}$,

* Plane Trig.

Or $\tan. \frac{1}{2}\overline{BD+AD} : \tan. \frac{1}{2}\overline{BC+AC} :: \tan. \frac{1}{2}\overline{BC-AC} : \tan. \frac{1}{2}\overline{BD-AD}$.

Q. E. D.

Schol.—The preceding proposition will be easily remembered, from its resemblance to the corresponding proposition in Plane Trigonometry.

PROPOSITION XIII. DAT.

If the ratios of the radius of any circle, to the trigonometrical lines corresponding to any given arc, were known, and conversely; then if any two of the parts of a right-angled spherical triangle were given, the remaining parts would also

hypotenuse and either of the adjoining angles were given, the sides and the remaining angle, would be given also.

Pl. XII.

Fig. 21. That is, if AC, and A are given, AB, BC, and C may be found.

1. Because the radius, the sine of AC, and the sine of A are given, *the sine of BC and consequently the arc BC, may be found.*

For (II) R : sine AC::sin. A : sin. BC; and a fourth proportional to any three straight lines may be found. (6. 12.)

Therefore the sine of BC may be found.

But as the radius, and the sine of BC are given, their *ratio* is given, and therefore by supposition the arc BC is given also.

It is to be remarked that any sine may be the sine of two arcs which are supplemental to each other, but as A is given, it is known of *what affection it is*, i. e. whether it is greater or less than a right-angle, and BC is of the same affection, (Sp. Geom. XIV.) therefore BC is given without ambiguity.

2. In a similar manner *AB, may be found.*
For (IV) R : cos. A : : tang. AC : tang. AB.

If AC be less than a quadrant, AB is of the same affection as A which is given. If AC be greater than a quadrant, AB, and A, are of different affections, in either case AB is given without ambiguity. (Sp. Geom. XIV. XV.)

3. Also, C *may be found.*
For (III) R : cos. AC : : tang. A : cot. C.

If AC is less than a quadrant, C is of the same affection as A; otherwise they are of different affections. In either case C is given unambiguously.
In the same manner, if AC and C are given, BC, AB, and B may be found.

II. If the hypotenuse and either of the sides are given the angles and the remaining sides are given.
That is, if AC, BC, are given, AB, A and C may be found.

1. *The side AB may be found.*
For (V.) cos. BC : cos. AC : : R : cos. AB.

If AC is less than a quadrant, AB is of the same affection as BC, if not, they are of different affections.

2. *The angle A may be found.*
For (II) sine AC : sine BC : : R : sin. A.
The angle A is of the same affection as BC.

3. *The angle* C *may be found.*
For (IV) tangent AC : tan. BC : : R : cos. C.

If AC is less than a quadrant, BC and C, are of the same affection, otherwise they are of different affections.

In the same manner, if AC and AB are given, BC, A and C may be found.

III. If one side and an adjacent angle are given, the hypotenuse and the remaining side and angle, are given.

That is, if BC and C are given, AC, AB and A may be found.

1. *The hypotenuse AC may be found.*

For (IV) cos. C : R::tan BC : tan. AC.

If BC and C, (both of which are given,) be of the same affection, AC is less than a quadrant, otherwise it is greater.

2. *The side AB may be found.*

For (I.) R : sin. BC::tan. C : tan AB.

AB is of the same affection as C.

3. *The angle A may be found.*

For (VI) R : cos. BC::sin. C : cos. A.

The angle A is of the same affection as BC.

In like manner, if AB and A are given, AC, BC and C may be found.

IV. If one side, and the angle opposite are given, the hypotenuse and the remaining side and angle are given.

That is, if BC and A are given, AC, AB and B may be found.

1. *The hypotenuse AC may be found.*

For (II) sin. A : sin. BC::R : sin. AC.

The hypotenuse may be either the arc AC, or its supplement C*d*, for both have the same sine.

The two triangles ABC, and BC*d* have each the parts given in this case. The hypotenuse therefore is given *ambiguously*. It is one of two given arcs. If the affection of either of the remaining parts of the triangle, should be given, the ambiguity would be removed.

2. *The side AB may be found.*

For (I) tan. A : tan. BC::R : sin AB.

The arc is ambiguous, being either AB or its supplement B*d*. The sine of both being the same.

3. *The angle C may be found.*

For, (VI) cos. BC : cos. A::R : sin. C, which is ambiguous, being either BCA, or BC*d*, which have the same sine.

In like manner if AB and C are given, AC, BC, and A may be found.

V. If the two sides are given, the hypotenuse and angles are given.

That is, if AB and BC are given, AC, A and C may be found.

1. *The hypotenuse AC may be found.*

For, (V) R : cos. AB::cos. BC : cos. AC, which is less than a quadrant, when AB and BC, are of the same affection, but greater than a quadrant when they are of different affections. (Sp. Geom. XV.)

2. *The angle A may be found.*

For, (I) sin. AB : R::tan. BC : tan. A; which is of the same affection as BC.

3. *In the same manner B may be found.*

For, (I) sin. BC : R::tang. AB : tan. C; which is of the same affection as AB.

VI. If the two angles are given, the Hypotenuse and the opposite sides are also given.

That is, if A and C are given, AC, AB, and BC may be, found.

1. *The hypotenuse AC may be found.*

For, (III) tan. A : cot. C::R : cos. AC.

When B and C are of the same affection, AC is less than a quadrant; but when they are of different affections, AC is greater than a quadrant. (Sp. Geom. XIV. XV.)

2. *The side AB may be found.*

For (VI) sin. A : cos. C::R : cos.AB; which is of the same affection as C.

3. *In the same manner BC may be found.*

For, (VI) sin. C : cos. A : : R : cos. BC, which is of the same affection as A.

PROPOSITION XIV. DAT.

The same being supposed as before, if any three of the parts of an oblique angled triangle be given, the other three are also given.

Fig 32. I. If two sides and the included angle are given, the remaining side and angles are also given.

That is, if AC and CB and C, are given, AB and A and B ; may be found.

1. *The angle B, may be found.*

For let fall the perpendicular AP, from the other unknown angle A, on BC,

Then, (IV) R : cos. C : : tan. AC : tan. CP ; which is therefore given,

And (X) sin. BP : sin. CP : : tan. C : tan B ; which is of the same affection as A when the perpendicular falls within the triangle, but if it falls without, B and A are of different affections.

In like manner, the angle A may be found ; or each of its parts CAP, and PAB, may be found ; as in case III, Prop. XIII Dat.

2. *The side AB may be found.*

For (IV) R : cos. C : : tan. AC : tan. CP,

And (IX) cos. CP : cos. BP : : cos. AC : cos. AB ; which is of the same affection as AC, when AP falls within the triangle, but different when it falls without. (Sp. Geom. XVI.)

Or AB, may be found, after B is found, by Prop. VII.

II. If two angles and the included line are given, the remaining sides and angle are given.

That is, if A, C, and AC, are given, AB, CB and B may be found.

1. *If AB may be found.*

For, letting fall the perpendicular as before,

Then (III) R : cos. AC : : tan. C : cot. CAP,

Therefore BAP may be found (=CAB − CAP,)

which is of the same affect as C, when AC is less than a quadrant.

and (XI) cos. BAP : cos. CAP : : tan. AC : tan. AB.

AB, is of the same affection as AC, when the perpendicular falls within the triangle, *i. e.* when $CAP < CKB$.

In like manner, BC may be found.

Or, after AB is found, by Prop. VII.

2. *The third angle B may be found.*

For, (III) R : cos. AC : : tan. C : cot. CAP.

and (VIII) sin. CAP : sin. BAP : : cos. C : cos. B; which is of the same affection as C, when AP falls within the triangle, but they are of different affections, when AP falls without.

Or, when AB is found, B may be found, by Prop. VII.

III. If two sides, and an angle opposite to one of them be given, the remaining angles and side are given.

That is, if AC and AB and C, be given, B, A, and BC may be found.

1. *The angle B opposite AC may be found.*

For (VII) sin. AB : sin. AC : : sin. C : sin. B.

When $AB + AC > Ce, C > ABe$,

But,, $AB + AC < Ce, C < ABe$, (Sp. Geom. XII. & Cor.)

This proposition may *sometimes* determine whether B is acute or obtuse; when it does not, B is ambiguous.

2. *The angle A is given.*

For, the perpendicular being drawn from A,

Then (III) R : cos. AC : : tan. C : cot. CAP.

And (XI) tan. AB : tan. AC : : cos. CAP : cos. BAP.

Therefore CAP and BAP being given, CAB is given, but it may be either acute or obtuse, as the perpendicular falls within or without the triangle.

It is therefore ambiguous.

3. *The third side BC is given.*

For, (IV) R : cos. C : : tan. AC : tan. CP.

And (IX) cos. AC : cos. AB : : cos. CP : cos. BP.

CB may therefore be found, but it is ambiguous, as AP, may fall within or without the triangle.

Or, after A is given, B*e* may be found, by Prop. VII.

IV. If two angles and a side opposite to one of them be given, the remaining sides and angle are given.

That is, if C, B and AC, be given, A, BC, and AB may be found.

1. *The side AB opposite C may be found.*

For, (VII) sin. B : sin. AC : : sin. C : sin. AB.

When $C+B>2$ right-angles $AC+AB>Ce$,

and „ $C+B<2$ „ „ $AC+AB<Ce$. (S. G. XII.)

This proposition will sometimes determine whether AB is greater or less than a quadrant, otherwise it is ambiguous.

2. *The side BC, adjacent to C and B may be found.*

For, (I) R : cos. C : : tan. AC : tan. CP,

and, (X) tan. B : tan. C : : sin. CP : sin. BP.

BP is ambiguous (see Prop. XIII, Dat. case IV.)

Therefore BC, is ambiguous both when the perpendicular falls within the triangle and when it falls without, so that BC might have four values, except that some of them are excluded by the impossibility of its being greater than a semi-circle.

3. *The remaining angle A may be found.*

For (III) R : cos. AC : : tan. C : cot. CAP.

And (VIII) cos. C : cos. B : : sin. CAP : sin. BAP.

A is ambiguous, in the same sense as BC.

V. If the three sides are given, the angles may be found.

That is, if AC, AB and BC, be given, A, B, and C may be found.

1. *The angle* C *may be found.*

For, (XII) tan. $\frac{1}{2}$BC : tan. $\frac{1}{2}$AC+AB : : tan. $\frac{1}{2}$AC − AB, : tan. $\frac{1}{2}$E = the difference between CP and BP.

When AP falls within the triangle.

Then (I) tan. AC : tan. CP : : R : cos. C.

In the same manner, the other angles may be found. But after C is found, the sides may be found by Prop. VII.

If AC+AB > C*e*, C+B > 2 right-angles, and the greater side, is opposite to the greater angle.

VI. If the three angles of a spherical triangle are given, the three sides are given.

The supplements of the angles, are the sides of the supplemental triangle, (Sp. Geom. X.) which are therefore given, therefore its angles are given by the last; but the supplements of those angles are the sides of the supposed triangle which therefore are given.

Schol. If the sides of a spherical triangle are reduced indefinitely, they approach continually to equality with their sines and tangents, and to coincidence with them, *i. e.* the *limit* of the ratio of the sides to their sines, or tangents, is a ratio of equality.

If they be considered as actually coinciding in their evanescent state, the triangle, becomes a plane triangle, and as the properties of the spherical triangle belong to it, whatever be the length of its sides, the propositions which have been demonstrated concerning spherical triangles, may be transferred to plane triangles, by substituting the sides of such plane triangles in the places of the sines or tangents, of the sides of the spherical triangle. It will be seen, that these results conform to the Propositions in Plane Trigonometry.

NAPIER'S RULES OF THE CIRCULAR PARTS.*

The rule of the *Circular Parts*, invented by Napier, is of use in spherical trigonometry, by reducing all the theorems employed in the solution of right-angled triangles to two. These two are not new propositions, but are merely enunciations, which, by help of a particular arrangement and classification of the parts of a triangle, include the first six propositions, with their corollaries, which have been demonstrated above. They are perhaps the happiest examples of artificial memory that is known.

Definition 1.—If in a spherical triangle, we set aside the right angle, and consider only the five remaining parts of the triangle, viz. the three sides and the two oblique angles, then the two sides which contain the right-angle, and the complements of the other three, namely, of the two angles and the Hypotenuse, are called the *Circular Parts*.

Fig. 13. Thus in the triangle ABC right-angled at A, the circular parts are AC, AB with the complements of B, BC, and C. These parts are called circular; because, when they are named in the natural order of their succession, they go round the triangle.

Def. 2.—When, of the five circular parts, any one is taken, for the middle part, then of the remaining four, the two which are immediately adjacent to it, on the right and left, are called the *adjacent parts;* and the other two, each of which is separated from the middle by an adjacent part, are called *opposite parts.*

* This account of Napier's Circular Parts, is taken principally from Playfair's *Appendix to Spherical Trigonometry.*

Thus, if AC, be reckoned the middle part, then AB and the complement of C, which are contiguous to it on different sides, are called *adjacent parts ;* and the complements of B and BC are the *opposite parts.* In like manner if AB be taken for the middle part, AC and the complement of B are the *adjacent parts,* and the complements of BC and C the *opposite.* If the complement of BC be the middle part, the complements of B and C are *adjacent,* AC and AB *opposite parts.*

PROPOSITION.

In a right-angled spherical triangle, the rectangle of *the radius* and *the sine* of *the middle part,* is equal to the rectangle of the *tangents* of the *adjacent parts ;* or to the rectangle of the *cosines* of the *opposite parts.*

The truth of the two theorems included in this enunciation may be easily proved, by taking each of the five circular parts in succession for the middle part, when the general proposition will be found to coincide with some one of the analogies already given for the resolution of the cases of right-angled spherical triangles.

If a spherical triangle have one side a quadrant, the supplemental triangle will be right-angled ; and as the sines, tangents, &c of an arc are the same as those of its supplement, *Quadrantal,* or *Rectilateral* spherical triangles, may evidently be solved like Rectangular Triangles. The same proposition of Napier, is applicable to both, if the following are taken as the *Circular Parts* in the former.

The Quadrant is in the place of the right-angle, and is not supposed to separate the circular parts, which are, the two angles adjacent to the Quadrant, the complements of the other two sides, and of the remaining angle.

SPHERICAL TRIGONOMETRY.

PART II.

Calculation of the sides and Angles of Spherical Triangles.

It has been said by Playfair, that, "Trigonometry is the application of numbers to express the relations of the sides and angles of triangles to each other." This is true of Trigonometry, as applied to the actual calculation of the sides and angles of triangles. The nature and use of this application of numbers will be evident, by remarking that the Data, Prop. XIII, and XIV. are founded on the supposition that the ratios of radius to the Trigonometrical lines belonging to any given arc, are given. These ratios however are not given, and cannot, except in a very few cases, be expressed by numbers.*

If however the circumference of the circle be divided into any number of equal parts, as 360, the ratios of radius to the trigonometrical lines, belonging to these arcs, may be calculated to any required degree of exactness. (*Days Trigonometry*, "*Computation* of *the canon*".) This is done, and numbers expressing the relations of their parts to radius, considered as unity, are arranged in tables.

By means of these, the sides and angles of spherical, as of plane triangles can be easily computed, from the preceding propositions.

* Cagnoli, Chap. VI.

I. RIGHT ANGLED SPHERICAL TRIANGLES.

CASE I.

Given the Hypotenuse (=55° 8',) and one side (=32° 12',) it is required to find the other parts. Fig. 21.

PROJECT THE TRIANGLE,* *at the circumference* of the Primitive, making (A*p*=) AC=55° 8' and (*fy*=*fz*=) BC=32° 12', (Prob. VIII.), on circles deseribed through AC*d*, and B *y*.

At the centre, make (*m n*=AP=) AC=Hyp. and (*xy*=) BC=side, on the oblique circle described through *m* and C. Fig. 22.

In the plane, from the point A, make (*yz*=*ym*=) AC= Hyp. on the oblique great circle described through *y* A and C, and BC, (on the oblique circle described through *x*C*d*)=given side. " 23.

1. *To find the other side*, AB.

By Circular Parts; R . cos. AC=BC . cos. AB.
Or, Prop. XIII.2.1 .; cos BC : R : : cos. AC : cos. AB.

Cos. BC=32° 12' av . co.	0.0725305.
R	10
Cos. AC=55° 8'	9.7571444.
Cos. AB=47° 30' 4"	9.8296749.

* The *Projections* are not given in full. To a person familiar with the Problems in Stereographic Projections, a mere reference to the steps of the process, it is presumed, will be sufficient.

The side AB is less than a quadrant, like BC, because the Hypotenuse is less than a quadrant.

2. *To find the angle* C.

Cir. Parts; R . cos . C=tan. BC . cot. AC,
Or prop. (XIII. 2. 3.) tan. AC : tan.BC : : R : cos.C.= 68° 58′ 30″. C<90°, like BC.

3. *To find the angle A.*

Cir. Parts; R . sin. BC=sin. AC . sin. A.
Or Prop. (XIII. 2. 2.) sin. AC : R : : sin BC : sin. A. 40° 30 5″. A like BC< 90°.

CASE II.

Given the Hypotenuse (=55° 8′) and one angle (=40′° 30′ 5 ″to find the remaining angle, and the sides.

Project the Triangle, making A=40° 30′ 5″ and AC= 55° 8′.

Fig, 21. 24. 25. The projections at the *circumference* and *centre* are evident. From a point, as A, *in the plane*, project the circle, making A=40° 30″ 5″, (Prob. IV.) From A cut off AC=Hypotenuse (VIII) project the circle *pCo*, through C; BC, AB are the required sides.

1. Cir. Parts; R . cos. 55° 8′=cot. 40° 30′ 5″ . cot. C. C<90°, like A, because AC<90°.

2. Sin. A . sin. AC=R . cos. BC. BC<90°. like A . because AC<90°.

3. R . cos. A=cot. AC . tan. AB.
AB<90 . like C; because AC<90.

CASE III.

Given one side (=47° 30′ 4″) and its opposite angle (=63° 58 30″) to find the other parts.

Project the triangle, at the circumference. Fig. 26.

Make AB=47° 30′ 4″=Draw the perpendicular diameters Bx, yz. Make ym=yn=63°.58′ 30″ and draw the small parallel circle, (Prob. II.) from l, as a pole, project the oblique circle ACqR. ACB, or RCx is the triangle required. Each of them contain the given parts.

At the centre. Cut off AB=given side (Prob. VIII.) Fig. 27. Project the circle dBe (Prob. III.); around x, its pole, describe the small circle, at the distance of the given angle. (Prob. II.) From h, where it intersects the primitive, as a pole, project the right circle *fad;* then ACB, or its supplement is the triangle.

From a point in the plane, as *A*, cut off AB, on a right Fig. 28. circle, equal to the given side. (Prob. VIII..) Project the oblique circle xBz, (Prob. III.) find its pole m, and around it, describe the small circle, at the distance of the given angle; v, its intersection with the circle xvz, whose pole is A, is the pole of a circle passing through A, making the required angle at C.

1. R . cos. C=cos. AB (= Rx) . sin. A or CRx its Fig.26. supplement. For the angle is ambiguous.

2. Tan. . AB (or Rx) . cot. C=R . sin. BC (or Cx.) The side is ambiguous. BC or its sup′ Cx.

3. R . sin. AB=sin. C . sin. AC, or CR . which is ambiguous.

CASE IV.

Given one side (=47° 30′ 4″), and the adjacent angle (40° 30′ 5″) to find the other parts.

Fig 21. 24. 25. *Project the Triangle*, making AB=47° 30′ 4″, and A=40° 30′ 5″. (Prob.IV.)

1. R . sin. AB=cot. A . tang. BC.
BC is of the same affection as A.

2. Cos. AB . sin. A=R . Cos. C.
C is of the same affection as CB.

3. R . cos. A=tang. AB . cot. AC.
The Hypotenuse is less then a quadrant, because AB and A are of the same affection.

CASE V.

Given two sides (=47° 30′ 4″ and 32° 12′) to find the other parts.

Fig. 21. 22. 23. *Project the Triangle*, making AB and BC equal to the given sides.

1. Cos. AB . cos. BC=R . cos. AC.
AC< 90, because AB and BC are alike.

2. R . sin. AB=tang. BC . cot. A.
A<90, like BC.

3. R . sin. BC=tan. AB . cot. C.
C<90, like AB.

CASE VI.

Given the two angles (=40° 30′ 5″ and 63° 58′ 30″) to find the other parts.

Project the Triangle. Fig. 29.

At the circumference, make A=40 30′ 5″.

Find y the pole of ACx, and around it describe the small circle, at the distance of 63° 58′ 30″, its intersection with the primitive in S, is the pole of the right circle BC, which makes the given angle at C, with ACx.

At the centre, make A=one of the given angles about x the pole of CAd, describe the small circle yz at the distance of the other given angle from x, their intersection is the pole of qBCl making the given angle at C. " 30.

From A, *a point in the plane of the Primitive;* make A= given angle, about m the pole of ACx, describe a small circle, at the distance of the other angle, intersecting ABp in p' which is the pole of dBCf' making C=the other given angle. " 31.

1. Cot. A . cot. C=R . cos. AC.
AC<90, because A and C are alike.

2. R . cos. C=sin. A . cos. AB.
AB<90, like opposite angle C.

2. R . cos. A=sin. C . cos. BC.
BC<90 like A.

OBLIQUE-ANGLED TRIANGLES.

CASE I.

Given two sides (=58°, and 79° 17′ 14″,) and the angle opposite one of them (=62°′ 34 6″,) it is required to find the other parts.

Fig. 32.

Project the Triangle, making AC=58° and describe the circle CB*e*, making C=62° 34′ 6″, and the small circle *kg*, at the distance of 79° 17′ 14″ from A, AB*d* will complete the triangle.

Through, *q*, the pole of CB*e* project the circle *Apqd*, perpendicular to B*e*.

1. Sin. AB : sin. C: :sin. AC : sin. B. i. e. sin. ABC, or AB*e*.
This case is ambiguous,

2. R . cos. C=cot. AC . tan. PC.
PC< 90, because C, and AC< 90.

3. R . cos. B=cot. AB . tan. BP.
PB< 90, as before, and PC+PB=CB.

4. Sin. AC : sin. B: :sin. BC : sin. A,
Which is ambiguous.

The projection gives B acute, and A obtuse, but with the same things given, another triangle might be projected, containing angles equal to the supplements of these.

That two spherical triangles may have two sides, and the angle opposite to one of them in each equal, while the remaining side and angles are unequal may be easily shown, as in fig. 6, where the triangles APB, APB′, have AP, AB, or AB′, and the angle P, in each equal. The projection of these two triangles is given Pl. XVI, fig. 2, where the ambiguity of A, B and BC is exhibited. (BC=PC∓PB,) and A, B, are either acute or obtuse.

CASE 2.

Given two sides (=58°. and 110°.) and the included angle, (62°. 34′ 6″) to find the other parts.

Project the Triangle ; making AC=58° C=62°. 34′ 6″ and CB=110°, by drawing the small circle *pBq*, at the distance of 180°.—110°.=70°. from *e* as its pole. Fig. 33.

Describe AB*d*, through.B, and AP*d* through the pole of C*e*.

1. R. cos.C=cot. AC . tan. PC ; <90°.
   Then BC—PC=PB.

2. (X) Sin. PB : sin. PC : : tan. C : tan. B.
   B<90°. like C, because AP is within the triangle.

3. Sin. B : sin. AC : : sin. C : sin. AB<90°.
   Because B and BP are both acute.

4. Sin. AC : sin. B : : sin. BC : sin. A .>90°.
   For BC+BA>180°,∴(S. G. XII)B A*e*<C.

CASE 3.

Given two angles (=50°, and 62° 34′ 6″) and the side opposite to one of them (79° 17′ 14″) to find the other parts.

Project the Triangle ; making BA =79° 17′ 14″, the angle B=50°, ; also describe the great circle AC*d*, making with BC an angle at C=62° 34′ 6″ Fig. 35.

Draw APD through the pole of BC.

1. R . cos. B=cot. AB . tan. PB$<$90°.

2. Tan. C : tan. B : sin. PB : sin. PB$<$90°.
For PB+PC=BC.

3. Sin. C : sin. AB : : sin. B : sin AC$<$90°.

4. Sin. AC : sin. B : : sin. BC : A $>$90°.
For AB+BC$>$180°.·.A$>$BC*d*.

CASE 4.

Given two angles (=50° and 62° 34′ 6″) and the side between them (=110°) to find the other parts.

Fig. 34. *Project the Triangle ;* making CB=110°. C=62° 34′ 6″, and B=50° ; also draw BP*d* through the pole of CA*e*.

1. R . cos. BC=cot. C . cot PBC$>$90°
Then PBC−ABC=PBA.

2. Sin. BC . sin. C=R . sin. PB$<$90°.

3. R. cos. PBA=tan. PB . cot. BA$<$90°.

4. R. cos. BA=cot. PBA . cot. PAB$<$90.
Then 180°−PAB=BAC$>$90.

5. R . cos. BA=cos. PB . cos. PA$<$90°.
Then PC−PA=AC$<$90.

CASE 5.

Given the three sides (=79° 17′ 14″ , 110° and 58°) to find the angles.

Fig. 33. *Project the Triangle ;* making AC=58°, AB=79° 17′ 14″ and BC=110°, by describing small circles *rg* and *pq*, at the distance of 79° 17′ 14″ and 70° from A and *e* as poles and then drawing ABd, CBe through the point of their intersection.

1. Tan. $\frac{1}{2}$BC : tan, $\frac{1}{2}\overline{AC+AB}$: : tan. $\frac{1}{2}\overline{AB-AC}$: tan. $\frac{1}{2}\overline{BP-PC}=\frac{1}{2}$D.
   Then $\frac{1}{2}$BC+$\frac{1}{2}$D=BP, and $\frac{1}{2}$BC − $\frac{1}{2}$D=PC.
   For the least segment, is adjacent to the least side.
(Sp. Geom. II Prop. XVII.)

2. The other parts are easily found as before.

CASE 6.

Given the three angles (=121°54′ 56″, 50° and 62° 34′ 6″) to find the angles.

The supplements of these angles, are the *sides* of the *supplemental* triangle, the angles of which are found, as in Case 5. Then the supplements of those angles are the sides of the given triangle.

Spherical Trigonometry is extensively applicable to the solution of questions connected with Geography, and especially in Trigometrical Surveying, and Geodesic operations.*

Its use and importance in astronomical investigations, is indicated by the declaration of M. de La Lande, himself an eminent Astronomer, "*La Trigonométrie Spherique est la véritable Science de l'Astronome.*"

The following questions will illustrate its application to each of these sciences.

1. Given the Latitude of St. Petersburg, 59° 66′ N. and its Longitude, 27° 59′ 30″ E. from Paris, also the Latitude and Longitude of Conception in South America, 36° 42′ 53″ S. and 75° W; required the *distance* of the two places, as measured on the arc of a great circle.†

In the triangle NAB two sides and the included angle are given; its solution therefore is according to Case 2, of Oblique angled Triangles. Pl. XV. Fig. 3.

* Hutton's Math. Vol. II. † Cagnoli (1150.)

2. Given the Longitude and Latitude of two stars, and the distance of a third star from each of them, required the Longitude and Latitude of the third Star.

Fig. 4. Let EL represent the *Ecliptic*,* PF, PH and PL arcs of great circles perpendicular to it. A and B the two stars whose places are given. and C the place of the third star.

Then in the Triangle APB, two sides and the included angle are given; therefore the remaining side and angles may be found.

Then in the triangle ABC, the three sides are given, whence the angles may be found.

Then in the Triangle PAC, two sides (PA and AC) are given, and the angle included by them. Therefore the remaining side PC, (which is the complement of the Latitude of C) and APC, (the difference of Longitude betwean A and C) may be found.

If the Declination and Right Ascension of each of the stars A and B were given, then EL would represent the Equator; and the solution would give the Declination and Right Ascension of the third Star.

* The Longitude of a heavenly body is measured by an arc of the *Ecliptic* and its Latitude is its *distance* from the same circle.

NOTES TO CONIC SECTIONS.

DEFINITIONS.

Conic Sections are treated differently by different authors; some defining them by *the sections of a cone*, and deriving their properties directly from the intersection of the cone and plane, while others define them as figures described in a plane, and derive their properties from their mechanical description. Each method has its advantages. If the latter, in some cases, renders the demonstrations more simple and easy; the former, which is adopted in this treatise, has the important advantage, of deriving the figures from solids whose properties have been demonstrated, and by operations which have become familiar in the Elements of Geometry.

"Thus far," says Newton, "I think I have expounded the construction of solid Problems by operations whose manual practice is most simple and expeditious. So the Ancients, after they had obtained a method of solving these Problems by a composition of solid places, thinking the Constructions by the Conick Sections as useless, by reason of the Difficulty of describing them, sought easier Constructions by the Conchoid, Cissoid, the Extension of Threads, and by any Mechanic Application of Figures.— "If the Ancients had rather construct Problems by Figures not received into Geometry at that Time, how much more ought these Figures now to be preferred which are received by many into Geometry as well as the Conick Sections.

However I do not agree to this new sort of Geometricians who receive all Figures into Geometry.-In my judgment, no Lines ought to be admitted into plain Geometry besides the right Line and the Circle, unless some Distinction of Lines might be first invented, by which a circular Line might be joined with a right Line, and separated from all the rest. But truly plain Geometry is not then to be augmented by the number of Lines. For all figures are plain that are admitted into plain Geometry, that is, those which the Geometers postulate to be described *in plano*."—" All these descriptions of the Conicks *in plano*, which the Moderns are so fond of, are foreign to Geometry. Nevertheless, the Conick Sections ought not to be flung out of Geometry. They indeed are not described Geometrically *in plano*, but are generated in the plane Superficies of a Geometrical Solid. A Cone is constituted geometrically, and cut by a Geometrical Plane. Such a segment of a Cone is a Geometrical Figure, and has the same place in solid Geometry, as the Segment of a Circle has in Plane, and for this reason its base, which they call a Conick Section, is a Geometrical Figure. Therefore a Conick Section hath a place in Geometry, so far as it is the Superficies of a Geometrical Solid; &c." *Universal Arith.* pp. 247-249.

Def. 8.—This definition was suggested by a communication in the *Journal of Science*, from Prof. Davies of the Military Academy at West Point. *Jour. Sci. Vol.* VI *page* 280.

ELLIPSE PROPOSITION XXVI.

The properties referred to, are the following.

PROPOSITION A.

If any line in the Ellipse pass through either the of the Foci, and a tangent be drawn through one of its extremities, a line drawn from the centre, parallel to the line passing through the Focus and intercepted by the tangent, is equal to the semi-transverse axis.

That is, $Ct = CA$. Fig. 26.

Bor, (IX Cor. 1.) $CF . CD = CA^2 - CA . EF$,

$\therefore \quad CA . EF = CA^2 - CF . CD$;

but, (VI) $CA^2 = CD . CT$; $\therefore CA . EF = CD . CT$
[$- CD . CF = CD . TF$,

sim. tri. $TC : Ct :: TF : FE :: TF . CD \ (= CA . FE) :$
[$FE . CD$,

$:: CA : CD$,

therefore, $Ct . CA = TC . CD = CA^2$.

$\therefore \quad Ct = CA$.

Q. E. D.

Cor. 1.—Hence $TF . CD = CA . FE$. *3d step of dem.*

Cor. 2.—$AB : kz : EH$ are continued proportionals.

For, draw the diameter LCIK, bisecting EH, and draw Ed parallel to it.

Then, (XXVI.) $Cd : Ck : Ct$, are continued proportionals,

That is $\quad IE : Ck : CA$, " " "

$\therefore \quad HE : zk : BA$, " " "

and $\quad zk^2 = HE . BA$.

PROPOSITION B.

The rectangle of the segments of any line in the Ellipse, passing through the Focus, is equal to the rectangle of one fourth of the Parameter, into the same line.

That is, $EF . FH = \frac{1}{4}P . EH$. Fig. 26.

For, (XXIV,) $Ck^2 : CA^2 :: EF . FH : AF . FB$;
But (VII. Cor. 3) $AF . FB = (Ca^2 =) CA . \frac{1}{2}P$,
Therefore, $Ck^2 : EF . FH :: CA^2 : CA . \frac{1}{2}P$,
But, $CA : \frac{1}{2}P :: CA . EI : \frac{1}{2}P . EI$,
And, $CA^2 : CA . \frac{1}{2}P :: CA . EI : \frac{1}{2} P . EI$.
$\therefore \; Ck^2 : EF . FH :: CA . EI : \frac{1}{2}P . EI$.
But, (XXVII. Cor. 2) $Ck^2 = EI . CA \therefore EF . FH = \frac{1}{2}P$.
[EI, or $\frac{1}{2}EH . \frac{1}{2}P$,
$\therefore \; EF . FH = \frac{1}{4}P . EH$.

Q. E. D.

Cor. 1.—As the proposition is applicable to all lines passing through the Focus,
$AF . FB = \frac{1}{4}P . AB$, as demonstrated. Prop. VII.
And $EF . FH = \frac{1}{4}P . EH$,
Therefore, $AB : EH :: AF . FB : EF . FH$. (See Parab. XIV. Cor. 7.)

Cor. 2.—When EH, is perpendicular to the Transverse Axis, it is a double ordinate to it, and since the rectangle of the abscisses at the Focus equals the square of the semi-conjugate, and EH then equals the parameter, and the rectangle EF . FH is the same as the square of EF, the proposition becomes
As the Transverse Axis
Is to the Parameter,
So is the square of the Semi-Conjugate
To the square of the Focal-Ordinate.

PROPOSITION C.

The rectangle of the two lines drawn from the Foci, to any point in the curve, is equal to the square of half the diameter conjugate to that which passes through the point.

Fig. 12. That is $FE . fE = Ce^2$,

Let ER be drawn perpendicular on eh,
Then, sim.. triang. $FE : FP :: EI : ER$,
And " " $fE : fp :: EI : ER$,
Therefore $FE . fE : FP . fp :: EI^2 : ER^2$.
But, (IX. Cor. 4.) $FP . fp = Ca^2$; and $EI^2 = CP^2 =$ (VII) [AC^2 ;
Therefore, $FE . fE : Ca^2 :: CA^2 : ER^2$;

But, (XVI.) $Ce . ER = AC . Ca$,
$\therefore$ $Ce : Ca :: CA : ER$,
And, $Ce^2 : Ca^2 :: CA^2 : ER^2$,
Therefore, $FE . fE = Ce^2$.

Q. E. D.

CURVATURE OF CONIC SECTIONS.

The application of Conic Sections to Physical Astronomy, which is one of their most important applications, requires an acquaintance with their *Curvatures*, and especially with *the method of reasoning* employed in treating of this part of Conic Sections. (*See Newtons Principia Lib.* I *Sect.* 1. *Cavallo's Philosophy, Introduction, Lemmas. Enfield's Phil. Central Forces. Lem.*) As the doctrine of the Curvature of Conic Sections, gives the student new and interesting views of their general properties, and is attended with no peculiar difficulties, a few elementary propositions are introduced in this part of the course.

PROPOSITION V.

"The angles at the base of an isosceles spherical triangle are *Symetrical* magnitudes, not admitting of being laid on one another, nor of coinciding, notwithstanding their equality. It might be considered as a sufficient proof that they are equal, to observe that they are each determined to be of a certain magnitude rather than any other, by conditions which are precisely the same, so that there is no reason why one of them should be greater than another. For the sake of those to whom this reasoning may not prove satisfactory, the demonstration below is given, which is strictly geometrical."

Playfair's Sph. Trig.

Pl. XVI. Fig. 1. Let ABC be a spherical triangle, having the side AB equal to the side AC; the spherical angles ABC and ACB are equal.

Let D be the centre of the sphere; join DB, DC, DA, and from A on the straight lines DB, DC, draw the perpendiculars AE, AF; and from the points E and F draw in the plane DBC the straight lines EG, FG perpendicular to DB and DC, meeting one another in G: Join AG.

Because DE is at right angles to each of the straight lines AE, EG, it is at right-angles to the plane AEG, which passes through AE, EG (4.2. Sup.); and therefore, every plane that passes throug DE is at right-angles to the plane AEG (17. 2. Sup.); wherefore, the plane DBC is at right-angles to the plane AEG. For the same reason, the plane DBC is at right-angles to the plane AFG, and therefore AG, the common section of the planes AFG, AEG is at right-angles (18. 2. Sup.) to the plane DBC, and the angles AGE, AGF are conseqnently right-angles.

But since the arch AB is equal to the arch AC, the angle ADB is equal to the angle ADC. Therefore the triangles ADE, ADF, have the angles EDA, FDA equal, as also the angles AED, AFD, which are right-angles; and they have the side AD common, therefore the other sides are equal, viz. AE to AF, (26. 1.), and DE to DF. Again, because the angles AGE, AGF are right-angles, the squares on AG and GE are equal to the square of AE; and the squares of AG and GF to the square of AF. But the squares of AE and AF are equal, therefore the squares of AG and GE are equal to the squares of AG and GF, and taking away the common square of AG, the remaining squares of GE and GF are equal, and GE is therefore equal to GF. Wherefore, in the triangles AFG, AEG, the side GF is equal to the side GE, and AF has been proved to be equal to AE, and the base AG is common, therefore, the angle AFG is equal to the angle AEG (8. 1.). But the angle AFG is the angle which the plane ADC makes with the plane DBC (4. def. 2. Sup.) because FA and FG, which are drawn in these planes, are at right-angles to DF, the common section of the planes. The angle AFG (3. def.) is therefore equal to the spherical angle ACB; and, for the same reason, the angle AEG is equal to the spherical angle ABC. But the angles AFG, AEG are equal. Therefore the spherical angles ACB ABC are also equal.

Q. E. D.

The converse of this proposition is thus demonstrated by Playfair.

Let ABC be a spherical triangle having the angles ABC ACB equal to one another; the sides AC and AB are also equal.

Let D be the centre of the sphere ; join DB, DA, DC, and from A on the straight lines DB, DC, draw the perpendiculars AE, AF ; and from the points E and F, draw in the plane DBC the straight lines EG, FG perpendicular to DB and DC, meeting one another in G ; join AG.

Then, it may be proved, as was done in the last proposition, that AG is at right angles to the plane BCD, and that therefore the angles AGF, AGE are right-angles, and also that the angles AFG, AEG are equal to the angles which the planes DAC, DAB make with the plane DBC. But because the spherical angles ACB, ABC are equal, the angles which the planes DAC, DAB make with the plane DBC are equal, (3. def.) and therefore the angles AFG, AEG are also equal. The triangles AGE, AGF have therefore two angles of the one equal to the two angles of the other, and they have also the side AG com mon, wherefore they are equal, and the side AF is equal to the side AE.

Again, because the triangles ADF, ADE are right-angled at F and E, the squares of DF and FA are equal to the square of DA, that is, to the squares of DE and DA ; now, the square of AF is equal to the square of AE, therefore the square of DF is equal to the square of DE, and the side DF to the side DE. Therefore, in the triangles DAF, DAE, because DF is equal to DE, and DA common, and also AF equal to AE, the angle ADF is equal to the angle ADE ; therefore also the arches AC and AB, which are the measures of the angles ADF and ADE, are equal to one another ; and the triangle ABC is isosceles.

Q. E. D.

ERRATA.

The following errors of the press should be corrected with a pen or pencil, before reading the book.

Page 10 line 4 fr. top, for XII Prop. 2, read Sup. I, Prop. IV.

47 Prop. IX	„ $fE+fE$	„ $FE+fE$,
51 line 13 fr. top	„ $\frac{1}{2}FG$	„ $\frac{1}{2}fG$,
52 „ 7 „ „	„ PAC	„ P . AC,
„ „ 21 „ „	„ $CF : \overline{CF+FT}$	„ $CF . \overline{CF+FT}$,
54 Prop: XIV	„ De^2 and $D'e^2$	„ DE^2 and $D'E^2$
55 line 5 fr. bot.	„ Ca,	„ CA,
„ “ 2 „ „	„ Gg	„ Cg,
57 „ 10 „ „	„ GMT,	„ CMT,
178 „ 2 fr. top,	„ $\frac{1}{2}BD-AD$,	„ $\frac{1}{2}\overline{BD-AD}$,
185 „ 2 „ „	„ $\frac{1}{2}AC+AB$ and $\frac{1}{2}$ AC − AB read $\frac{1}{2}\overline{AC+AB}$ and $\frac{1}{2}\ \overline{AC-AB}$.	

204 for Prop. V, read Prop. IV.

ERRATA.

[Of the following Errata, some were obviously owing to the press, and many to inaccuracies of the manuscript, which had been transcribed, and was written originally in haste, and not very legibly A considerable number of them are unimportant, but it was thought adviseable to make the list as complete as possible.]

PAGE	Line	Place		For		Read
12	margin,	- - - -	for	6 6	read	6.8.
15	24th line from top	-	"	Fg^2	"	fg^2.
17	10 " "	bottom,	insert	*that of,* after *equal to.*		
20	9 " "	top,	for	AF	read	$AF.^2$
–	12 " "	do.	"	2AF	"	–2AF.
23	last line,	- -	"	(VII)	"	(VI).
27	5 "	from top,	"	*circle*	"	*curve.*
–	27 " "	do.	"	II Cor.	"	(XI Cor. 4.)
29	11 " "	bottom,	"	(II Cor. 1)	"	(XI Cor. 4.)
–	2 " "	do.	"	(VIII)	"	(XI)
30	6 " "	top,	"	2CM : TP	"	2CM.TP.
34	- -	margin,	"	Fig. 6. Pl. VII	"	Pl. VIII. Fig. 14.
35	1 & 2 " "	bottom,	"	*ad.ab*	"	*ad db.*
38	14 " "	top,	"	*ek.kl*	"	*ik.kl.*
46	18 " "	do.	"	(III.	"	(II.
–	22 " "	do.	"	FF^2	"	Ff^2.
47	" "	do.	"	*f*E+*f*E	"	FE+FE.
49	16 " "	do.	"	TF	"	T*f*.
50	12 " "	do.	"	OEP	"	OE*f*.
51	8 " "	do.	"	*fep,*	"	*f*E*p*.
–	13 " "	do.	"	½FG	"	½*f*G.
–	4 " "	bottom,	"	*semi-diameter*	"	*diameter.*
–	– " "		"	*f*PK	"	*p*PK.
52	7 " "	top,	"	PAC	"	P.AC.
–	21 " "	do.	"	$CF:\overline{CF+FT}$	"	$CF.\overline{CF+FT}$.
54	10 & 11 " "	do.	"	$De^2\ D'e^2$	"	$DE^2\ D'E^2$.
55	- -	margin,	"	Fig. 14.	"	Fig. 13.
–	5 " "	bottom,	"	Ca	"	CA.
–	2 " "	do.	"	Gg	"	Cg.
56	5 " "	top,	"	De^2	"	de^2.
57	10 " "	bottom,	"	GMT	"	CMT.
60	8 " "	do.	"	+	"	:
64	- -	top,	"	HK	"	*k*K.
–	2 & 3 " "	do.	"	DT	"	BT.
–	- -	margin,	"	Fig 24.	"	Fig. 25.
–	6 " "	top,	"	+.	"	:
75	- -	do.	"	CL	"	CA.
–	3 " "	do.	"	CD	"	CT.
77	- -	margin,	"	Fig. 10.	"	Fig. 9.
78	6 " "	top,	"	BHF	"	BH*f*.
–	15 " "	do.	"	(III.	"	(II.

Page 78	17th	line	from	top,	for	$CA^2 - CF^2$	read	$CF^2 - CA^2$.
-	2	"	"	bottom,	insert *their squares* after *are equal.*			
81	13	"	"	top,	for	FE*f* -	read	FEP.
82	-	-		bottom,	"	*semi-diameter*	"	*diameter.*
-	-	-		do.	"	*f*PK -	"	*p*PK.
87	5	"	"	top,	"	De^2 -	"	*de^2*.
88	15	"	"	do.	"	CHN -	"	CHA.
91	5	"	"	do.	"	CA^2 -	"	Ca^2
-	11	"	"	do.	"	*he eh* -	"	*he.ek.*
-	-	-		margin,	"	Fig. 14. -	"	Fig. 16.
92	4	"	"	top,	"	AEK -	"	CEK.
96	3,5 & 6	"	"	bottom,	invert the terms of the following propositions—viz: MT is less than MA; CM is less than MA; CM^2 is less than MA^2			
99	6	"	"	top,	"	*diameter*	read	*radius.*
-	11 & 22	"	"	do.	"	CB^2 -	"	Cr^2.CF.
-	19	"	"	do.	insert $\therefore Cr^2$.FA		after	CV^2.
-	4	"	"	bottom,	note, omit *varies as* FC^2.			
102	9	"	"	do.	insert	*the cube of*	after	*inversely as.*
-	1,2&3	"	"	do.	insert	*inversely,*	after	*varies,* and
					for	HC -	read	Hc.
110	13	"	"	top,	"	BA -	"	CA.
112	9	"	"	do.	"	CA -	"	C*a*.
113	-	-		do.	"	FNC -	"	FVC.
114	13	"	"	bottom,	"	NP -	"	NF.
-	11	"	"	do.	"	NEA -	"	NFA.
-	7	"	"	do.	"	ER -	"	ET.
117	9	"	"	do.	"	CA : C*a*	"	C*a* : C*a*′.
119	6	"	"	bottom,	"	LIR -	"	LIK.
-	-	"	"	do.	"	AK*h* -	"	A*k*H.
-	2	"	"	do.	"	EKG -	"	EK*g*.
135	4	"	"	do.	"	PE*p* -	"	PF*p*.
139	20	"	"	do.	"	ABC -	"	ABD.
140	3	"	"	do.	"	AEC -	"	EAC.
-	4	"	"	do.	"	AC -	"	EC.
-	16	"	"	do.	"	C, B & A	"	C, A & B.
-	18	"	"	do.	"	DF, FE & ED	"	AB, BC & CA.
143	23	"	"	do.	"	$<$ -	"	$>$
-	24	"	"	do.	"	CEA -	"	CED.
144	-	-		margin,	dele Fig. 12.			
148	18	"	"	top,	for	FG	--	EG.
149	3	"	"	do.	"	Cor. 7 -	"	Cor. 11.
-	8	"	"	do.	"	BDC -	"	ADC.
153	9	"	"	bottom,	"	l*if* -	"	P*if.*
155	14	"	"	do.	"	HI*i* -	"	*h*I*i*.
157	-	-		margin,	"	Pl. XIV.	"	Pl. XVI.
160	13	"	"	top,	"	*axis of the eye,*	"	*plane of projection.*
165	14	"	"	do.	insert	*and* H *to* P	after	*reduce* P *to* G.
-	15	"	"	do.	for	Y, P & G	read	P, Y & *p*.
166	9	"	"	bottom,	"	TV -	"	TH, and after VY insert, *through* X *where* EH *intersects the great circle, whose pole is* E. (Cor. 3.)
171	10	"	"	top,	"	AB, AD, AC,	read	DB.DA, & DC.
172	5	"	"	do.	"	ABC -	"	tang. ABC
174	5	"	"	bottom,	"	Cos. A. -	"	Cos. CA

ERRATA.

176	5th line from bottom,		for tang. BC	read tang. DC.
178	2	" " top.	" $\frac{1}{2}$BD—AD	" $\frac{1}{2}\overline{BD—AD}$
177	11	" " do.	" +	" :
-	15	" " do.	" $\frac{1}{1}$	" $\frac{1}{2}$
-	18	" " do.	" BB	" BC.
-	4	" " bottom,	" *rad*	" *rad*² .
-	5	" " do.	" *cot*	" *.cot.*
181	20	" " top,	" B	" C.
185	2	" " do.	" $\frac{1}{2}$AC+AB & $\frac{1}{2}$AC—AB,	read $\frac{1}{2}\overline{AC+AB}$ & $\frac{1}{2}\overline{AC—AB}$.
189	7	" " do.	for AP	read *Ap.*
-	11	" " do.	" *y*AC	" *y*A*f.*
-	15	" " do.	" BC	" *Cos* BC.
191	13	" " do.	" *fad*	" *f*A*d.*
195	6	" " do.	" Pl. XVI	" Pl. XV.
196	2	" " do.	" PB	" PC.
204	-	- top,	" TRIGONOMETRY	" GEOMETRY,
		and	" PROP. V	" PROP. IV,

The assertion at the top of page 144 is obviously false—the converse of it is true. Also, in the corollary, at the bottom of the next page, the assertion implied in the last clause of each sentence, is not *necessarily* true but of one side, and one angle.

M. R. D.

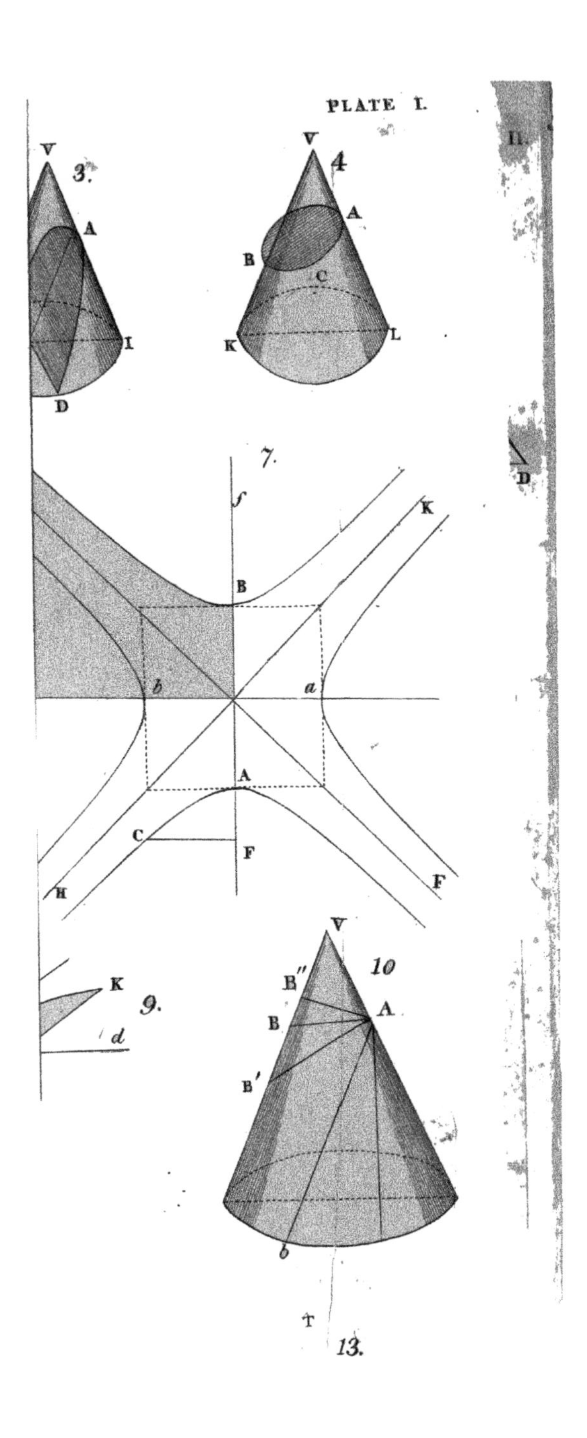
V
3.
A
I
D
V
4
A
B
C
K
L
7.
f
K
B
b
a
A
C
F
H
F
V
10
B″
A
B
B′
b
K
9.
d
13.

PLATE I.

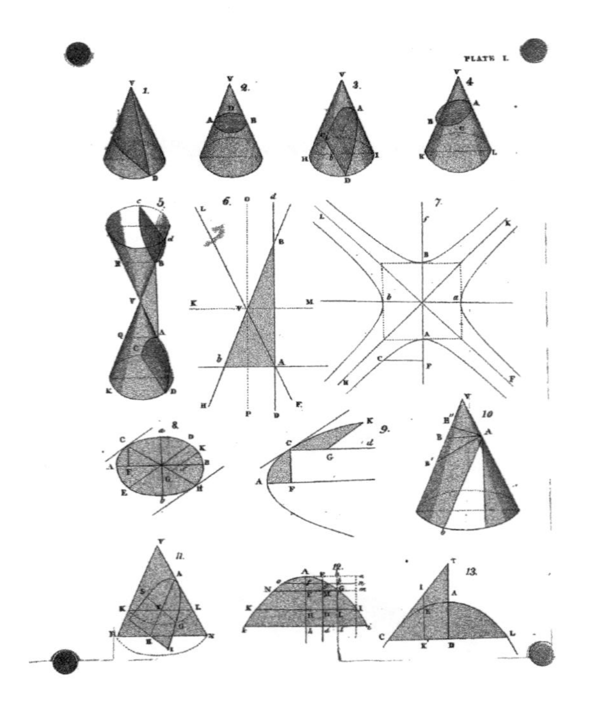

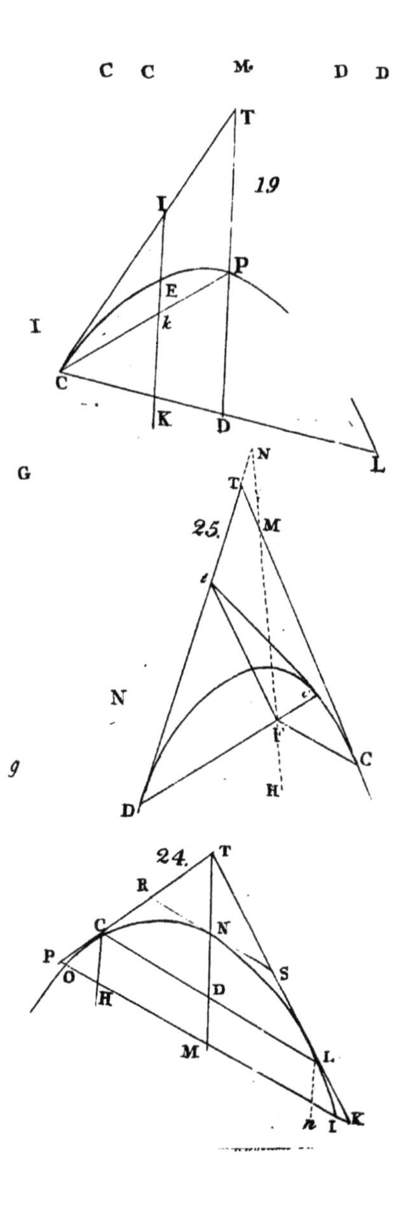
C C M. D D
T
19
L
P
E
I
k
C
K
D
L
G
N
T
25.
M
N
F
C
g
H
D
24.
T
R
C
N
P
S
O
D
H
M
L
n
I
K

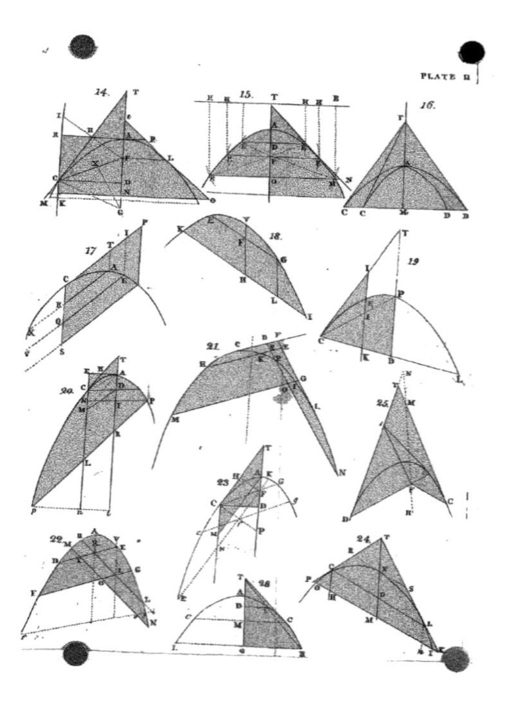
PLATE II
14.
15.
16.
17
18.
19
20.
21.
22
23
24.
25.
26

PLATE III.

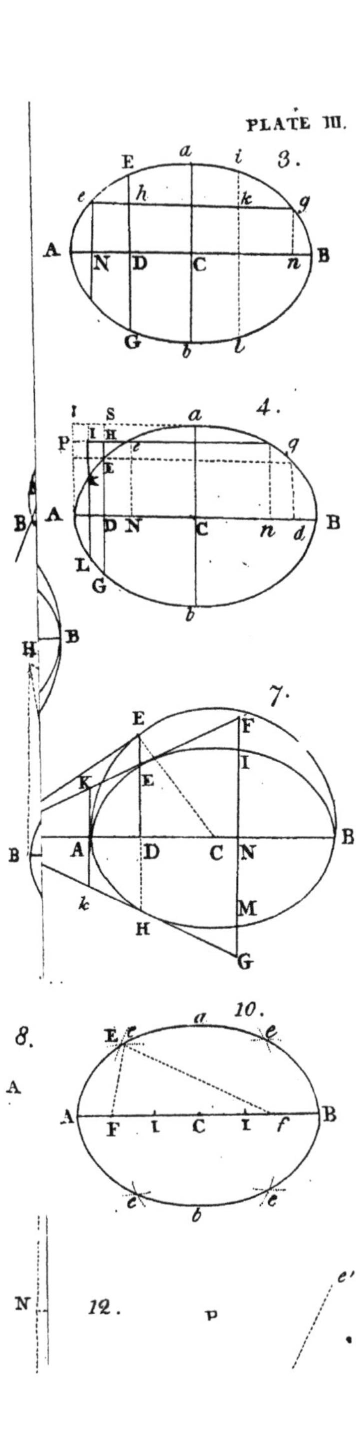

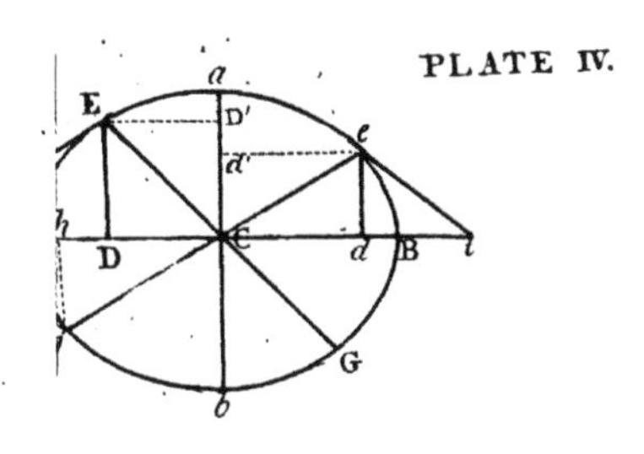
PLATE IV.
a
E
D'
e
d'
h
D
C
d
B
t
G
b

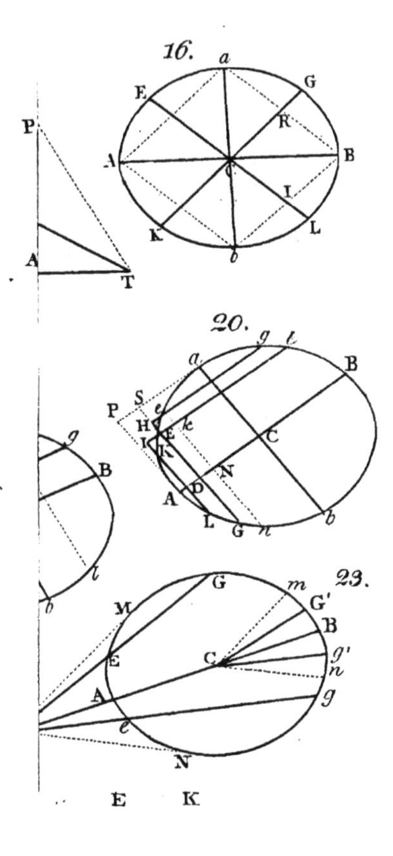
16.
a
E
G
R
P
A
B
C
I
K
L
b
A
T
20.
g
t
a
B
S
P
H
e
E
k
I
K
C
N
D
A
L
G
n
b
g
B
t
b
23.
G
m
M
G'
B
C
g'
E
n
A
g
e
N
E
K

k

B

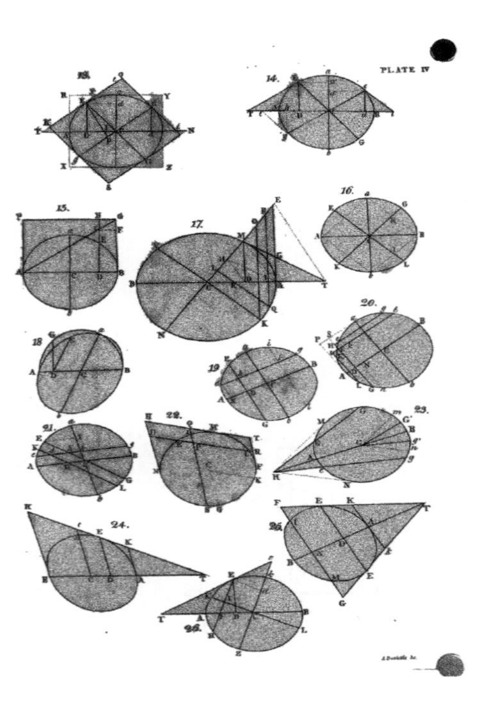
PLATE IV
13.
14.
15.
16.
17.
18
19
20.
21.
22.
23.
24.
25.
26.

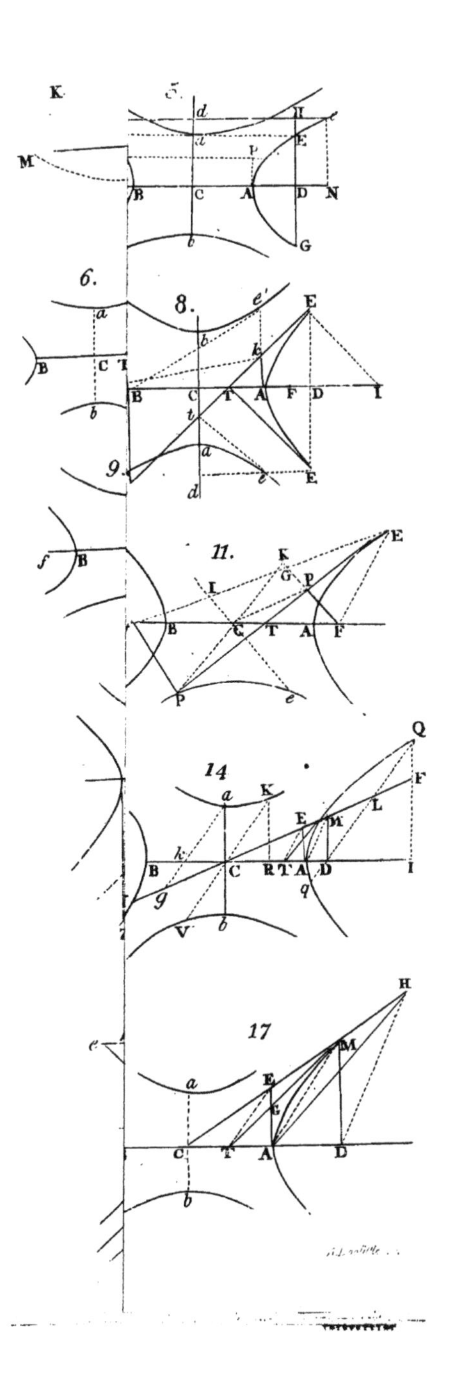

5.
6.
8.
9.
11.
14
17

PLATE V

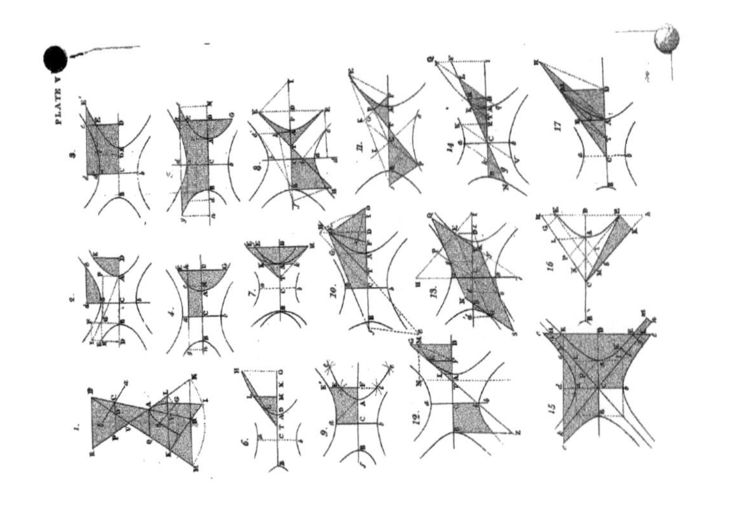

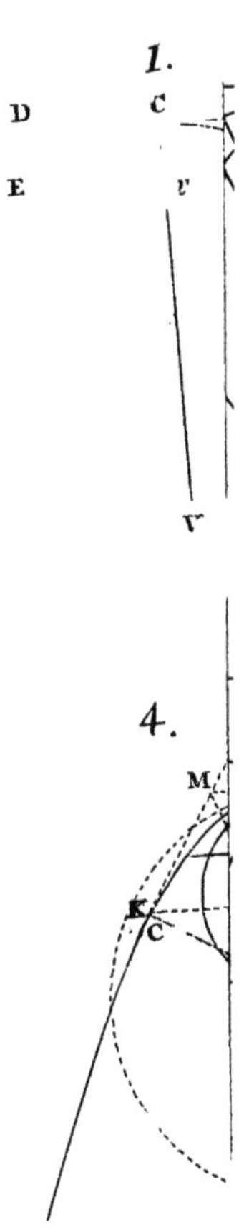
1.
D
C
E
F
V
4.
M
K
C

6.
H
B

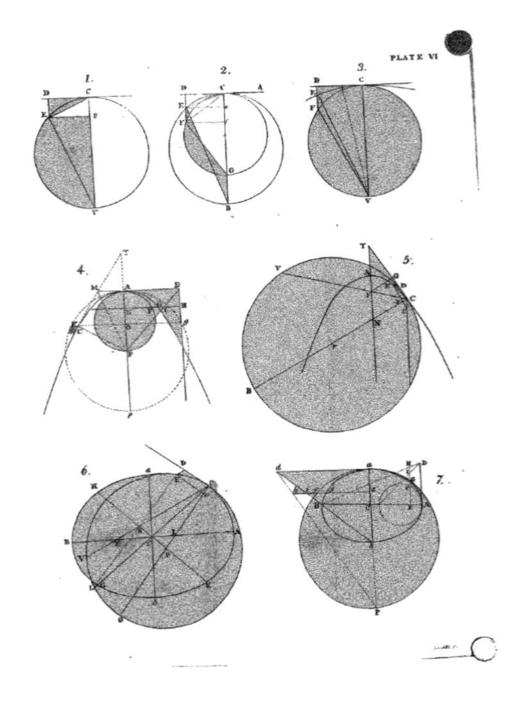
PLATE VI
1.
2.
3.
4.
5.
6.
7.

PLATE VII.

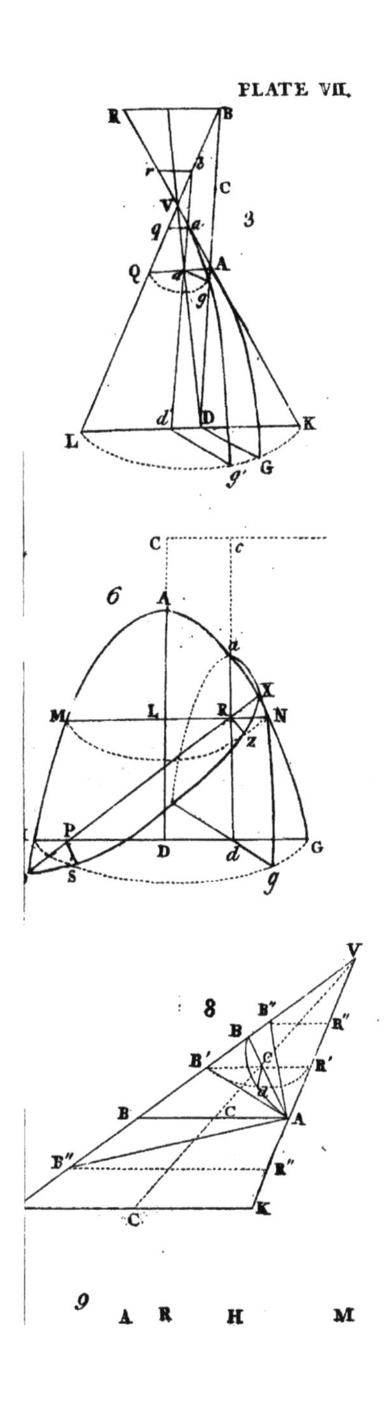

R
B
r
3
C
V
q
a
3
Q
A
g
d
D
K
L
g'
G
C
c
6
A
a
X
M
L
R
N
z
P
D
d
G
S
g
V
8
B''
B
R''
c
B'
R'
a
B
C
A
B''
R''
C
K
9
A
R
H
M

PLATE VII.

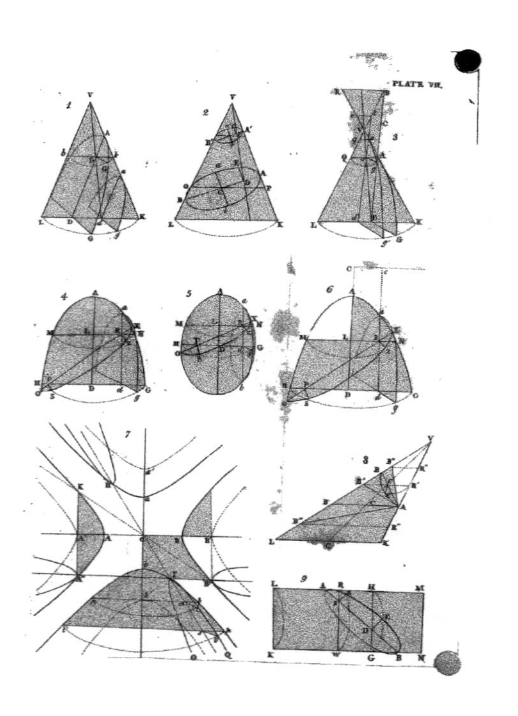

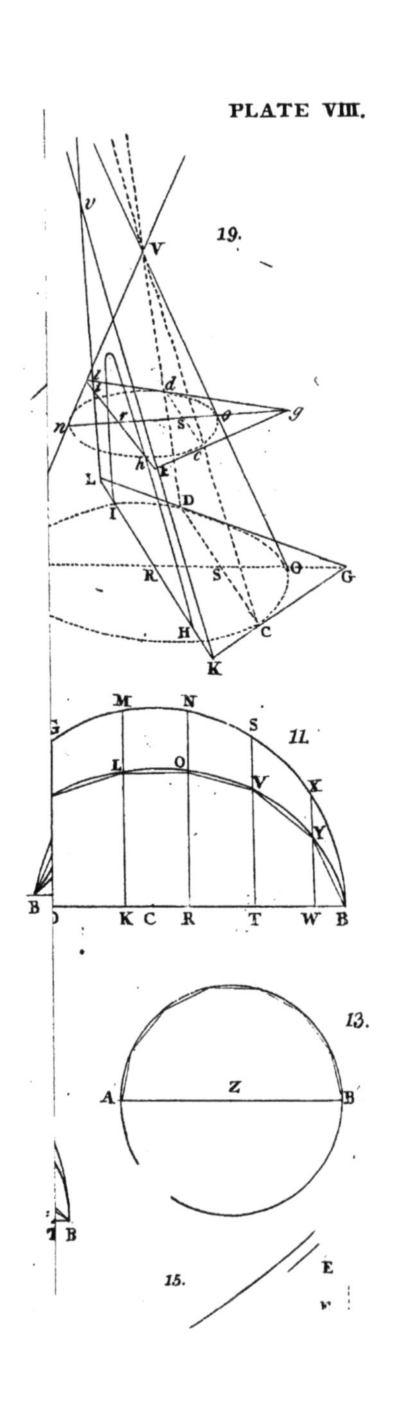
19.
v
V
d
g
n
r
s
o
h
c
L
D
I
R
S
O
G
H
C
K
M
N
S
11.
L
O
V
X
Y
B
K
C
R
T
W
B
13.
A
Z
B
B
15.
E

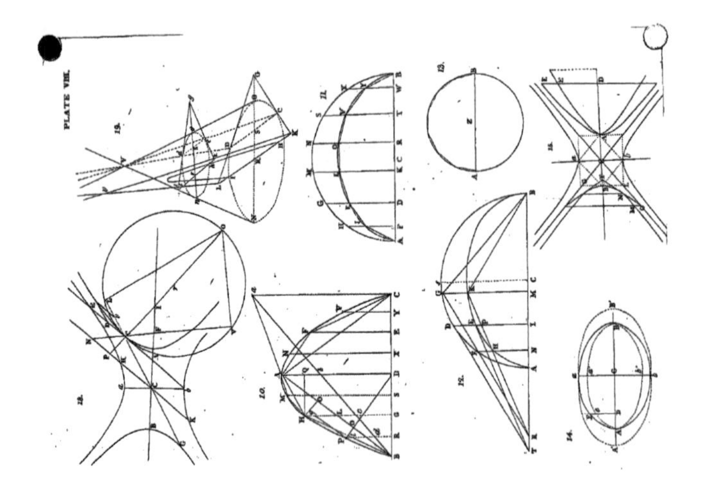
PLATE VIII.

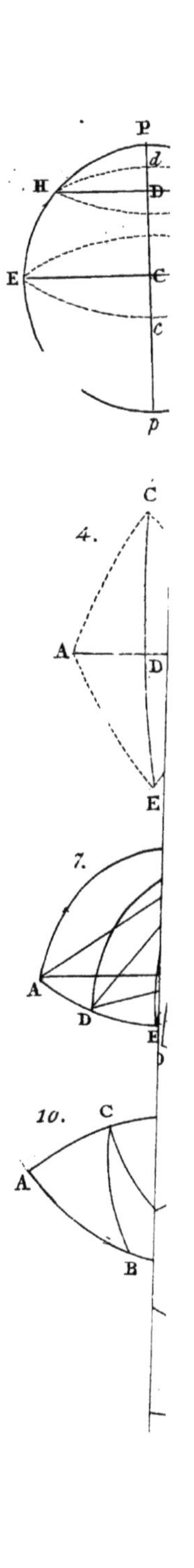
P
d
H
D
E
C
c
p
C
4.
A
D
E
7.
A
D
E
10.
C
A
B

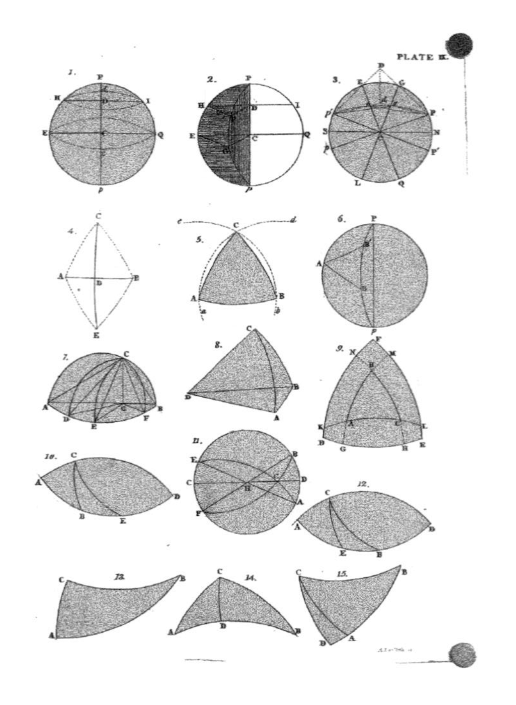
PLATE II.

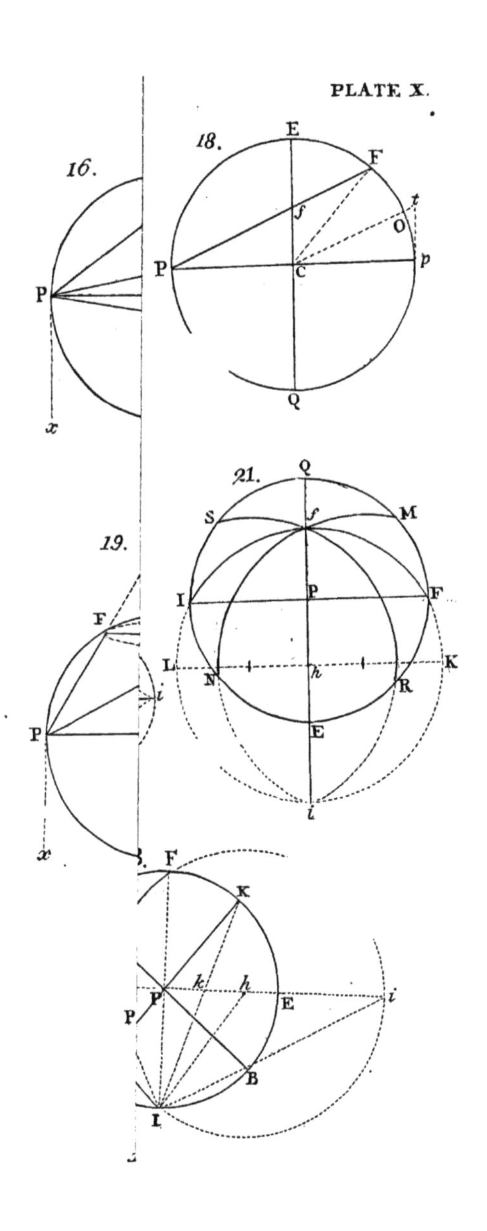
16.
P
x
18.
E
F
f
t
O
P
C
p
Q
19.
F
i
P
x
21.
Q
S
f
M
I
P
F
L
K
N
h
R
E
i
F
K
P
k
h
E
i
P
B
I

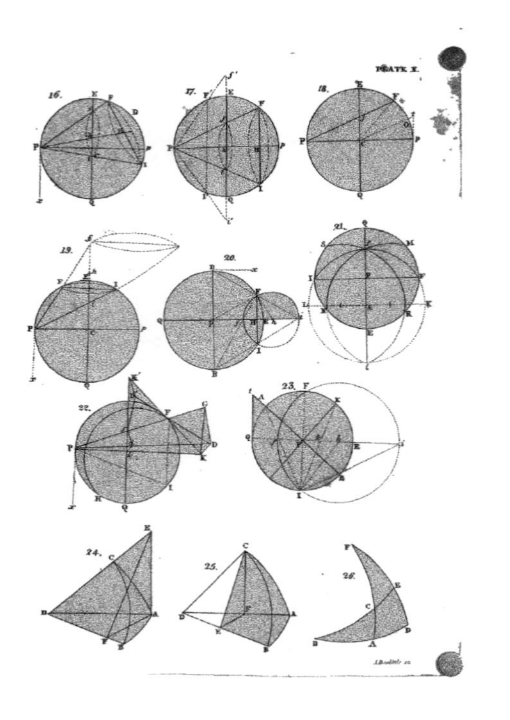
PLATE X.
16.
17.
18.
19.
20.
21.
22.
23.
24.
25.
26.

/G

B

B

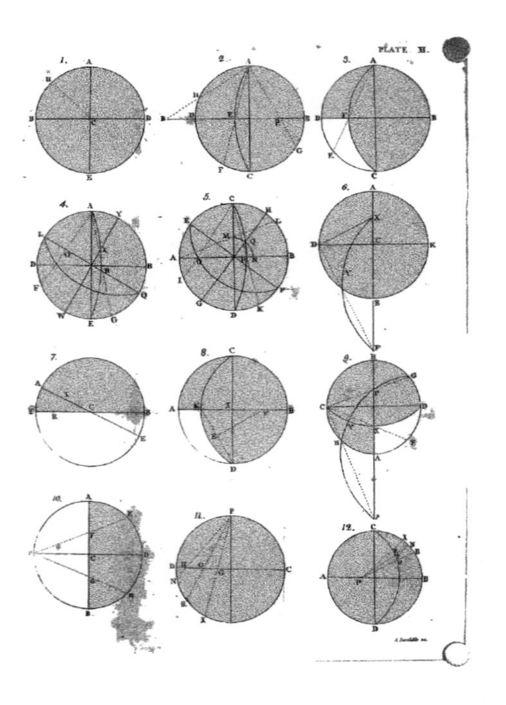
PLATE II.
1.
2.
3.
4.
5.
6.
7.
8.
9.
10.
11.
12.

PLATE XII.

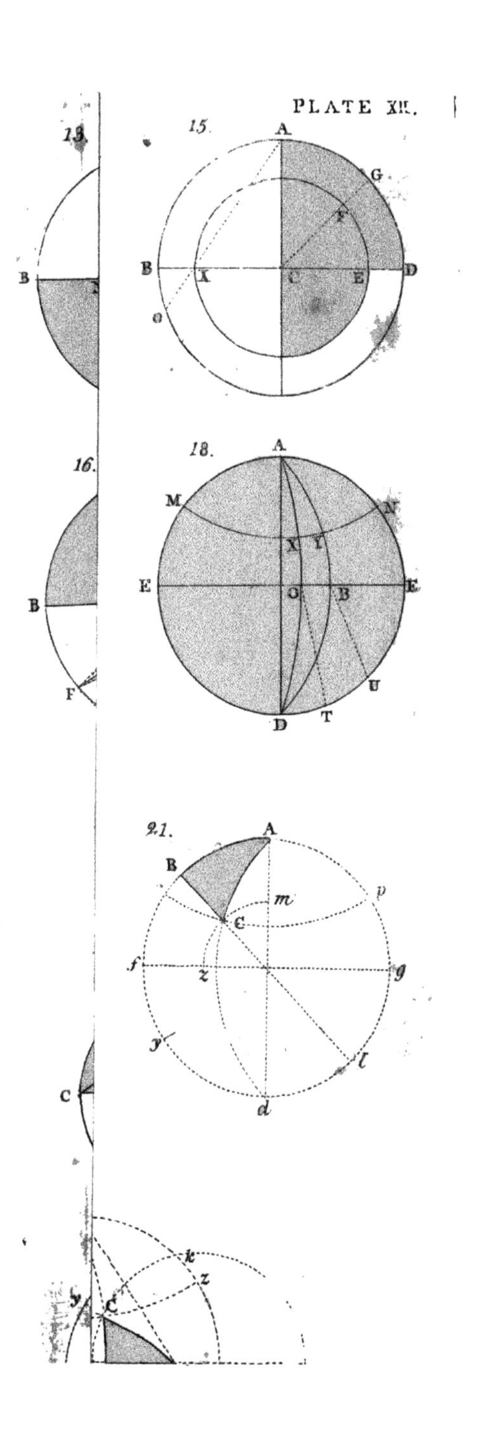

13.
B
15.
A
G
F
B
X
C
E
D
O
16.
B
F
18.
A
M
N
X
Y
E
O
B
E
U
D
T
21.
A
B
m
p
C
f
z
g
y
l
d
C
k
z
y
C

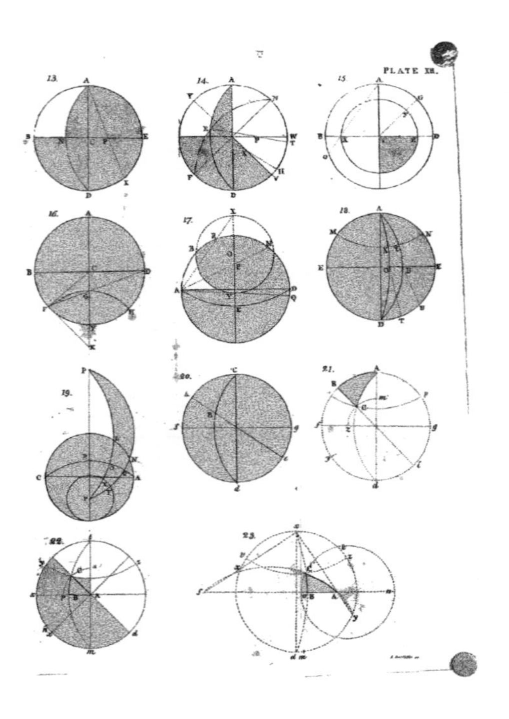
PLATE XII.
13.
14.
15
16.
17.
18.
19.
20.
21.
22.
23.

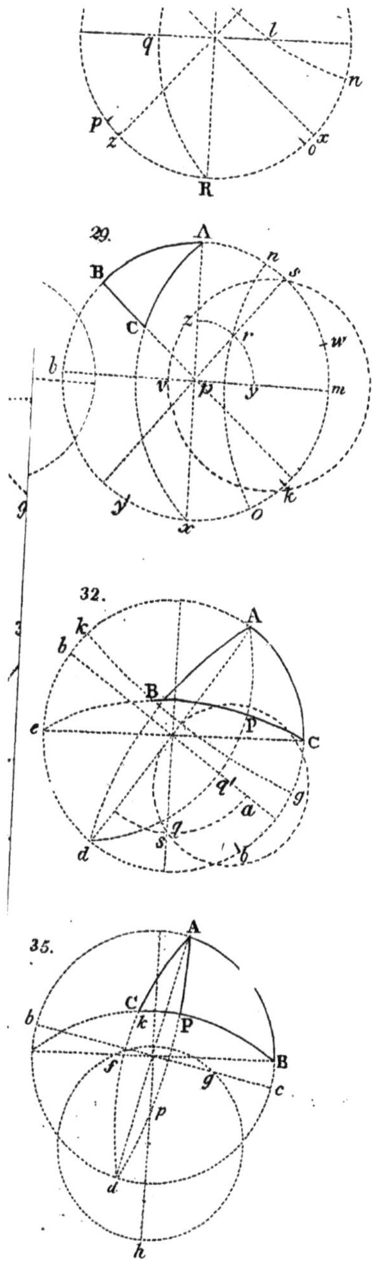

A Doolittle sc

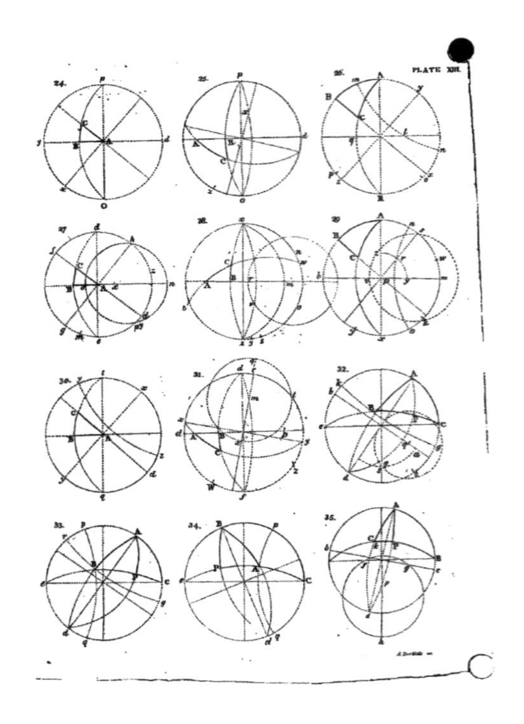
PLATE XIII.
24.
25.
26.
27
28.
29
30.
31.
32.
33.
34.
35.

PLATE XIV

E

E · ·

PLATE XIV

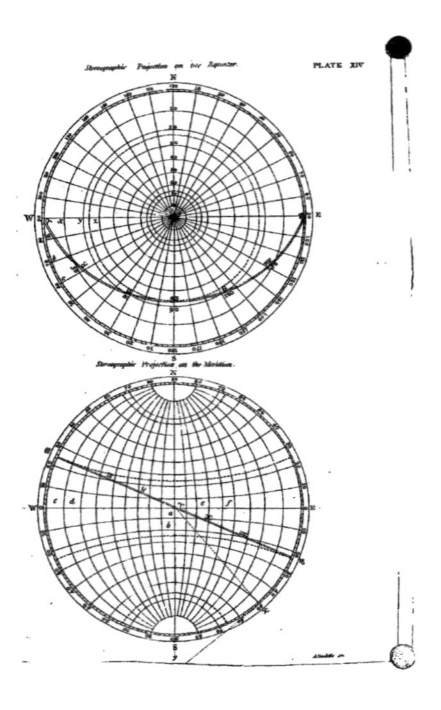

Stereographic Projection on the Equator.

Stereographic Projection on the Meridian.

H

C

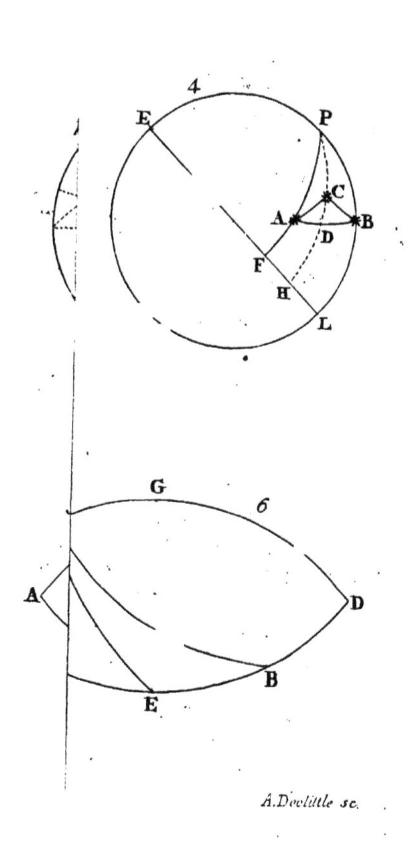

PLATE XV

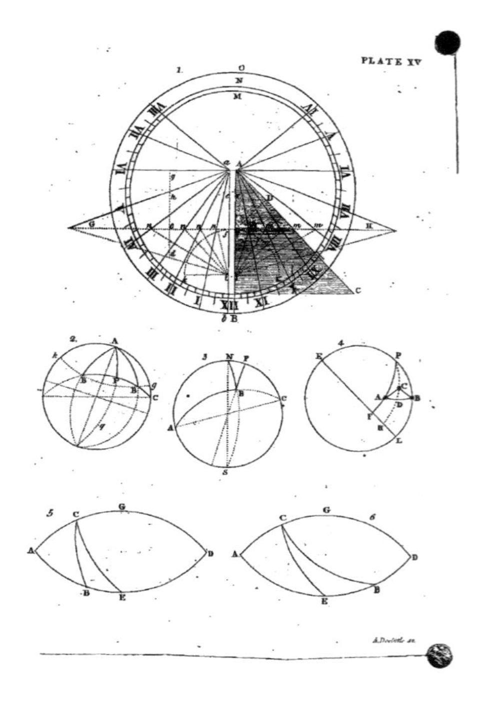
1.
O
N
M
A
a
B
b
C
D
G
H
XII
XI
X
IX
VIII
VII
VI
V
IIII
I
II
III
2.
A
B
F
C
3
N
P
B
C
A
S
4
E
P
A
C
B
D
F
L
5
C
G
A
D
B
E
6
C
G
A
D
E
B

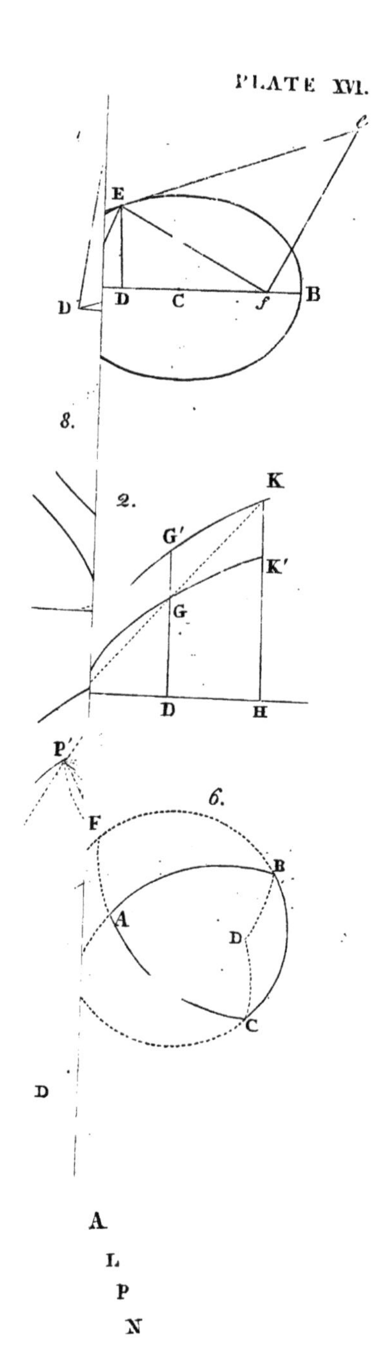
PLATE XVI.
e
E
D
D
C
f
B
8.
2.
K
G′
K′
G
D
H
P′
6.
F
B
A
D
C
D
A
L
P
N

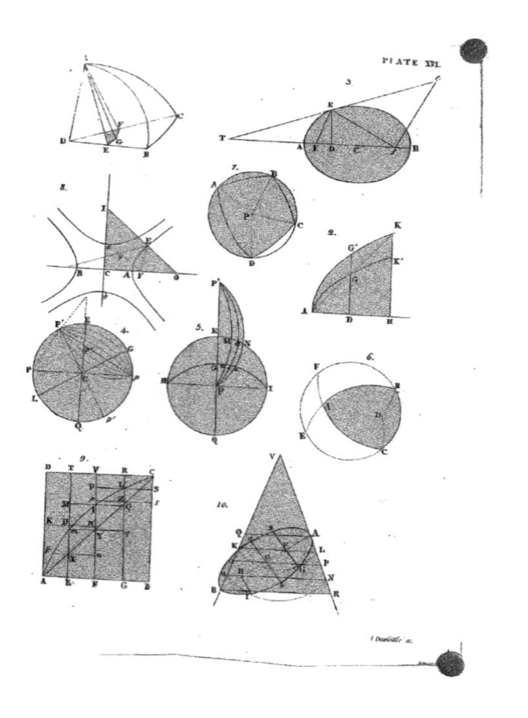
PLATE

Lightning Source UK Ltd.
Milton Keynes UK
UKHW020031160219
337399UK00010B/550/P

9 781330 165331